STUDENT
ACTIVITY MANUAL
LAUREL TECHNICAL SERVICES

COLLEGE ALGEBRA

A View of the World Around Us

DAVID WELLS
LYNN SCHMITT-TILSON

PRENTICE HALL, Upper Saddle River, NJ 07458

Acquisition Editor: *Sally Denlow*
Manufacturing Buyer: *Alan Fischer*
Production Editor: *Dawn Blayer*
Special Projects Manager: *Barbara A. Murray*
Supplement Cover Manager: *Paul Gourhan*
Supplement Cover Designer: *PM Workshop Inc.*
Supplement Editor: *April Thrower*

Simon & Schuster/A Viacom Company
Upper Saddle River, NJ 07458

Printed in the United States of America

10 9 8 7 6 5 4 3 2 1

ISBN 0-13-573940-3

Prentice-Hall International (UK) Limited, *London*
Prentice-Hall of Australia Pty. Limited, *Sydney*
Prentice-Hall Canada, Inc., *Toronto*
Prentice-Hall Hispanoamericana, S.A., *Mexico*
Prentice-Hall of India Private Limited, *New Delhi*
Prentice-Hall of Japan, Inc., *Tokyo*
Simon & Schuster Asia Pte. Ltd., *Singapore*
Editora Prentice-Hall do Brasil, Ltda., *Rio de Janeiro*

TABLE OF CONTENTS

TIPS FOR SURVIVAL

What are these? There are general tips on how to use this textbook to learn the mathematics of algebra.

- Read the Nutshell for each chapter before you begin the chapter, after you finish, the chapter and before each exam.

  This is what you should get from each reading: the 1st reading will give you an overview of the chapter and place it into a framework with the chapters you already read.

  When you read it the 2nd time, look at the Nutshell as an outline. You can then fill in the details yourself to reinforce the connections discussed in the text.

  The 3rd reading will help you remember important points.

- Go through this text with pencil and paper in hand!!!!
- Understanding mathematics takes commitment, time, patience and personal involvement.
- Most of us cannot learn mathematics in a cram session the night before an exam. The understanding comes gradually over time as you listen, ask, ask, listen, write, state, ask again, listen again, write, state, etc. It is a process of study time alone and study time with others.
- Take advantage of your classmates! Your study time together can be valuable. Your explanations to each other are often *right on target.*
- Think of your graphing calculator as your "side-kick", not as the trail boss. That means, view the graphing calculator as an "aide" to your understanding of concepts. It should not replace understanding with automatic finger tip exercises.
- Neatness counts in mathematics! Do not skip steps. If you have everything written down it is easier to review, check the results, or explain it to someone else.
- Go back to page 21, now and again, to refresh your memory about the Polya strategies. You might want to copy them on a separate sheet of paper and keep them handy.
- When you are asked to "explain", don't worry too much about "saying it right" the first time. Use your own words as best you can. The ability to explain clearly and concisely doesn't come right away. It takes work, time, and practice. Write several drafts of your explanation. Read each one a few hours after you have written it. Rewrite it, and revise it if it isn't clear to you. Better yet, have a classmate read it.
- Sometimes our questions are meant to give you intellectual space, that is, not just to have you mimic what you have heard or read. How do you approach a questions like these? Approach them by noticing the main points and mentally organizing them in a broad view of what you have read over the last several pages, rather than looking for an example that is just like the problem.
- Multiple views are the backbone of critical thinking. There is always *another* way to think through a problem!
- Don't be limited by old habits. What usually gets in the way of learning is not what you *don't* know but what you know already.

Here are the solutions to the odd problems and some hints on how to use them. The hints are listed for each chapter at the beginning of the exercises for that chapter.

- Most of us, after looking at a solution, have the normal reaction, "Gee, I would have gotten that!" The fact is we probably wouldn't have. Don't fool yourself into believing that you can solve a given type of problem until you can solve several without looking at the solution manual.

- The solution to a problem may take from less than a minute to more than an hour. If you can't solve a problem right away, think about it, put it aside, think about it some more.

- When you finally decide to look at a solution from this manual, put a piece of paper over it and uncover it line by line. You will probably be able to "pick up the ball" at some point and carry the solution through to the end on your own.

- If you have worked through a problem and suspect that your solution is correct but at odds with ours, please check with someone. Even though all who worked on this text looked at it carefully, there is still a possibility that some errors have occurred. If this happens, please let us know and accept our apologies.

David Wells
Department of Mathematics
Penn State University
New Kensington, PA 15068

412-339-6049

dmw8@psu.edu

Lynn Tilson
517 Arcadia Park
Lexington, KY 40503

606-277-1149

Survival tips for studying for an exam:

- Do enough exercises so you can *easily* do all of the basic drill problems.
- To do all the drill problems *easily* means to solve them and *resolve* them until they become second nature. Treat mathematics as A SECOND LANGUAGE. IT IS!
- Do enough exercises so that you feel comfortable with the concepts and can solve similar but new problems that you haven't seen before.
- You might use the problems in the student manual to test yourself. This will give you extra practice and increase your conceptual understanding.
- The night before the exam SHOULD BE for only a quick review of the main ideas and skills for each section. Then it's time for a good night's sleep!!

Global list for doing exercises throughout the textbook

- Once in a while you might find the answer to a problem through a "flash of insight". This certainly counts as a solution, but you should also find a systematic solution that you can explain to someone else.
- Sometimes when you make a table, try using tractions instead of decimals. Then use the strategy of *Looking for a Pattern* to answer questions.
- Be sure to notice that some exercises are springboards for the next.
- Some exercises ask for your opinion. Remember that your opinions need to be based on a combination of facts (mathematical and otherwise) and experience. Pay attention to *all* the information provided.
- Be sure to complete the Chapter Review exercises at the end of each chapter. These will help you to know the meanings of all new terms introduced in the chapter. They will also help you understand all of the chapter's main ideas.
- Many exercises will ask you to confirm a graphical result analytically or support an analytical result graphically. Solving a problem both ways provides insight that you might miss with either solution alone.

NUTSHELLS

Nutshell for Chapter 1 - MODELING AND PROBLEM SOLVING

What's familiar?

- This book is about algebra. It is about solving equations, drawing graphs, and simplifying expressions.
- It builds on algebraic skills you already have. Appendix A reviews the most important skills. They are also listed as prerequisites at various points in the book.

What's new?

- This book is different, in many ways, from mathematics textbooks you may have used in the past. Section 1-1 introduces you to its structure. "Some Suggestions for Feeling at Home with This Book" (pages 5-7 will help you take advantage of its unique features.
- Section 1-1 explores algebra as more than solving equations, drawing graphs, and simplifying expressions. It discusses ways to create mathematical models based on actual problems and the tools with which to solve them.
- Section 1-2 contains a systematic discussion of tables, equations, and graphs as mathematical models and how to construct them.
- Section 1-3 describes a comprehensive process for solving problems and introduces you to several specific problem-solving strategies.

List of hints for exercises in each chapter.

Chapter 1

- These exercises represent the various models presented in this chapter. They require careful reading. You might want to refer back to the text.
- Several exercises such as Section 1-1, Exercises 4 and 6 ask for your opinion.
- There is a concentration of *Problem Solving* exercises in this chapter. Spend a good deal of time here. It will serve you well as you proceed to other chapters.

Name:____________________

Date:____________________

Chapter 1 Additional Exercises

For Problems 1–4, use the graph shown. (x,y)

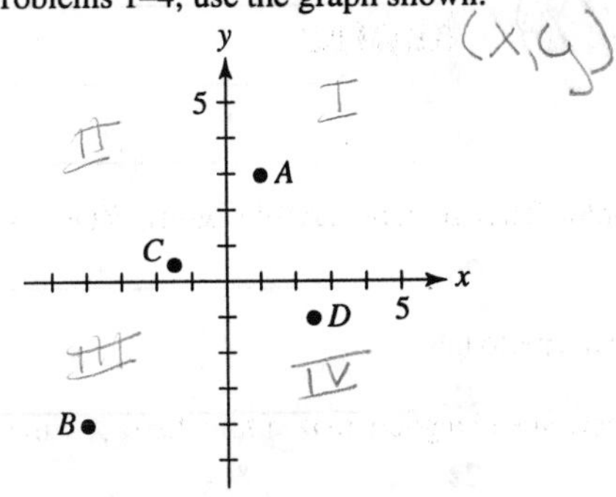

1. What are the coordinates of point A? — **1.** (1,3)

2. What are the coordinates of point B? — **2.** (−4,−4)

3. What are the coordinates of point C? — **3.** (−1.5,−0.5)

4. What are the coordinates of point D? — **4.** (2.5,−1)

5. Graph $A\left(-\frac{7}{2}, 4\right)$ and $B(2, 3)$ on the same coordinate system. — **5.**

y−7 = 2x + 5

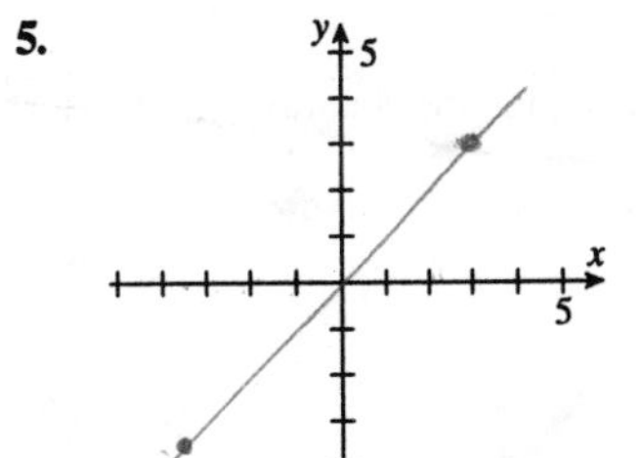

6. Find the y-intercept of the graph of $y-7=2x+5$. — **6.** __________

7. Find the y-intercept of the graph of $5y-8=x-20$. — **7.** __________

8. Find the x-intercept of the graph of $6-y=3x+8$. — **8.** __________

9. Find the x-intercept of the graph of $y+7=\frac{x}{4}+9$. — **9.** __________

10. Find the slope of the graph of $5y+11=7x-2$. — **10.** __________

11. Rewrite the equation $6x+11=\frac{y}{2}+4$ in slope-intercept form. — **11.** __________

12. $y=-4x+13$. Find y when $x=-5$. — **12.** __________

13. $8x+19=\frac{x}{3}+4y$. Find y when $x=15$. — **13.** __________

$m=\frac{y_2-y_1}{x_2-x_1}$

In 14–19, find the slope-intercept form of the equation with the given properties. Round all answers to two decimal places.

14. Passes through (3.6, 0.6) and (5.1, 1.2) — **14.** __________

15. Passes through (5, 0.03) and (9, 0.11) — **15.** __________

16. Passes through (−5.97, 0.36) and (4.2, −1.77) — **16.** __________

Chapter 1 Additional Exercises *(cont.)*

Name:____________________

17. Passes through (7.11, 2.5) with slope 1.07 — **17.** ________________

18. Passes through (4.20, 3.39) with slope 4.8 — **18.** ________________

19. Passes through (5.37, 8.116) with slope –8.6 — **19.** ________________

20. Find the y-intercept of the line passing through (6.11, 4.109) and (17.06, 9.88). Round the answer to two decimal places. — **20.** ________________

21. Find the x-intercept of the line passing through (–3.26, 9.98) and (11.93, 12.06). Round the answer to two decimal places. — **21.** ________________

22. Graph the equation $y = \frac{3}{5}x - \frac{8}{3}$. — **22.**

23. Graph the equation $y = 3 - \frac{1}{4}x^2$. — **23.**

24. Use the equation $y = 9x - 5$ to complete the table. — **24.**

y=9(1)-5
9-5
13=9x-5
-9x=18
22=9x-5
-9x=-3
x=3

x	y
1	4
2	13
3	22
8	

25. Use the equation $y = \frac{\sqrt{x+4}}{3}$ to complete the table. — **25.**

x	y
0	
	$\frac{4}{3}$
32	
	7

26. Sketch a possible graph of the relationship between x and y. — **26.**

x	y
–4	1
–1	–2
2	1
3	$\frac{10}{3}$

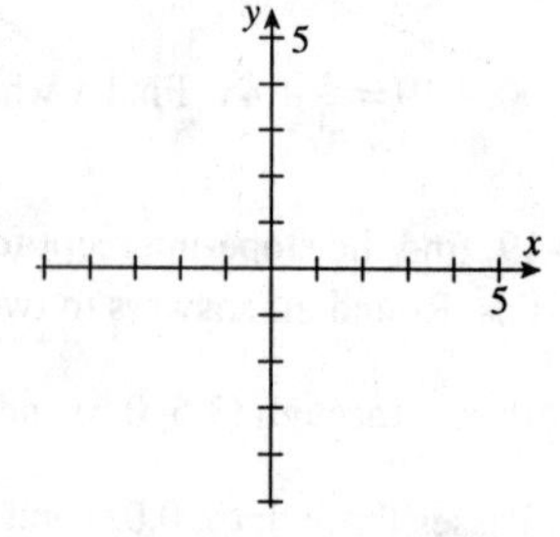

Chapter 1 Additional Exercises *(cont.)*

Name:____________________

27. Sketch a possible graph of the relationship between x and y.

x	y
0	−6
2	−2
4	2
6	6

27.

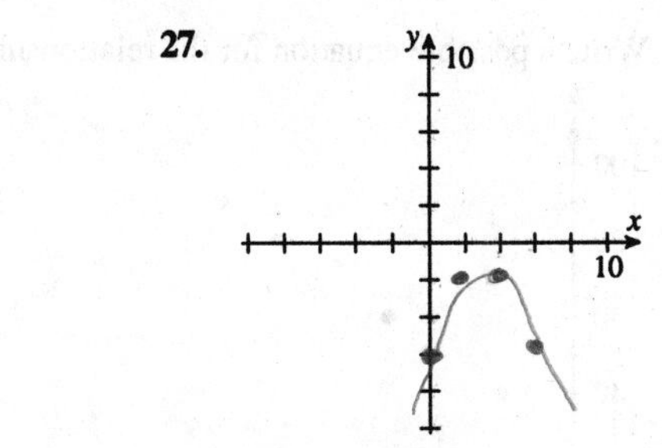

28. Use the graph to complete the table.

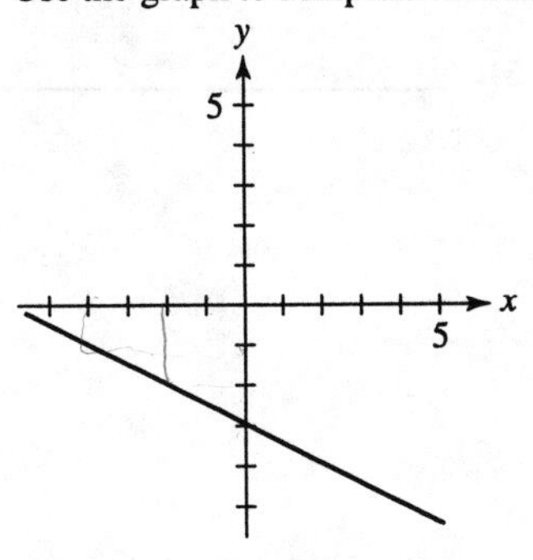

28.

x	y
−4	-1
-2	−2
0	−3
2	-4

29. Use the graph to complete the table.

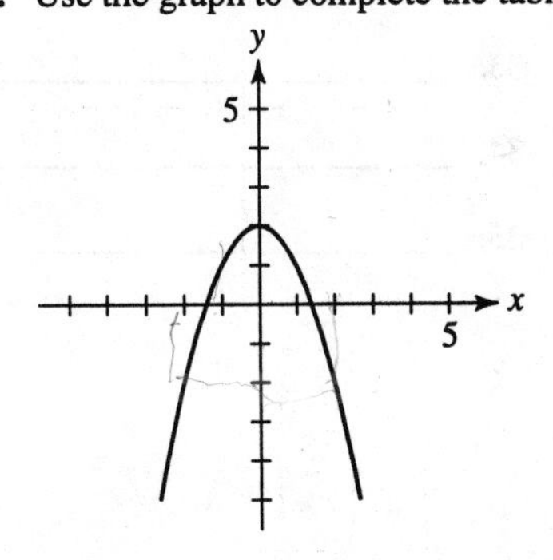

29.

x	y
−2	2
-2	2
−1	2
2	

30. Write a possible equation for the relationship shown in the graph.

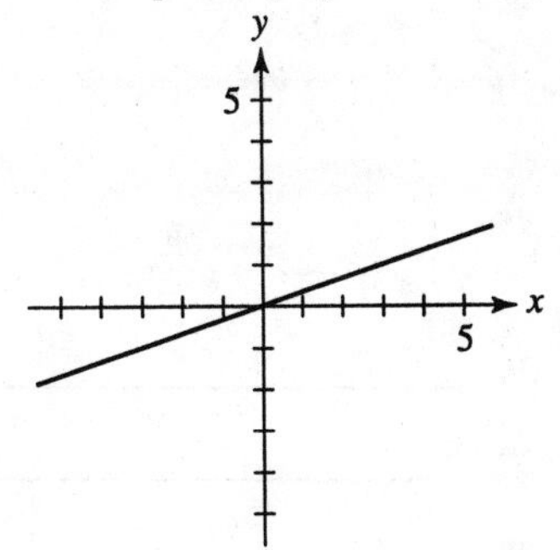

30. ____________________

Chapter 1 Additional Exercises *(cont.)*

Name:________________

31. Write a possible equation for the relationship shown in the graph.

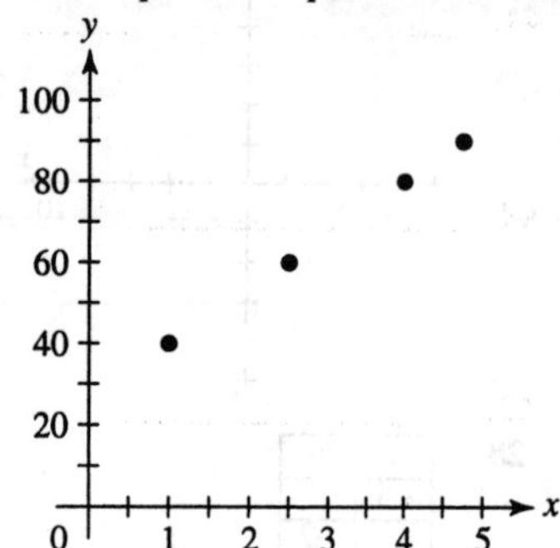

31. ________________

For Problems 32–35, use the table shown.

a	b	c	d
5	6	–3	–5
9	5	0	–6
17	3	6	–8
21	2	9	–9

32. Write a possible equation defining a in terms of b. **32.** ________________

33. Write a possible equation defining c in terms of a. **33.** ________________

34. Write a possible equation defining b in terms of d. **34.** ________________

35. Write a possible equation defining d in terms of c. **35.** ________________

For Problems 36–39, use the graph shown.

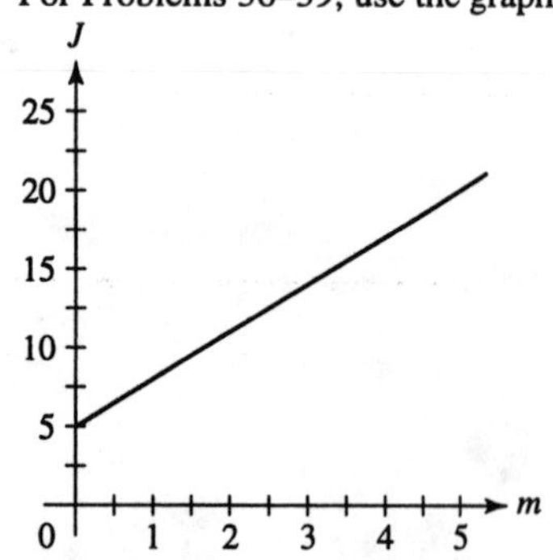

36. Estimate the value of J for $m = 3$. **36.** ________________

37. Estimate the value of J for $m = 5$. **37.** ________________

38. Predict the value of J for $m = -9$. **38.** ________________

39. Predict the value of J for $m = 13$. **39.** ________________

40. A current of I amperes through a circuit is generating a voltage V given by $V = 300I + 12$. If $I = 0.25$ A, what is V? **40.** ________________

Chapter 1 Additional Exercises *(cont.)*

Name:____________________

41. The speed of a falling body, v (m/s), at time t (s), is given by $v = -9.81t + 11.35$. What is v when $t = 2.71$ s?

41. ____________________

42. In the ocean, the water pressure P, in kilopascals, existing h meters below the surface is given by $P = 10.06h$. What is the pressure at a depth of 85 m?

42. ____________________

43. A college has 9750 students enrolled. Two years ago the enrollment was 8626. If the growth is linear, what was the enrollment five years ago?

43. ____________________

44. In a psychological test, people are able to remember an average of 0.9 words from a list they have seen for 0.3 seconds, and 3.2 words from a list seen for 0.4 seconds. Give a realistic estimate for the number of words remembered from a list seen for 0.6 seconds.

44. ____________________

45. A sum of money is deposited in an account where it earns simple interest. The account contains \$1673.00 after three years and \$1855.00 after five years. How much was in the account to start, and what is the annual interest rate?

45. ____________________

46. The Olympic record for women's 1500-m speed skating was 2 min 25.2 s in 1960 and 2 min 22.4 s in 1968. Assuming a linear change in the record, what will it be in the year 2000?

46. ____________________

47. A 15-lb mix of foods A and B must contain exactly 24.5 units/lb of Vitamin Z. A pound of A contains 20 units of Z, and a pound of B contains 40 units of Z. How much of A should be used in the mix?

47. ____________________

48. A shop making model windmills has monthly fixed costs of \$54,200, in the form of rent, utility bills, and so forth. Each windmill costs \$7.13 in materials and labor to make. Write the shop's total monthly costs, C, in terms of w, the number of windmills produced.

48. ____________________

49. A piece of machinery is worth \$12,500 new and \$400 at the end of its useful life, which is 8 years. Assuming linear depreciation, what is the machine worth after 6 years?

49. ____________________

50. According to the graph, what is the cost of three hours of computer time? Of eight and a half hours?

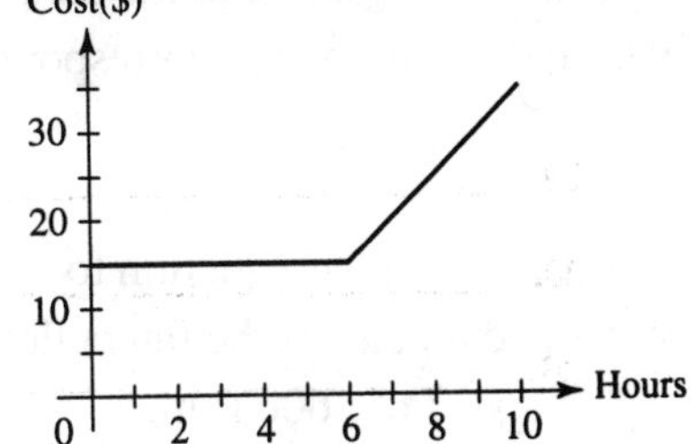

50. ____________________

A Mathematical Looking Glass

More Pulleys

(to follow Section 1-2)

Figure 1-4 from **Pulleys** (page 9) is reproduced below. If John keeps the diameter of M (the motor pulley) constant at 5 inches and varies the diameter of F (the *fan pulley*), then increasing the diameter causes the fan to turn more *slowly*.

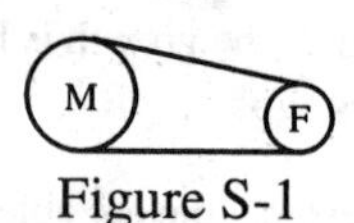

Figure S-1

Table S1 is a numerical model of the relationship between the fan pulley diameter and the fan speed. Fan speeds are rounded to the nearest 0.1 rpm.

Fan pulley diameter F in inches	Fan speed S in rpm
3	2833.3
4	2125
5	1700
8	1062.5
10	850
12	708.3

Table S1

Exercises

1. (Creating Models) Write an equation to describe this relationship by looking for a pattern in the data. (*Hint*: Multiply each diameter by the corresponding fan speed.)

2. (Interpreting Mathematics) Use your equation to determine the required diameter of the fan pulley if the fan is to turn at a speed of 3600 rpm.

3. Sketch a graph of this relationship with pulley diameter as the independent variable and fan speed as the dependent variable. Remember that pulley diameters and fan speeds must be positive.

4. (Interpreting Mathematics) Use your graph to determine the required diameter of the fan pulley if the fan is to turn at a speed of 3600 rpm.

5. (Writing to Learn) Compare the processes used to calculate the results in Exercises 2 and 4.

   a. Which yielded the more accurate answer?
   b. Which process did you find easier to use, and why? (There are no "right" or "wrong" answers. The important part of your answer is the "why", which reflects your judgment.)

6. (Making Observations) Use Table S1 to decide which of the following causes the greater decrease in fan speed.

   a. Increasing the diameter from 3 to 5 inches

   b. Increasing the diameter from 10 to 12 inches

7. (Writing to Learn) Describe how you could complete Exercise 6 by looking at your graph, without subtracting fan speeds. (*Hint*: Compare the steepness of two different segments of the graph.)

A Mathematical Looking Glass

Testing for HIV

(to follow Section 1-3)

In 1987 the Illinois state legislature debated a bill to enact a program of mandatory premarital testing for the human immunodeficiency virus (HIV), which causes autoimmune deficiency syndrome (AIDS). Part of the debate focused on the effectiveness of the test.

We can identify four aspects of effectiveness, embodied in the following four questions.

- Among all infected people tested, what percentage will have the correct test result? If this number is high, the test has high **sensitivity**.

- Among all uninfected people tested, what percentage will have the correct test result? If this number is high, the test has high **selectivity**.

- Among all people with positive tests, what percentage will have the correct test result? (An incorrect result in this case is called a **false positive**.)

- Among all people with negative tests, what percentage will have the correct test result? (An incorrect result in this case is called a **false negative**.)

Although we may seem to be repeating ourselves here, that is not the case. In Exercises 1-4, you will see that the four percentages referred to here can all be different.

The legislators knew the name of the test being proposed (the ELISA test), and had access to data relating to its performance. In particular, they had access to Table S2.

	infected persons	uninfected persons	total
positive tests	1325	7648	8973
negative tests	23	3816372	3816395
total	1348	3824020	3825368

Table S2

Source: Paul D. Cleary et al, *Compulsory Premarital Screening for the Human Immunodeficiency Virus*, JAMA, v. 258, no. 13, 10/2/87

Table S2 was not compiled from actual experience, but contains predicted data based on a study done by a group of researchers at Harvard University. It represents an estimate of annual test results if every applicant for a marriage license in the United States were tested. The estimates were derived from the researchers' knowledge of the ELISA test's sensitivity and selectivity. For example, it's sensitivity is 98.3%, so if 1348 infected people were tested, 1325 (98.3% of 1348) would have correct test results.

Exercises

1. (Interpreting Mathematics) The ELISA test has a selectivity of about 99.8%. How is this fact reflected in Table S2?

2. (Interpreting Mathematics) Among all persons with positive tests, what percentage would have correct test results according to Table S2?

3. (Interpreting Mathematics) Among all persons with negative tests, what percentage would have correct test results according to Table S2?

4. (Writing to Learn) Which of the percentages in Exercises 1-3 was lowest? What are some practical consequences of this result? In light of your results, what judgment would you make about the value of this testing program?

5. (Writing to Learn) Describe the four steps in the Polya problem-solving process as you used them in completing Exercises 1-4.

At first glance, the results do not seem to be reasonable in the context of the problem. How can a test with both high sensitivity and high selectivity yield such a high percentage of false positives? A careful look at Table S2 reveals the answer. The 7648 false positive test results represent only a small percentage of the number of uninfected people, but a large percentage of the positive test results. This pattern stems from the rarity of HIV in the general population. Thus, you can conclude that in testing for any rare disease, any test is likely to yield a high percentage of false positives.

The cost of the Illinois program was estimated to be about $312,000 per infected individual identified, compared with less than $2000 per individual through other state programs of counseling and testing. In addition, many Illinois residents chose to apply for marriage licenses in neighboring states, resulting in a 22.5% decrease in the number of licenses applied for in Illinois, and a corresponding loss of revenue. (For further details, see Bernard J. Turnock and Chester J. Kelly, *Mandatory Premarital Testing for Human Immunodeficiency Virus*, JAMA, vol 261, no. 23, 6/16/89.) Had the Illinois legislature chosen to consult some mathematicians before instituting the program, they could have saved a great deal of needless expense.

Chapter 1

Section 1-1

1. The slope of the line is $\frac{0.100-0.133}{2.5-0.7} = -\frac{0.036}{0.18} = -0.2$. In point-slope form, the equation of the line is $y-0.100 = -0.2(t-2.5)$. Simplifying, and rounding the coefficients to three decimal places, $y = -0.02t + 0.150$

3. Since $t = 0$ at the time of the accident, Bill's BAC was 0.091.

5. **a.** The left portion of the graph would become less steep. It would still pass through (0.7, 0.136), since that point corresponds to the reading taken at 2:52. Therefore, the y-intercept on the graph would be higher than the one shown in Figure 1-2.

   b. Bill's friend's testimony favors the prosecution.

Section 1-2

1. Responses will vary.

3. If $S = 3600$ rpm, then $3600 = 340M$
   $M = \frac{3600}{340} \cong 10.59$ inches

5. The fan speed appears to be about 3700 or 3800 rpm. This result is less precise than that obtained from the equation, but is consistent with it.

7. (Sample response) The equation provides a more precise answer than the graph, but the graph provides an estimate more quickly than the equation. The table does not provide a direct answer to either question.

9. (Sample response) I would rather have access to Table 3 because it provides reasonably precise answers to the questions that customers would ask most often.

11. Estimates will vary, but should be close to 360 parts per million (ppm) for 2000, and 420 ppm for 2050. The estimate for the year 2000 should be more trustworthy, because it assumes that the graph will show no surprising changes over a period of only 15 years beyond 1985.

13. **a.**

year	CO_2 concentration (parts per million)
1960	315
1970	325
1980	335
1985	342

Answers may vary slightly.

   b. 360 (answers may vary slightly)

   c. 420 (answers may vary slightly)

15. **a.** $y = 10 - x$

   b.

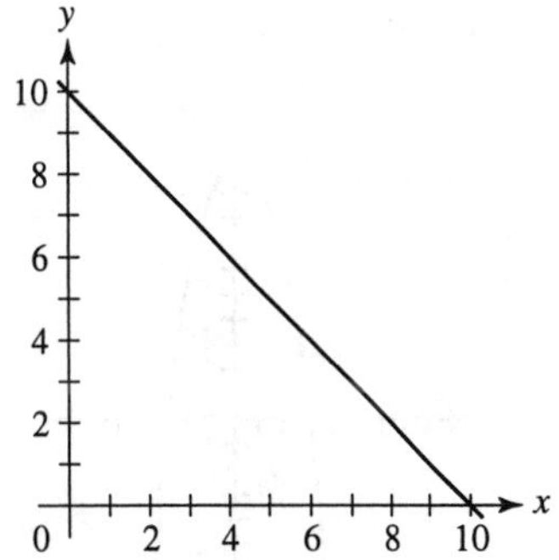

17. **a.** $y = \frac{12}{x}$

   b.

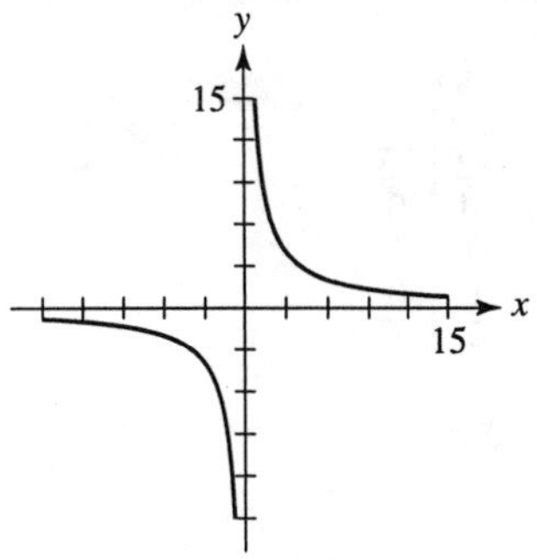

19. a.

x	y
−4	14
−2	12
0	10
2	8
4	6

b.

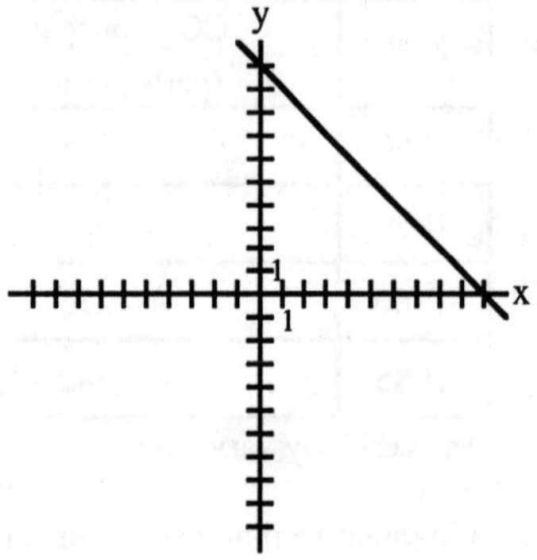

21. a.

x	y
−4	−6
−2	6
0	10
2	6
4	−6

b.

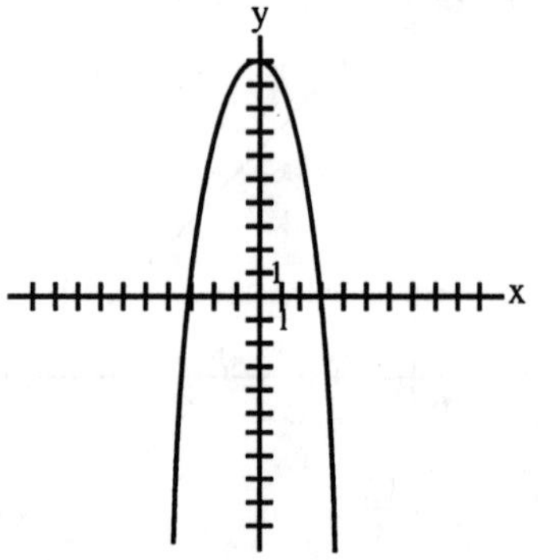

23. a.

x	y
−2	−3
−1	0
0	1
1	0
2	−3

b. $y \cong -24$

25. a.

x	y
−2	0
−1	−1
0	−2
1	−1
2	0

b. $y \cong 3$

27. a. (Sample response) The information is incomplete because the table does not show the life expectancy at every age.

b. (Sample response) The information is approximate because the values of E have almost certainly been rounded.

c. (Sample response) The information is explicit because the values of the variables can be read directly from the table.

29. a. (Sample response) The information is incomplete because it shows sea levels only for about the past 50,000 years.

b. (Sample response) The information is approximate because sea levels must be estimated from the graph.

c. (Sample response) The information is hidden because sea levels must be read from the graph.

31. a.

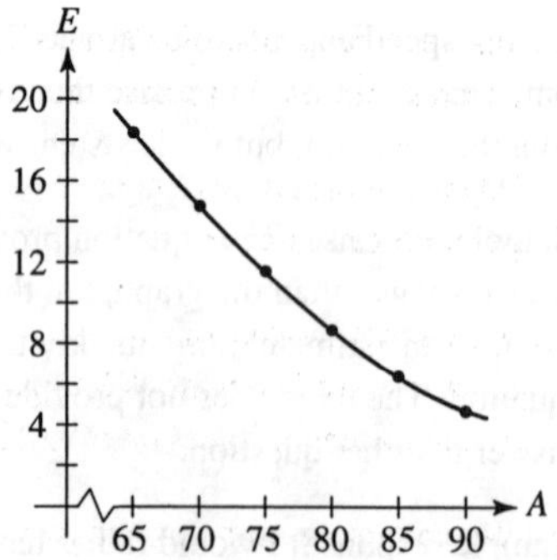

b. About 2 years (Answers will vary.)

33. a.

D	r
200	2.0
300	3.1
400	4.2
500	5.3
600	6.4

b.

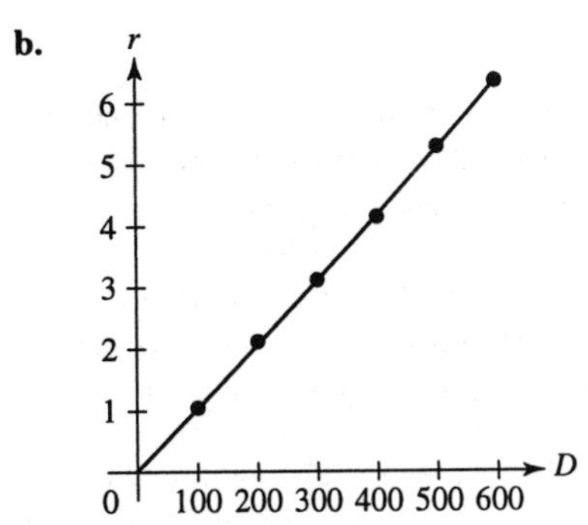

35. Responses will vary.

37. Responses will vary.

Section 1-3

1. Responses will vary.

3. When $C = 360$:
If $A = 25C - 5625$, then $A = 25(360) - 5625 =$ 3375 ft^2
If $A = 15C - 2025$, then $A = 15(360) - 2025 =$ 3375 ft^2

5. a. We cannot answer this question. We could answer if we knew the dollar amounts spent on prescription drugs in either 1990 or 1991.

b. Physician care experienced the smallest percentage increase.

c. We cannot answer this question. We could answer if we knew the dollar amounts spent on each category in either 1990 or 1991.

d. We cannot answer this question. We could answer if we knew the dollar amounts spent on each category in either 1990 or 1991.

7. Each of the fifteen masters played each of the other fourteen, so that twice the number of games is equal to $(15)(14) = 210$, and the total number of games is 105. (The result of 210 represents twice the number of games because it counts a single game between players A and B as both a game for A and a game for B.)

9. Fill the larger container, pour 3 liters from it into the smaller container, then empty the smaller container. There are now 2 liters in the larger container. Pour that into the smaller container. Then once again fill the larger container. There are now 5 liters in the larger container, and 2 liters in the smaller one. Pour from the larger to the smaller until the smaller one is full. There are now 4 liters in the larger container.

11. According to Slats, there are 60 animals in the enclosure. If E represents the number of elephants, then $60 - E$ represents the number of ostriches. The total number of legs is $4E$ (from the elephants), plus $2(60 - E)$ (from the ostriches). Thus,

$$\begin{aligned} 4E + 2(60 - E) &= 194 \\ 4E + 120 - 2E &= 194 \\ 2E &= 74 \\ E &= 37 \end{aligned}$$

There are 37 elephants.

13.

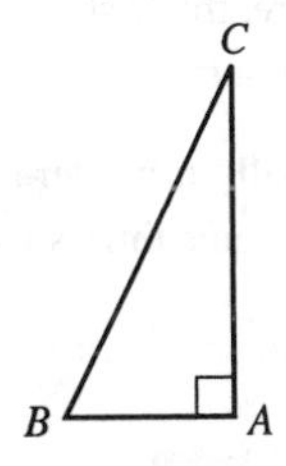

In the triangle above, the antenna is represented by AC, and the wire by BC. The base AB is half of a diagonal across the roof. According to Exercise 12, its length is 40.31 feet. By the Pythagorean Theorem,

$$\begin{aligned} (AB)^2 + (AC)^2 &= (BC)^2 \\ (40.31)^2 + (15)^2 &= (BC)^2 \\ 1849.90 &\cong (BC)^2 \\ BC = \sqrt{1849.90} &\cong 43.01 \text{ feet} \end{aligned}$$

15. If p represents the number of loaves of pumpernickel Robert will buy, then some additional entries in the table are:

	number of loaves	price per loaf	total cost
pumpernickel	p	\$1.75	\$1.75p
day-old	$14 - p$	\$0.50	\$0.50$(14 - p)$
total	14		\$17

From the last column of the table,
$1.75p + 0.50(14 - p) = 17$
$1.75p + 7 - 0.50p = 17$
$1.25p = 10$
$p = 8$
Robert will buy 8 loaves of pumpernickel, and 6 loaves of day-old bread.

17. We don't know how you feel, but we would rather ride in the first elevator. The steep portion of the second graph indicates a long drop in a very short time. Our guess is that the cable on the second elevator snapped as it reached the second floor.

19. The first graph is at least as high as the second one at every time, indicating that the first car is always going at least as fast as the second car, and usually faster. Therefore, the first car has traveled farther during the hour.

The two cars are going equally fast after an hour, since both graphs show the same time-speed coordinates after an hour.

21. Responses will vary.

NUTSHELLS

Nutshell for Chapter 2 - FUNCTIONS

What's familiar?

- We continue to view quantitative relationships from numerical, analytical, and graphical perspectives.
- You gain experience in using the Polya strategies and process (from Chapter 1) by creating mathematical models, analyzing them and interpreting the results.

What's new?

- We use the word, *function*, to describe a quantitative relationship in which each input value generates a unique output value.
- Section 2-1 discusses how you can tell whether a table, an equation, or a graph describes a function.
- Section 2-1 also explains why the concept of a function is fundamental to the study of algebra.
- Section 2-2 shows you how to view a function as a *process* leading from input to output. You are introduced to a notation that makes it natural to think about them this way.
- Section 2-3 looks at graphs of functions called *dynamically*. You will take a walk along the graph from left to right. You will notice which x-values (*domain*) and y-values (*range*) you encounter on the way. This will tell you whether the graph is going uphill (*increasing*) or downhill (*decreasing*).
- Section 2-3 discusses special functions called *sequences*. These functions have only positive integer x-values. You will learn how sequences are the same as other functions and how they are different.
- Section 2-4 gives examples of how to use the graphical models of functions as a tool for solving equations.
- Section 2-4 suggests that the tools for solving graphical models depends on the assumption that the graphs are unbroken curves, i.e., *continuous*. We return to this idea in Sections 3-4 and 8-2.

List of hints for exercises in each chapter.

Chapter 2

- Know the list of terms at the end of this chapter well, These terms are used frequently throughout the textbook and understanding them will help you do many exercises.
- This chapter is where we first ask you to use your graphing calculator extensively for particular exercises. Do it! Make sure you know how to graph, zoom, and trace on your calculator.
- Be brave enough to explore functions on your calculator. Use your trace and zoom keys to formulate and answer questions beyond those posed in the exercises.

Name: ______________________

Date: ______________________

Chapter 2 Additional Exercises

1. Is y a function of x? Assume that the table is complete.

x	y
3	1
7	11
3	−3
−1	5

1. ______________________

2. Is y a function of x? Assume that the table is complete.

x	y
4	2
2	7
7	−2
−2	4

2. ______________________

3. Given $x^3 + |y| = -18$, is y a function of x?

3. ______________________

4. Given $x^2 - 9 = \sqrt{y}$, is y a function of x?

4. ______________________

5. Is y a function of x?

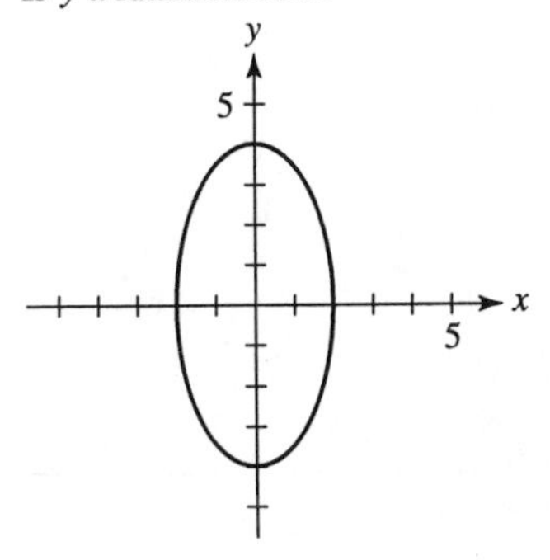

5. ______________________

6. Use function notation to say that the result of applying process h to the number 0.65 is the number −7.7.

6. ______________________

For problems 7–20, use $f(x) = x^3 - 36$ and $g(x) = \sqrt{x} - 10$.

7. Find $f(3)$.

7. ______________________

8. Find $g(-9)$.

8. ______________________

9. Write an expression for $(f + g)(x)$.

9. ______________________

10. Find $(f + g)(5)$.

10. ______________________

11. Write an expression for $(f - g)(x)$.

11. ______________________

12. Find $(f - g)(16)$.

12. ______________________

13. Write an expression for $(fg)(x)$.

13. ______________________

14. Find $(fg)(9)$.

14. ______________________

15. Write an expression for $\left(\frac{f}{g}\right)(x)$.

15. ______________________

Chapter 2 Additional Exercises *(cont.)*

Name:____________________

16. Find $\left(\frac{g}{f}\right)(-25)$.

16. ____________________

17. Find $\left(\frac{g}{f}\right)(4)$.

17. ____________________

18. Write an expression for $(g \circ f)(x)$.

18. ____________________

19. Find $(f \circ g)(81)$.

19. ____________________

20. Find $(g \circ f)(5)$.

20. ____________________

21. What is the domain of the function $f(x) = \sin\left(\frac{1}{x}\right) - x$?

21. ____________________

22. What is the range of the function $f(x) = \sqrt[4]{x-28} + \frac{3}{7}$?

22. ____________________

23. Give the domain of $f(x)$, making a reasonable guess about the portion of the graph not shown.

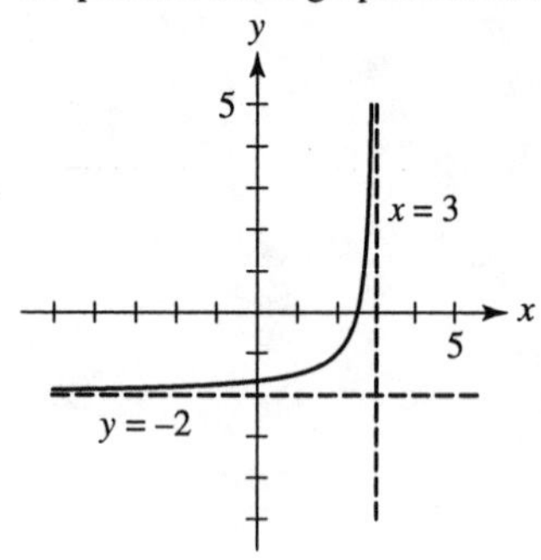

23. ____________________

24. Give the range of $f(x)$, making a reasonable guess about the portion of the graph not shown.

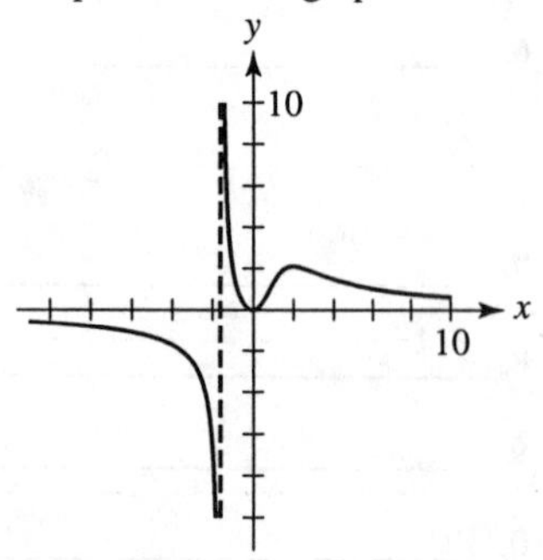

24. ____________________

25. Give the range of $f(x)$, assuming that the table is complete.

x	y
8	2
6	6
0	5
4	5

25. ____________________

26. If $f(x) = \frac{1}{x-10}$ and $g(x) = \sqrt{x}$, what is the domain of $(f \circ g)(x)$?

26. ____________________

Chapter 2 Additional Exercises *(cont.)*

Name:____________________

27. What is the range of $h(x)=\frac{x^4}{16}-1$ if the domain of h is restricted to (–8, 12]?

27. ____________________

28. If $g(x)=\sqrt{x-7}$, what restricted domain will lead to a range of (14, 21]?

28. ____________________

29. Write the first five terms of the sequence for which $a_k=\frac{k+4}{3}$.

29. ____________________

30. Write the first five terms of the sequence for which $a_k=k^2+\frac{k}{4}-13$.

30. ____________________

31. Write the first five terms of the sequence for which $a_1=4$ and $a_k=3a_{k-1}$ for $k\geq 2$.

31. ____________________

32. Write the first five terms of the sequence for which $a_1=-6$, $a_2=5$ and $a_k=a_{k-2}-2a_{k-1}$ for $k\geq 3$.

32. ____________________

33. Using the recursive definition $a_1=8$ and $a_k=a_{k-1}+\frac{1}{3}$ for $k \geq 2$, write an explicit formula for the sequence.

33. ____________________

34. Identify the interval(s) on which $f(x)$ is increasing. Make a reasonable guess about the portion of the graph not shown.

34. ____________________

35. Solve graphically the equation $x+5=\frac{3}{4}x^2$. Give the solution(s) to two decimal places.

35. ____________________

36. Solve graphically the equation $x^2-2=7x^3$. Give the solution(s) to two decimal places.

36. ____________________

37. Solve graphically the inequality $\sqrt{x+1}>-4x^2+9$. Give the solution to two decimal places.

37. ____________________

38. Solve graphically the inequality $x^3-5x\leq -x^3+4x-3$. Give the solution to two decimal places.

38. ____________________

Chapter 2 Additional Exercises *(cont.)*

Name:____________________

39. The minimum number of dice rolls required to score a total of 100 points is a function of the number of dice. Complete the table for 1, 2, 3, and 4 dice.

39.

dice	min. rolls
1	
2	
3	
4	

40. A piece of cardboard can be used to make a cube-shaped box up to 42 cm tall. The volume of the box is a function of its height. What is the range of the volume function?

40. ____________________

41. A computer program keeps track of three variables, "rep," "jcur," and "nprob." nprob is a function of jcur according to the formula nprob = -jcur + 5, and jcur is a function of rep according to the formula jcur = sqrt(rep) + 9. Write a formula describing nprob as a function of rep.

41. ____________________

42. The amount of grass seed needed to seed a lot is a function of the lot's size. Write a general formula if it takes 5.8 bags of seed to cover a 1/2-acre lot.

42. ____________________

43. Two radioactive substances, A and B, decay at different rates. The remaining mass, in grams, is a function of time t in seconds: A's mass is $e^{-0.055t}$ grams, while B's mass is $e^{-0.323t}$ grams. What is ratio of A's remaining mass to B's remaining mass, as a function of time?

43. ____________________

44. The amount of fuel an aircraft needs to travel between two cities is a function of the number of passengers aboard. What would be wrong with modeling this function as Fuel = $C - k \times$ Passengers?

44. ____________________

45. At constant temperature, an ideal gas's pressure and volume are related by the formula $PV = k$. Given this fact, is the pressure a function of the volume, the volume a function of the pressure, both, or neither?

45. ____________________

46. Suppose that a new car loses around 17% of its value the moment it leaves the sales lot and loses 9% of its remaining value each year after that. How much is a car that sold for $24,000 worth after four years? Round your answer to the nearest $50.

46. ____________________

47. The conversion from degrees Kelvin (K) to degrees Fahrenheit (F) involves subtracting 273.15, then multiplying by $\frac{9}{5}$, then adding 32. Express this process as a formula using function notation.

47. ____________________

48. A cannonball is shot straight into the air. It lands after 33.5 seconds. Over what interval was height as a function of time increasing?

48. ____________________

49. A basketball team's first basket is a three-point field goal. After that, every basket is a two-point field goal. Write a recursive formula and an explicit formula for the team's score after k baskets.

49. ____________________

50. A company's profit function is $P(x) = -0.03x^2 + 173x - 4200$, where x is in units sold per month and P is in dollars. Find the interval of integer values of x over which $P >$ \$200,000.

50. ____________________

A Mathematical Looking Glass

ECG

(to follow Section 2-1)

When blood flows through your heart, it enters the upper into the lower chambers (ventricles). The ventricles then contract, pumping the blood to the rest of your body. The contractions are triggered by two electrical impulses from your nervous system.

The time interval between the two electrical impulses is critical. If the second impulse comes too soon after the first, the ventricle contracts before it is filled. If it comes too late, the flow of blood into the body is delayed. In either case, your body is not getting the amount of blood it needs. You may become dizzy or light-headed, or experience a heart attack in extreme cases. Consequently, cardiac patients in hospitals often have electrodes attached to their skin, which send the impulses to a monitoring device. This device produces a graphical model, called an electrocardiogram (ECG), of the relationship between time and the electrical charge in the patient's heart.

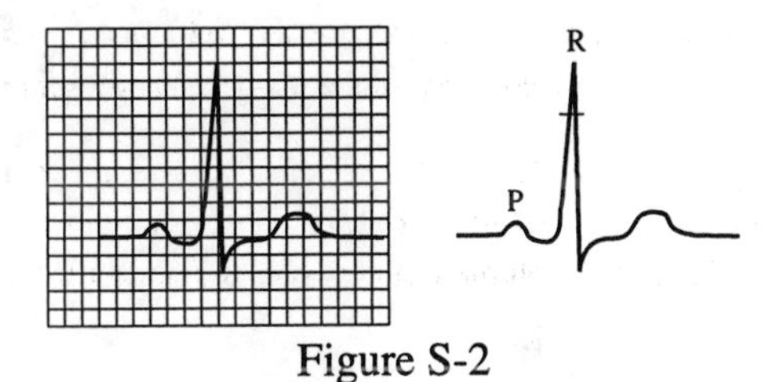

Figure S-2

Figure S-2 is a portion of an ECG made at Allegheny General Hospital, where Dave's wife, Alice, works as a nurse. It conveys important information about the time intervals between impulses, and about the relative sizes of the impulses. A centimeter of horizontal distance (one square) represents 0.2 seconds. A centimeter of vertical distance is usually referred to simply as "a centimeter", since the people who read the graphs are interested only in the relative sizes of the impulses.

In Figure S-2, the point P corresponds to the first impulse, which produced a small electrical charge. The point R corresponds to the second impulse, which produced a larger charge. The horizontal distance between P and R, sometimes called the PR-interval, represents the time between the impulses. Howard Leonhard, a monitor technician at Allegheny General, uses calipers to measure the PR-interval for each heartbeat. Howard looks for significant changes in the length of the interval. A pattern of abnormal intervals indicates that the patient needs assistance.

Exercises:

1. (Writing to Learn) Figure S-3 shows a portion of an ECG set in a coordinate plane, with t measured in seconds and C measured in centimeters.

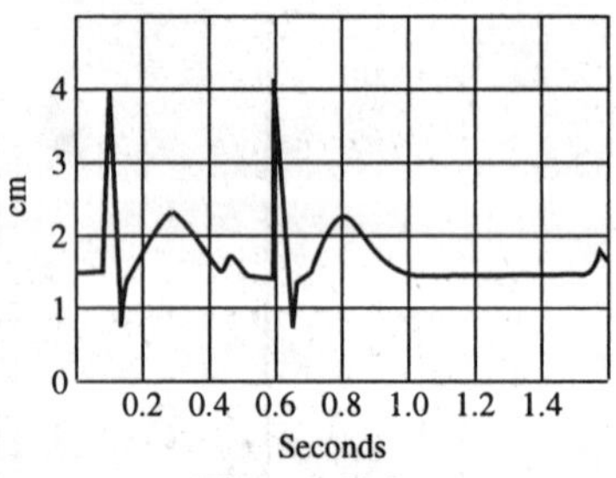

Figure S-3

   a. Explain why the graph shows that the electrical charge C is a function of the time t.

   b. Explain why the graph shows that t is not a function of C.

2. a. Use Figure S-3 to complete the following table.

t	C
0.0	
0.5	
1.0	
1.5	
2.0	

b. (Writing to Learn) How does the table support the conclusion that C is a function of t?

c. (Writing to Learn) Could you conclude that C is a function of t from the table alone? Why or why not?

3. (Writing to Learn)

a. Which model of the ECG would be more helpful to Mr. Leonhard in measuring the PR-interval, the graph in Figure S-3 or the table in Exercise 2?

b. Do you think Mr. Leonhard could measure the PR-interval more effectively if he had access to an equation for the ECG? Why or why not?

A Mathematical Looking Glass
Gas Station
(to follow Section 2-2)

J. C. Baker is ninetysomething years old, and runs a gas station in Gassaway, WV. In a typical month, he sells about 50,000 gallons of gasoline. His profit varies according to the price he must pay to buy his gasoline, but for the purpose of illustration, let's assume that he makes a profit of \$0.009 per gallon on the sale of regular gas, and \$0.014 per gallon on premium. These numbers are typical of actual profits on gasoline sales by retail dealers.

Gassaway is an isolated community, and until recently, J. C.'s station was the only one in town, so his total sales were about the same each month. The only way he could increase his profit was to sell more premium gasoline, and less regular. One possible course of action would be to promote his premium gasoline. In order to determine whether the additional profit would be worth the expense, he would need to analyze the relationship between premium sales and profit.

Exercises

1. (Interpreting Mathematics) Let S represent the amount of premium gasoline that J. C. sells each month, in thousands of gallons.

   a. Tell what physical quantities are represented by each of the following expressions: $14S$, $50 - S$, $9(50 - S)$.

   b. Explain why J. C.'s monthly profit in dollars can be expressed as a function $f(S) = 14S + 9(50 - S)$

2. a. (Creating Models) Let R represent thousands of gallons of regular gasoline sold by J. C. each month. Write an equation to express S as a function $g(R)$.

   b. (Creating Models) Write an equation to express J. C.'s monthly profit as a function $h(R)$.

   c. (Making Observations) If $f(S)$ is defined as in Exercise 1, verify that $h = f \circ g$.

   d. (Writing to Learn) In the language of gasoline and dollars, explain why $h = f \circ g$.

Chapter 2

Section 2-1

1. a. The independent variable if F.

   b. Yes, since each Celsius temperature C corresponds to a unique Fahrenheit temperature F.

3. a. Two rows in the table have a first entry of 8, and different second entries, so y is not a function of x.

   b. No matter what other rows are added to the table, y is not a function of x.

5. $4x+2y=9$
   $2y=9-4x$
   $y=\dfrac{9-4x}{2}$
   For any value of x, the value of $\dfrac{9-4x}{2}$ is uniquely determined, so y is a function of x.

7. $x^4+y=1$
   $y=1-x^4$
   For any value of x, the value of $1-x^4$ is uniquely determined, so y is a function of x.

9. a. $4x+2y=9$
   $x=\dfrac{9-2y}{4}$
   Therefore, x is a function of y.

   b. $0.2x=3-0.1y$
   $x=\dfrac{3-0.1y}{0.2}=15-0.5y$
   Therefore, x is a function of y.

   c. $x^4+y=1$
   $x^4=1-y$
   $x=\pm\sqrt[4]{1-y}$
   Therefore, x is not a function of y.

   d. $x^3+y^2=1$
   $x^3=1-y^2$
   $x=\sqrt[3]{1-y^2}$
   (No $\pm$ is needed, since every real number has only one real cube root.) Therefore, x is a function of y.

11. Since no vertical line intersects the graph more than once, y is a function of x.

13. Many vertical lines intersect the graph more than once, so y is not a function of x.

15. 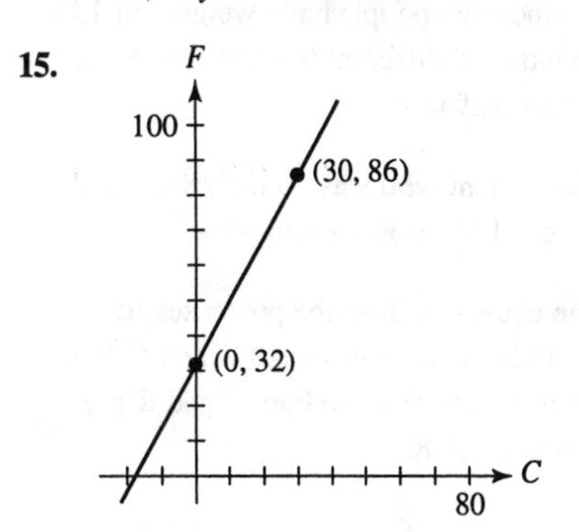

    No vertical line intersects the graph more than once, so F is a function of C. This conclusion agrees with the one obtained in Exercise 10.

17. The equation $x=y^5+y+1$ represents y as a function of x, since no vertical line intersects its graph more than once.

19. a. No two rows in the table have the same first entry, so y is a function of x.

    b. More information is needed to tell whether y is a function of x.

21. a. No two rows in the table have the same first entry, so y is a function of x.

    b. More information is needed to tell whether y is a function of x.

23. $y-2x^5+3x=0$
    $y=2x^5-3x$
    Therefore, y is a function of x.

25. This equation cannot be solved for y.

27. Every value of y corresponds to the value $x = 6$, so y is not a function of x.

29. Since no vertical line intersects the graph more than once, y is a function of x.

31. Many vertical lines intersect the graph more than once, so y is not a function of x.

33. a. Since two people have a height of 70 inches and different weights, w is not a function of h. Since two people have weights of 155 pounds and different heights, h is not a function of w.

b. The student who was 70 inches tall and weighed 155 pounds withdrew.

35. a. The equations describe processes for obtaining unique values of F and C from given values of K, so both F and C are functions of K.

b.

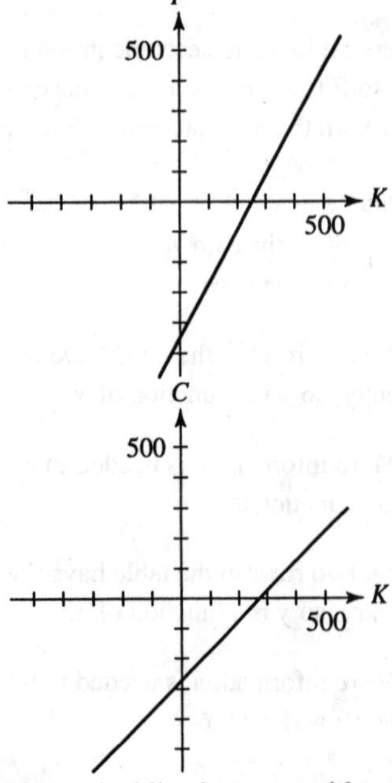

No vertical line intersects either graph more than once, so both F and C are functions of K.

c. $F = \frac{9}{5}K - 459.7$ $\qquad C = K - 273.2$

$\frac{9}{5}K = F + 459.7$ $\qquad K = C + 273.2$

$K = \frac{5}{9}(F + 459.7)$

K is a function of both F and C.

37. Responses will vary.

39. (Sample response) Many tables of physical relationships, such as those found in newspapers and magazines, are not complete tables of a relationship. If such a table fits a function, you need more information before concluding that it was actually generated by a function.

41. Most responses will build on the bulleted paragraphs on page 29.

43. The correspondence can be put into the form of a table:

Artist	Work
Michelangelo	David
Picasso	Guernica
Van Gogh	Self-portrait
Jefferson	Monticello

Since every work corresponds to only one artist, and vice versa, the correspondence is both a function from A to B, and a function from B to A.

Section 2-2

1. $H(7) = 5(7) - 4 = 31$

$H(0) = 5(0) - 4 = -4$

$H(A) = 5A - 4$

$H(A - 1) = 5(A - 1) - 4 = 5A - 9$

3. $H(7) = \frac{7}{7+1} = \frac{7}{8}$

$H(0) = \frac{0}{0+1} = 0$

$H(A) = \frac{A}{A+1}$

$H(A-1) = \frac{A-1}{(A-1)+1} = \frac{A-1}{A}$

5. a. $g(41) = \frac{5}{9}(41-32) = 5$

This represents the Celsius temperature corresponding to 41°F.

b. $g(-40) = \frac{5}{9}(-40-32) = -40$

This says that –40°F and –40°C are the same temperature.

c. $g(86) = 30$

7. $(f+g)(x) = (x^2-2)+(4-x) = x^2-x+2$

$(f-g)(x) = (x^2-2)-(4-x) = x^2+x-6$

$(fg)(x) = (x^2-2)(4-x)$

$= -x^3+4x^2+2x-8$

$\left(\frac{f}{g}\right)(x) = \frac{x^2-2}{4-x}$

9. $(f+g)(x) = \sqrt{t-1}+\sqrt{t^2-1}$

$(f-g)(x) = \sqrt{t-1}-\sqrt{t^2-1}$

$(fg)(x) = \left(\sqrt{t-1}\right)\left(\sqrt{t^2-1}\right) = \sqrt{(t-1)(t^2-1)}$

$= \sqrt{(t-1)(t-1)(t+1)} = |t-1|\sqrt{t+1}$

$\left(\frac{f}{g}\right)(x) = \frac{\sqrt{t-1}}{\sqrt{t^2-1}} = \sqrt{\frac{t-1}{t^2-1}} = \sqrt{\frac{t-1}{(t-1)(t+1)}}$

$= \sqrt{\frac{1}{t+1}}$ if $t \neq 1$.

11. If $p(x) = 16 - 0.005x$, then

$p(2500) = 3.50$

$p(2100) = 5.50$

$p(1700) = 7.50$

$p(1300) = 9.50$

$p(900) = 11.50$

13. $(U \circ f)(h) = U[f(h)] = U\left(\frac{2h}{h-100}\right)$

$= 3\left(\frac{2h}{h-100}\right)+100 = \frac{6h}{h-100}+100$

$= \frac{6h}{h-100}+\frac{100(h-100)}{h-100} = \frac{106h-10{,}000}{h-100}$

14. $(u \circ v)(x) = u[v(x)] = u(5x) = 2(5x)+1$

$= 10x+1$

$(v \circ u)(x) = v[u(x)] = v(2x+1) = 5(2x+1)$

$= 10x+5$

15. $(u \circ v)(x) = u[v(x)] = u(x^2) = 8-4(x^2)$

$= 8-4x^2$

$(v \circ u)(x) = v[u(x)] = v(8-4x) = (8-4x)^2$

$= 64-64x+16x^2$

17. $(u \circ v)(x) = u[v(x)] = u\left(\frac{1}{x}\right) = \left(\frac{1}{x}\right)^2 - 3\left(\frac{1}{x}\right)$

$= \frac{1}{x^2}-\frac{3}{x}\ \left(\text{or } \frac{1-3x}{x^2}\right)$

$(v \circ u)(x) = v[u(x)] = v(x^2-3x) = \frac{1}{x^2-3x}$

19. $x \rightarrow x^2 \rightarrow x^2+3 \rightarrow 5(x^2+3)$

$f(x) = 5(x^2+3)$

21. $x \rightarrow x+3 \rightarrow 5(x+3) \rightarrow [5(x+3)]^2$

$f(x) = [5(x+3)]^2$

23. the multiplying-by-six function, the subtracting-two function, the square-root function

25. the square-root function, the multiplying-by-six function, the subtracting-two function

27. $T(50) = 2$

29. a. h represents the process of multiplying the input by 20, and adding 13 to the result.

b. $h(4) = 20(4) + 13 = 93$

$h(0) = 20(0) + 13 = 13$

$h(-0.5) = 20(-0.5) + 13 = 3$

c. $h(3Z) = 20(3Z) + 13 = 60Z + 13$

d. $(h+g)(x) = 20x+13+x = 21x+13$

$(h-g)(x) = 20x+13-x = 19x+13$

$(hg)(x) = (20x+13)(x) = 20x^2+13x$

$\left(\frac{h}{g}\right)(x) = \frac{20x+13}{x}$

e. $(h \circ f)(x) = h[f(x)] = h(x+4)$

$= 20(x+4)+13 = 20x+93$

$(f \circ h)(x) = f[h(x)] = f(20x+13)$

$= 20x+13+4 = 20x+17$

31. a. h represents the process of subtracting 4 times the square of the input from 100 times the input.

b. $h(4) = 100(4) - 4(4)^2 = 336$
$h(0) = 100(0) - 4(0)^2 = 0$
$h(-0.5) = 100(-0.5) - 4(-0.5)^2 = -51$

c. $h(3Z) = 100(3Z) - 4(3Z)^2 = 300Z - 36Z^2$

d. $(h+g)(x) = 100x - 4x^2 + x = 101x - 4x^2$
$(h-g)(x) = 100x - 4x^2 - x = 99x - 4x^2$
$(hg)(x) = (100x - 4x^2)(x) = 100x^2 - 4x^3$
$\left(\frac{h}{g}\right)(x) = \frac{100x - 4x^2}{x} = 100 - 4x$ if $x \neq 0$.

e. $(h \circ f)(x) = h[f(x)] = h(x+4)$
$= 100(x+4) - 4(x+4)^2$
$= -4x^2 + 68x + 336$
$(f \circ h)(x) = f[h(x)] = f(100x - 4x^2)$
$= (100x - 4x^2) + 4 = -4x^2 + 100x + 4$

33. a. h represents the process of dividing the input by the sum of the input and 1.

b. $h(4) = \frac{4}{4+1} = \frac{4}{5}$
$h(0) = \frac{0}{0+1} = 0$
$h(-0.5) = \frac{-0.5}{-0.5+1} = -1$

c. $h(3Z) = \frac{3Z}{3Z+1}$

d. $(h+g)(x) = \frac{x}{x+1} + x$
$= \frac{x}{x+1} + \frac{x(x+1)}{x+1} = \frac{x + x(x+1)}{x+1}$
$= \frac{x^2 + 2x}{x+1}$
$(h-g)(x) = \frac{x}{x+1} - x$
$= \frac{x}{x+1} - \frac{x(x+1)}{x+1} = \frac{x - x(x+1)}{x+1}$
$= \frac{-x^2}{x+1}$
$(hg)(x) = \left(\frac{x}{x+1}\right)(x) = \frac{x^2}{x+1}$
$\left(\frac{h}{g}\right)(x) = \frac{x}{x+1} \div x = \frac{x}{x+1} \cdot \frac{1}{x}$
$= \frac{1}{x+1}$ if $x \neq 0$

e. $(h \circ f)(x) = h[f(x)] = h(x+4)$
$= \frac{(x+4)}{(x+4)+1} = \frac{x+4}{x+5}$
$(f \circ h)(x) = f[h(x)] = f\left(\frac{x}{x+1}\right)$
$= \frac{x}{x+1} + 4 = \frac{x}{x+1} + \frac{4x+4}{x+1} = \frac{5x+4}{x+1}$

35. a. h represents the process of doubling the input, adding 10 to the result, and taking the square root of that result.

b. $h(4) = \sqrt{2(4)+10} = \sqrt{18} = 3\sqrt{2}$
$h(0) = \sqrt{2(0)+10} = \sqrt{10}$
$h(-0.5) = \sqrt{2(-0.5)+10} = \sqrt{9} = 3$

c. $h(3Z) = \sqrt{6Z+10}$

d. $(h+g)(x) = \sqrt{2x+10} + x$
$(h-g)(x) = \sqrt{2x+10} - x$
$(hg)(x) = x\sqrt{2x+10}$
$\left(\frac{h}{g}\right)(x) = \frac{\sqrt{2x+10}}{x}$

e. $(h \circ f)(x) = h[f(x)]$
$= h(x+4) = \sqrt{2(x+4)+10} = \sqrt{2x+18}$
$(f \circ h)(x) = f[h(x)]$
$= f\left(\sqrt{2x+10}\right) = \sqrt{2x+10} + 4$

37. a. h represents the process of associating each input with itself.

b. $h(4) = 4$
$h(0) = 0$
$h(-0.5) = -0.5$

c. $h(3Z) = 3Z$

d. $(h+g)(x) = x + x = 2x$
$(h-g)(x) = x - x = 0$
$(hg)(x) = (x)(x) = x^2$
$\left(\frac{h}{g}\right)(x) = \frac{x}{x} = 1$ if $x \neq 0$

e. $(h \circ f)(x) = h[f(x)] = h(x+4) = x+4$
$(f \circ h)(x) = f[h(x)] = f(x) = x+4$

39. $(v \circ u)(P) = v[u(P)] = v[(2P-6)^3 - 5(2P-6)^2 + 4(2P-6) - 7]$
$= 0.5[(2P-6)^3 - 5(2P-6)^2 + 4(2P-6) - 7] + 3$
$= 0.5[(8P^3 - 72P^2 + 216P - 216) - 5(4P^2 - 24P + 36) + 4(2P-6) - 7] + 3$
$= 0.5(8P^3 - 92P^2 + 344P - 427) + 3$
$= 4P^3 - 46P^2 + 172P - 210.5$
$(u \circ v)(Q) = u[v(Q)] = u(0.5Q + 3)$
$= \{[2(0.5Q+3) - 6]^3 - 5[2(0.5Q+3) - 6]^2 + 4[2(0.5Q+3) - 6] - 7\}$
$= \{[(Q+6) - 6]^3 - 5[(Q+6) - 6]^2 + 4[(Q+6) - 6] - 7\}$
$= Q^3 - 5Q^2 + 4Q - 7$

41. $x \to x^3 \to x^3 - 5 \to \frac{x^3 - 5}{2}$
$f(x) = \frac{x^3 - 5}{2}$

43. $x \to x - 5 \to \frac{x-5}{2} \to \left(\frac{x-5}{2}\right)^3$
$f(x) = \left(\frac{x-5}{2}\right)^3$

45. the multiplying-by-three function, the taking-the-reciprocal function, the adding-one function

47. the multiplying-by-three function, the adding-one function, the taking-the-reciprocal function

49. a. $(q \circ p)(A)$ does not make sense in the physical context.

b. $(p \circ q)(D)$ represents the air pressure in pounds per square inch at a distance of D miles west of St. Louis.

51. a. $\left(\frac{f}{g}\right)(1) = \frac{f(1)}{g(1)} = \frac{3}{0}$, which is undefined.

b. $(f \circ g)(4) = f[g(4)] = f(8)$, which is undefined.

53. a. $u(T+5) = \frac{9(T+5)}{5} + 32 = \frac{9T+45}{5} + 32$
$= \frac{9T}{5} + \frac{45}{5} + 32 = \frac{9T}{5} + 41$

b. The Fahrenheit temperature is
$\frac{9(23)}{5} + 41 = 82.4°$.

c. $v(T+5) = (T+5) + 273.2 = T + 278.2$

d. The Kelvin Temperature is
$(23) + 278.2 = 301.2°$.

55. $(u \circ c)(T)$ represents the Fahrenheit temperature of the water after T seconds, and $(u \circ c)(60)$ represents the Fahrenheit temperature of the water after 1 minute.

57. a. $f(P) = P - (15\% \text{ of } P) = P - 0.15P = 0.85P$

b. $g(P) = P - 1000$

c. $(f \circ g)(P) = f[g(P)] = f(P - 1000)$
$= 0.85(P - 1000) = 0.85P - 850$
$(g \circ f)(P) = g[f(P)] = g(0.85P)$
$= 0.85P - 1000$
Regardless of the original price, the value of $(g \circ f)(P)$ is always \$150 less than the value of $(f \circ g)(P)$, and therefore results in the lower price. The composition $g \circ f$ is obtained by applying the discount first.

59. The cost of producing x cakes per week is $1.50x$ dollars for ingredients, plus \$212.50 of Chris' salary, so Periwinkle's cost function is
$C(x) = 1.50x + 212.50$.
The revenue from the sale of x cakes is $10x$ dollars, so Periwinkle's revenue function is
$R(x) = 10x$.
Periwinkle's profit function is
$P(x) = R(x) - C(x) = 10x - (1.50x + 212.50)$
$= 8.50x - 212.50$.

Section 2-3

1. domain $= (-\infty, \infty)$; range $= [-5, \infty)$

3. domain $= (-\infty, \infty)$; range $= (0, 1]$

5. The function is never undefined, so its domain is $(-\infty, \infty)$. To find the range analytically, solve for x:

$y = 3x + 24$
$y - 24 = 3x$
$x = \frac{1}{3}(y - 24)$

The resulting expression in y is never undefined, so the range of the original function is $(-\infty, \infty)$.

7. $Q(x)$ is undefined when

$3x + 24 < 0$
$x < -8$

The domain is therefore $[-8, \infty)$.
To find the range analytically, observe that $\sqrt{3x+24} \geq 0$ for any x in the domain of Q, so that $Q(x) \geq 0$. The range of Q is therefore $[0, \infty)$.

9. **a.**

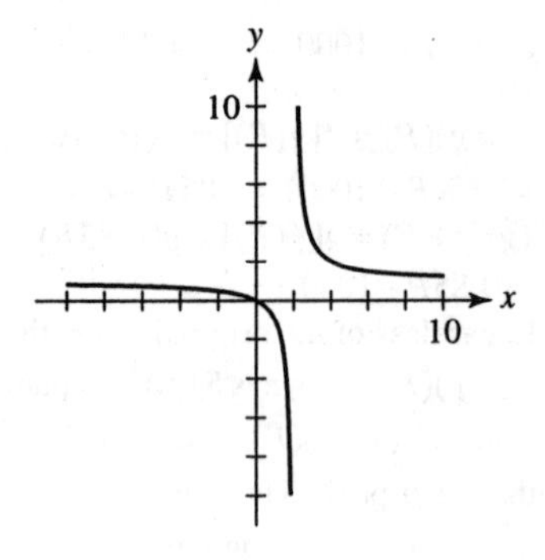

$f(x) = \frac{x}{x-2}$, $[-10, 10]$ by $[-10, 10]$

b.

x	$f(x)$
0	0
1	−1
2	undefined
3	3
4	2
5	$\frac{5}{3}$

c. The domain of f is $(-\infty, 2) \cup (2, \infty)$. The limitation $x \neq 2$ shows up in the equation as a factor of $x - 2$ in the denominator, in the graph as a vertical asymptote at $x = 2$, and in the table as an undefined y-value opposite $x = 2$.

11. The range is $(1, 9)$.

13. The range is $(-\infty, 2]$.

15. **a.** 0.000000078

b. 0.00078

c. 0.49

d. 0.4999999999

17. If the domain is restricted to $[0, \infty)$, the graph indicates a range of $[0, 0.5)$. This agrees with the conclusion of Exercise 16c.

19. 2, 4, 6, 8, 10, 12, 14, 16, 18, 20

21. 2, 0, −4, −10, −18, −28, −40, −54, −70, −88

23. 1, 2, 6, 24, 120, 720, 5040, 40320, 362880, 3628800

25. 1024, 0, 512, 256, 384, 320, 352, 336, 344, 340

27. **a.** 3, 7, 11, 15, 19, 23, 27, 31, 35, 39

b. $a_n = 4n - 1$, $a_{1000} = 3999$

29. The function is increasing on $(-\infty, 2)$, and decreasing on $(2, \infty)$.

31. The function is increasing on $(-\infty, \infty)$.

33. domain = $\{0, 1, 2, 3\}$; range = $\{0, 1, 2, 3\}$

35. domain = $\{-2, 0, 2, 4, 6\}$;
range = $\{-1, 2, 5, 8, 11\}$

37. domain = $(-\infty, \infty)$; range = $(-\infty, 2]$

39. domain = $[-5, 5]$; range = $[0, 3]$

41. The domain and range are each $(-\infty, \infty)$

43. The function is undefined when

$x - 2 = 0$
$x = 2$

Its domain is therefore $(-\infty, 2) \cup (2, \infty)$. The graph indicates a range of $(-\infty, 3) \cup (3, \infty)$.

45. The domain is $(-\infty, \infty)$.
To find the range, solve for x:
$z = 3x + 600$
$3x = z - 600$
$x = \frac{1}{3}(z - 600)$
Since the expression in z is defined for all values of z, the range is $(-\infty, \infty)$.

47. The function is undefined when $Z - 2 < 0$, that is, when $Z < 2$. The domain is therefore $[2, \infty)$.
To find the range, observe that $\sqrt{Z-2} \geq 0$ for all Z in the domain of the function, so that $2 + \sqrt{Z-2} \geq 2$. The range is therefore $[2, \infty)$.

49. The range is $[-1, 3]$.

51. The range is $[2, \infty)$.

53. The domain is $(-\infty, \infty)$.
To find the range, solve for x:
$y = x^2$
$x = \pm\sqrt{y}$
Since $\sqrt{y}$ is defined only when $y \geq 0$, the range is $[0, \infty)$.

55. The function is undefined when $x^4 = 0$, that is when $x = 0$. The domain is therefore $(-\infty, 0) \cup (0, \infty)$.
To find the range, observe that $x^4 > 0$ for any x in the domain the function, so that $\frac{1}{x^4} > 0$ as well. The range is therefore $(0, \infty)$.

57. The function is undefined when $1 - x^2 < 0$, that is, when $x^2 > 1$. This occurs when $x < -1$ or $x > 1$. The domain of the function is therefore $[-1, 1]$.
To find the range, observe that $\sqrt{1-x^2} \geq 0$ for any x in the domain. Also $\sqrt{1-x^2} \leq \sqrt{1} = 1$ for any x in the domain, since $1 - x^2$ cannot exceed 1. The range is therefore $[0, 1]$.

59. The domain is $(-\infty, \infty)$. Since $x^4 \geq 0$ for any x, the range is $[0, \infty)$.

61. The radical is undefined when $1 - x^2 < 0$, that is, when $x^2 > 1$, or equivalently, when $x < -1$ or $x > 1$. The denominator is zero when $1 - x^2 = 0$, that is, when $x = \pm 1$. The domain is therefore $(-1, 1)$.
A graph indicates a range of $[1, \infty)$.

63. The domain is $[-1, 4]$. A graph indicates a range of $[0, 65]$.

65. The pulley diameter fan speed must both be positive. Furthermore, the fan speed can be no more than 3600 rpm, so the pulley diameter must be no more than $\frac{3600}{340} \cong 10.58$ inches. The domain is therefore $(0, 10.58]$, and the range is $(0, 3600]$.

67. Absolute zero is –273.2°C and –459.7°F, so the domain is $(-273.2, \infty)$, and the range is $(-459.7, \infty)$.

69. The length of the pasture must be positive, and must be less than 2000 feet. The domain is therefore $(0, 2000)$.

A graph indicates a range of $(0, 1{,}000{,}000)$.

71. 4, 3, 2, 1, 0, –1, –2, –3, –4, –5

73. 1, 8, 27, 64, 125, 216, 343, 512, 729, 1000

75. 1, 0.1, 0.01, 0.001, 0.0001, 0.00001, 0.000001, 0.0000001, 0.00000001, 0.000000001

77. 1, –1, 1, –1, 1, –1, 1, –1, 1, –1

79. The function is increasing on $(-\infty, \infty)$.

81. The function is increasing on $(-\infty, -1)$, and $(2, \infty)$, and decreasing on $(-1, 2)$.

83. $\frac{(\sqrt{5}+1)^2 - (\sqrt{5}-1)^2}{2^2\sqrt{5}} = 1$
$\frac{(\sqrt{5}+1)^4 - (\sqrt{5}-1)^4}{2^4\sqrt{5}} = 3$
$\frac{(\sqrt{5}+1)^6 - (\sqrt{5}-1)^6}{2^6\sqrt{5}} = 8$

85. **a.** If $a_{n+1} = a_n + a_{n-1}$, then $a_4 = 3$.
If $a_{n+1} = a_1 + a_2 + \cdots a_{n-1}$, then $a_4 = 4$.

b. If $a_{n+1} = a_n + 1$, then $a_4 = 4$.
If $a_{n+1} = a_n + a_{n-1}$, then $a_4 = 5$.

c. If $a_{n+1} = (n+1)a_n$, then $a_4 = 24$.
If $a_{n+1} = a_{n-2}$, then $a_4 = 1$.

d. If $a_{n+1} = a_{n-1} - a_n$, then $a_4 = -3$.
If $a_{n+1} = (a_n)^2 + (a_{n-1})^2$, then $a_4 = 5$.

87. Responses will vary.

89. Many graphs are possible.

91. **a.** The domain of f is $(-\infty, \infty)$.

b. The domain of g is $(-\infty, \infty)$.

c–e. $(f+g)(x) = (3x-2)+(x^2+x-6)$
$= x^2+4x-8$
$(f-g)(x) = (3x-2)-(x^2+x-6)$
$= -x^2+2x+4$
$(fg)(x) = (3x-2)(x^2+x-6)$
$= 3x^3+x^2-20x+12$
The domain of each is $(-\infty, \infty)$.

f. $\left(\frac{f}{g}\right)(x) = \frac{3x-2}{x^2+x-6}$
The expression is undefined when
$x^2+x-6=0$
$(x+3)(x-2)=0$
$x=-3, 2$
The domain of $\frac{f}{g}$ is therefore
$(-\infty, -3)\cup(-3, 2)\cup(2, \infty)$.

g. $(f\circ g)(x) = f[g(x)] = f(x^2+x-6)$
$= 3(x^2+x-6)-2 = 3x^2+3x-20$
The domain of $f\circ g$ is $(-\infty, \infty)$.

h. $(g\circ f)(x) = g[f(x)]$
$= g(3x-2) = (3x-2)^2+(3x-2)-6$
$= (9x^2-12x+4)+(3x-2)-6$
$= 9x^2-9x-4$
The domain of $g\circ f$ is $(-\infty, \infty)$.

93. **a.** The function f is undefined when $x<0$. The domain of f is therefore $[0, \infty)$.

b. The function g is undefined when $4-x<0$, that is, when $x>4$. The domain of g is therefore $(-\infty, 4]$.

c–e. $(f+g)(x) = \sqrt{x}+\sqrt{4-x}$
$(f-g)(x) = \sqrt{x}-\sqrt{4-x}$
$(fg)(x) = \sqrt{x}\cdot\sqrt{4-x} = \sqrt{x(4-x)}$
Each is undefined when either radicand is undefined, that is, if $x<0$ or $x>4$. The domain of each is therefore $[0, 4]$.

f. $\left(\frac{f}{g}\right)(x) = \frac{\sqrt{x}}{\sqrt{4-x}} = \sqrt{\frac{x}{4-x}}$
The function $\frac{f}{g}$ is undefined when either radicand is undefined. It is also undefined when $4-x=0$, that is, when $x=4$. The domain of $\frac{f}{g}$ is therefore $[0, 4)$.

g. $(f\circ g)(x) = f[g(x)] = f(\sqrt{4-x})$
$= \sqrt{\sqrt{4-x}} = \sqrt[4]{4-x}$
The function $f\circ g$ is undefined when $4-x<0$, that is, when $x>4$. Its domain is $(-\infty, 4]$.

h. $(g\circ f)(x) = g[f(x)] = g(\sqrt{x}) = \sqrt{4-\sqrt{x}}$
The function $g\circ f$ is undefined when $x<0$. It is also undefined when $4-\sqrt{x}<0$, that is, when $\sqrt{x}>4$, or equivalently, when $x>16$. The domain of $g\circ f$ is therefore $[0, 16]$.

95. **a.** The function f is undefined when $x<0$. Its domain is $[0, \infty)$.

b. The function g is undefined when $x-2=0$, that is, when $x=2$. Its domain is $(-\infty, 2)\cup(2, \infty)$.

c–e. $(f+g)(x)=\sqrt[4]{x}+\frac{1}{x-2}$

$(f-g)(x)=\sqrt[4]{x}-\frac{1}{x-2}$

$(fg)(x)=\frac{\sqrt[4]{x}}{x-2}$

Each is undefined whenever $f(x)$ or $g(x)$ is undefined. The domain of each is $[0, 2)\cup(2, \infty)$.

f. $\left(\frac{f}{g}\right)(x)=\frac{\sqrt[4]{x}}{\frac{1}{x-2}}=\sqrt[4]{x}(x-2)$ if $x\neq 2$.

The domain of $\frac{f}{g}$ is $[0, 2)\cup(2, \infty)$.

g. $(f\circ g)(x)=f[g(x)]=f\left(\frac{1}{x-2}\right)=\sqrt[4]{\frac{1}{x-2}}$

The function $f\circ g$ is undefined when $x=2$. It is also undefined when $\frac{1}{x-2}<0$, or equivalently, when $x-2<0$, that is, when $x<2$. The domain of $f\circ g$ is $(2, \infty)$.

h. $(g\circ f)(x)=g[f(x)]=g(\sqrt[4]{x})=\frac{1}{\sqrt[4]{x}-2}$

The function $g\circ f$ is undefined when $x<0$. It is also undefined when $\sqrt[4]{x}-2=0$, that is, when $\sqrt[4]{x}=2$, or equivalently, when $x=16$. The domain of $g\circ f$ is $[0, 16)\cup(16, \infty)$.

97. The domains of $f+g$, $f-g$, and fg each consist of the x-values common to the domains of f and g. The domain of $\frac{f}{g}$ consists of those x-values common to the domains of f and g, for which $g(x)\neq 0$.

99. In Exercise 93, the domain of $g\circ f$ is $[0, 16]$. This is contained in the domain of f, which is $[0, \infty)$. A graph indicates that the range of $g\circ f$ is $[0, 2]$. This is contained in the range of g, which is $[0, \infty)$.

Section 2-4

1. The third x-intercept on the graph of $y=x^3-2x^2-10x+8$ occurs at about (0.73, 0), so the solution is $x\cong 0.73$.

3. The x-intercepts on the graph of $y=x^2+x-10$ occur at about (–3.70, 0) and (2.70, 0), so the solutions are $x\cong -3.70,\ 2.70$.

5. The x-intercept on the graph of $y=x^3-34x^2+408x-1728$ occurs at about (12, 0). Since $(12)^3-34(12)^2+408(12)-1728=0$, the solution is exactly $x=12$.

7. The inequality is true when the graph of $y=x^2+x-10$ is on or below the x-axis. The solution is [–3.70, 2.70].

9. The inequality is true when the graph of $y=x^3-34x^2+408x-1728$ is on or above the x-axis. The solution is $[12, \infty)$.

11. The x-intercepts on the graph of $y=x^2-4x-5$ occur at (–1, 0) and (5, 0), so the solutions of the equation are $x=-1, 5$. The inequality is true when the graph is below the x-axis, so its solution is (–1, 5).

13. The x-intercept on the graph of $y=x^3-4$ occurs at about (1.59, 0), so the solution of the equation is $x\cong 1.59$. The inequality is true when the graph is on or above the x-axis, so its solution is about $[1.59, \infty)$.

15. The graph of $y=x^2-2x+2$ has no x-intercepts, so the equation has no solution. The inequality is true when the graph is above the x-axis, so its solution is $(-\infty, \infty)$.

17. The x-intercepts on the graph of $y=25x^2-x^4$ occur at (–5, 0), (0, 0), and (5, 0), so the solutions of the equation are $x=-5, 0, 5$. The inequality is true when the graph is on or above the x-axis, so its solution is [–5, 5].

19. The x-intercepts on the graph of $y=|2x-5|-3$ occur at (1, 0) and (4, 0), so the solutions of the equation are $x=1, 4$. The inequality is true when the graph is on or below the x-axis, so its solution is [1, 4].

21. a.

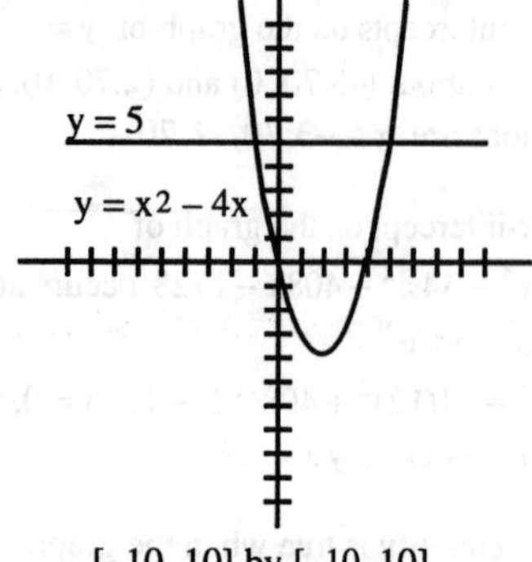

[–10, 10] by [–10, 10]

b. The two graphs intersect at a particular x-value exactly when the corresponding y-value is the same on both graphs. Since $y = x^2 - 4x$ on the first graph, and $y = 5$ on the second graph, the two graphs intersect at exactly those points for which $x^2 - 4x = 5$.

c. The solutions are $x = -1, 5$. This result agrees with that in Exercise 11.

23. See Exercise 12 for the solution.

25. See Exercise 14 for the solution.

27. The graph of $y = x^2$ intersects the graph of $y = x + 2$ at (–1, 1) and (2, 4), so the solutions of the equation are $x = -1, 2$. The inequality is true when the graph of $y = x^2$ is below the graph of $y = x + 2$, so the solution is (–1, 2).

29. a. Periwinkle's break-even point is the solution to the equation $P(x) = 0$. The x-intercept on the graph of $y = P(x)$ occurs at (25, 0), so Periwinkle's can break even by selling 25 cakes per week.

b. The graph of $y = P(x)$ is above the x-axis in $(25, \infty)$, so Periwinkle's can make a profit by selling more than 25 cakes per week.

NUTSHELLS

Nutshell for Chapter 3 - LINEAR FUNCTIONS

What's familiar?

- You have probably seen linear functions before. You may not have *called* them functions, so read Section A-6 in the Basic Algebra Review to refresh your memory.
- We continue to look at functions dynamically.
- We continue our practice of exploring mathematics in a physical context.
- Section 3-1 focuses on the familiar idea of *slope*. This discussion looks at slope from numerical, analytical and graphical perspectives.
- Section 3-2 introduces you to the fact that arithmetic sequences are linear functions and therefore have some of the same properties.
- Section 3-4 show you how to use your graphical equation-solving methods from Chapter 2 to solve absolute value equations and methods from Chapter 2 to solve absolute value equations and inequalities. Review the analytical methods in Section A-9, also.
- Section 3-4 returns to the idea of continuity.

What's new?

- When we take the dynamic view of functions, the *slope* becomes the *rate* at which a linear function increases or decreases. You will discover what physical information this *rate of change* can provide.
- Section 3-3 provides insight as to why you might want to use a linear function to model a nonlinear relationship.
- Section 3-4 introduces you to functions that are put together from linear pieces (*piecewise* linear functions). They are discussed from the three perspectives: analytical, numerical, and graphical.
- Section 3-4 discusses how to *transform* the basic graph of *linear absolute value* function, $y = |x|$, to the graph of $y = a|x - h| + b$ by shifting, flipping, compressing, and/or stretching it. The idea of obtaining graphs by transforming a basic graph will recur several times in later chapters.

List of hints for exercises in each chapter.

Chapter 3

- Prerequisites are especially important in this chapter. You should be familiar with Sections A-6 through A-9.
- Focus on understanding that *slope* is *average rate* of change and how this is expressed numerically, graphically and analytically.
- Pay particular attention to the techniques for graphical transformations of absolute value functions. These techniques are repeated several times throughout the textbook.
- Many problems in Section 3-2 and 3-3 ask you to create models and solve problems. Be sure you apply the Polya strategies from Chapter 1.

Name:______________

Chapter 3 Additional Exercises

Date:______________

1. Find $\frac{\Delta y}{\Delta x}$ for the values of x and y in the second and third rows of the table.

x	y
–4	$\frac{5}{3}$
–2	3
5	8
9	12

1. ______________

2. Use the linear function $f(x) = -\frac{5}{3}x + 6$ to fill in the table.

2.

x	$f(x)$
–6	
0	
9	
16	

3. Decide whether or not the table fits a linear function.

x	y
–5	$\frac{1}{3}$
–2	1
8	6
11	$7\frac{1}{3}$

3. ______________

4. Calculate the average rate at which $f(x)$ changes with respect to x over the interval [4, 8].

x	$f(x)$
2	–5
4	20
5	26
8	–12

4. 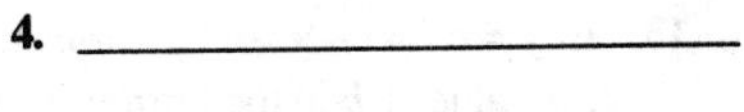

5. Find the slope of the line passing through (–65, 42) and (19, –13).

5. 

6. Graph a linear function passing through the points $\left(3, \frac{3}{2}\right)$ and $\left(-\frac{7}{2}, -\frac{3}{4}\right)$.

6. 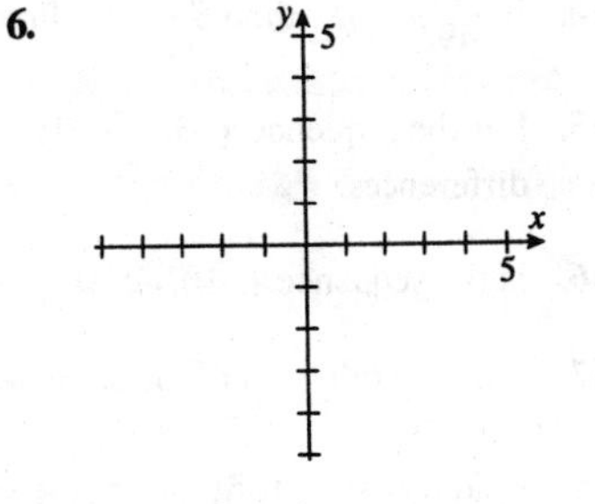

Chapter 3 Additional Exercises *(cont.)*

Name:_______________

7. Find the rate of change in y with respect to x in the graph.

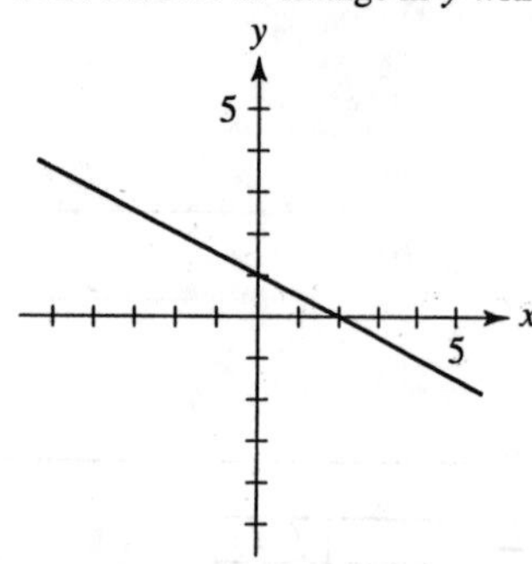

7. _______________

8. Write an equation for a linear function whose graph passes through the points $\left(\frac{3}{5}, 1\right)$ and $\left(\frac{8}{5}, \frac{2}{3}\right)$.

8. _______________

9. Rewrite the linear equation $13x - 27 = \frac{4+y}{9}$ in $y = mx + b$ form and find the slope.

9. _______________

10. Finish the table, given that it fits a linear function.

10.

x	y
4	5
	8
−2	9
11	

11. If possible, rewrite $\frac{y-4}{6} = x - \frac{2}{3}$ as an equation stating that y is directly proportional to x.

11. _______________

12. If $4y = 12x$, what is the constant of proportionality in x's direct variation with y?

12. _______________

13. If $y = c + 4x - 9$ and y varies linearly with x, what is the value of b in the form $y = mx + b$?

13. _______________

14. If $\frac{t^2}{49S} = 7t^5$ then S varies directly as ________.

14. _______________

15. For the sequence {−3, 5, 6, 19, ...}, find the first three first differences.

15. _______________

16. Is the sequence {−10, −5, 0, 5, 10, ...} arithmetic?

16. _______________

17. Find the 6th term of the sequence {17.6, 16.7, 15.8, 14.9, ...}.

17. _______________

18. Find a general formula for the sequence $\left\{\frac{5}{3}, \frac{13}{4}, \frac{29}{6}, \frac{77}{12}, 8, \ldots\right\}$.

18. _______________

19. Find a_{115} for the sequence whose formula is $a_n = -3 + \frac{1}{5}n$.

19. _______________

20. Find a general formula for a_n, given that $a_1 = -17$ and the constant difference is $\frac{2}{5}$.

20. _______________

Chapter 3 Additional Exercises *(cont.)*

Name:____________________

21. Find the constant difference of the sequence whose formula is $a_n = 11 - 42n$.

21. ____________________

22. Use the table to estimate *f*(17) by constructing a linear equation.

x	$f(x)$
5	12.7
10	20.3
15	22.3
20	23.8

22. ____________________

23. Find the average rate of change over the interval [–3, 6] of the function $f(x) = 2x^2 - \frac{x}{3} + 1$.

23. ____________________

24. Find the average rate of change over the interval [–4, 0] of the function whose graph is shown.

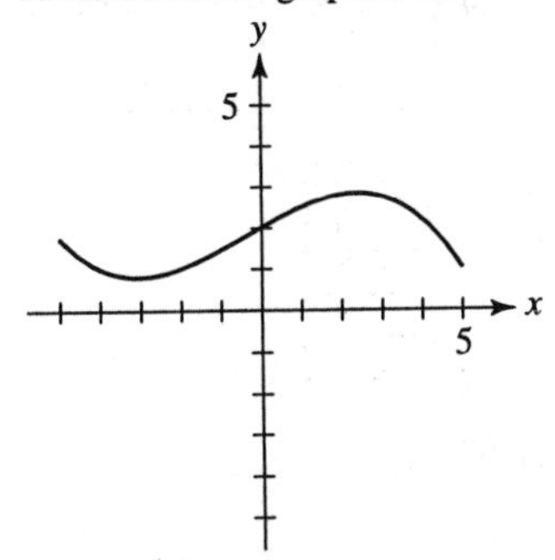

24. ____________________

25. Approximate the function $f(x) = 0.01x^2 + \frac{x}{5} - 2$ over the interval [11.5, 12.2] using a linear function. Round all constants to two decimal places.

25. ____________________

26. For the function $f(x) = x^2 + 2x - 5$, find the difference between $\frac{\Delta f(x)}{\Delta x}$ for the interval [–3, –1] and $\frac{\Delta f(x)}{\Delta x}$ for the interval [5, 7].

26. ____________________

27. Graph the piecewise linear function

$$f(x) = \begin{cases} 2x + 1 & \text{if } 0 \le x \le 2 \\ -2 - \frac{x}{2} & \text{if } -2 \le x < 0 \end{cases}$$

27.

y 5

x 5

28. Complete the table for the piecewise defined function

$$f(x) = \begin{cases} 2x - 16 & \text{if } 80 \le x < 120 \\ x - 2 & \text{if } 120 \le x \end{cases}$$

28.

x	$f(x)$
	138
90	
120	
	140

Chapter 3 Additional Exercises *(cont.)*

Name:____________________

29. Give a formula definition of the piecewise linear function whose graph is shown:

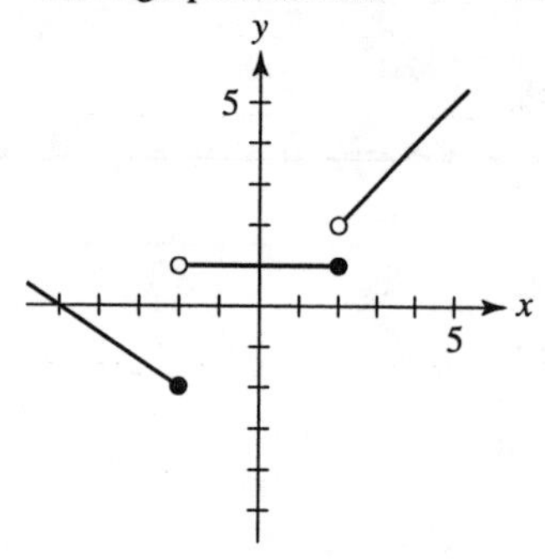

29. ____________________

30. Graph the function $f(x) = -3|x-2|+4$.

30.

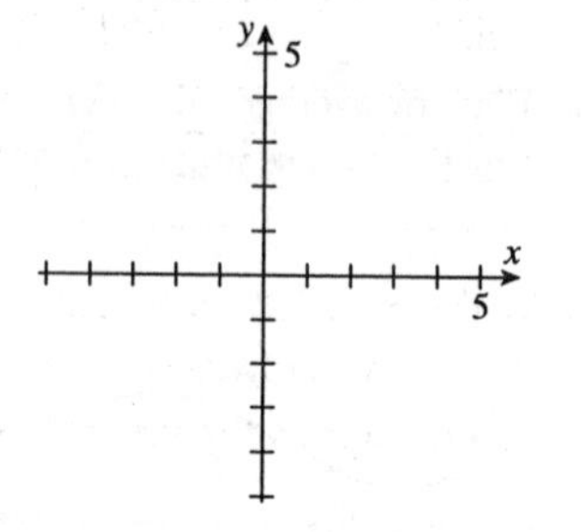

31. For the function $f(x) = -\frac{|x+11|}{4} + 4$, name the vertex and axis of symmetry, and state whether the function opens up or down.

31. ____________________

32. Find a formula for the linear absolute value function whose graph is shown:

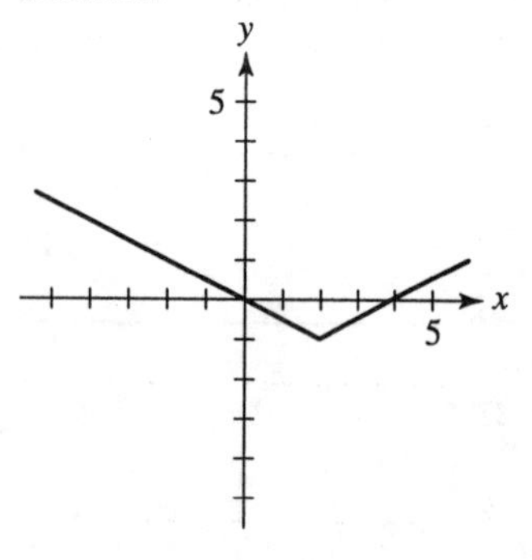

32. ____________________

33. Solve $|5x-17| \le 8$.

33. ____________________

34. Give the interval(s) on which $f(x) = \frac{|x-\frac{2}{5}|}{43} + \frac{22}{7}$ is increasing and the interval(s) on which it is decreasing.

34. ____________________

35. If temperature decreases with altitude at a linear rate of 6.7°F per 1000 ft, find the change of temperature as one descends from 6750 ft to 5210 ft above sea level. Round your answer to the nearest tenth of a degree.

35. ____________________

Chapter 3 Additional Exercises *(cont.)*

Name:____________________

36. A vacuum cleaner takes 2 min to get out and set up, and each room takes 11 min to vacuum. Putting the vacuum cleaner away takes another 4 min. Write the time a vacuuming job takes as a linear function of the number of rooms vacuumed.

36. ____________________

37. A bank account starts out with $400 in it and earns simple annual interest. After 6 years the account contains $502. Write the amount in the account as a function of the years since the initial deposit.

37. ____________________

38. The momentum of a sliding 7-kg block of ice is a function of its speed. Part of the function is shown in the table. Is the function linear?

Speed (m / s)	Momentum (kg • m / s)
3	21
9	63
13	91
22	154

38. ____________________

39. With a hose pumping 8 gal a minute, how long would it take to fill a cylindrical reservoir with a 6-m radius and a height of 11 m? Round your answer to the nearest hour. ($264 \text{ gal} = 1 \text{ m}^3$.)

39. ____________________

40. The gravitational force between two masses, m and M, a fixed distance apart varies directly with the product mM. If the force is 6.670×10^{-7} newtons when $m = M = 100$ kg, what is the force when $m = 375$ kg and $M = 512$ kg?

40. ____________________

Use the following scenario for problems 41–43. A family went on a 1299-mile trip by car, driving for 8.5 hours the first day, 9.5 hours the second, and 5 hours the third. At the end of the first day, the family had traveled 476 mi; at the end of the second, the family had traveled a total of 989 mi.

41. What was the difference between the average speed on day 1 and the average speed on day 2?

41. ____________________

42. What was the average speed over days 2 and 3 combined? Round to two decimal places.

42. ____________________

43. Represent the total miles traveled by time t as a piecewise linear function, where t represents total driving hours up to that point.

43. ____________________

44. Given the table, calculate the rate, in cents per minute, at which cost increases as a function of the length of a phone call.

time	cost
4 min	$0.96
9 min	$1.76
16 min	$2.88
22 min	$3.84

44. ____________________

45. The weight a certain weight lifter lifts in the clean-and-jerk event, divided by 2, is always right around the lifter's body weight of 100 kg—give or take 2.7 kg. What range of weights does the athlete lift?

45. ____________________

Chapter 3 Additional Exercises *(cont.)*

Name:____________________

46. The temperature near a fireplace lit at $t = 0$ is 23.2°C after 6 min, 25.4°C after 13 min, and 29.1°C after 25 min. Use linear approximation to estimate the temperature at $t = 22$ min. Round to the nearest tenth of a degree.

46. ____________________

Use the following scenario for problems 47–48. Achilles and the tortoise are racing toward the finish line. At $t = 26$ min, Achilles is within 210 ft of the tortoise, and at $t = 29$ min, Achilles is within 72 ft of the tortoise.

47. Write a linear equation expressing the tortoise's lead over Achilles as a function of time.

47. ____________________

48. At what time does Achilles catch up with the tortoise? Round to the nearest second.

48. ____________________

Use the following scenario for problems 49–50. In a planted field, 9 seedlings appear one day. The next day 45 seedlings appear, and the day after that, 81.

49. If the trend continues arithmetically, how many seedlings will appear on the sixth day?

49. ____________________

50. How many seedlings will have appeared after one week?

50. ____________________

A Mathematical Looking Glass
Seismographs
(to follow Section 3-2)

During an earthquake, the shifting of the earth at a location called the **epicenter** produces shock waves that travel through the ground, sometimes causing tremors at great distances from the epicenter. These waves are of two types, called **primary** and **secondary waves**. They travel at different speeds, and are therefore separated by increasingly greater distances as they move away from the epicenter. Thus at greater distances from the epicenter, there will be a greater time difference between the onset of the primary and secondary waves. See Figure S-4a.

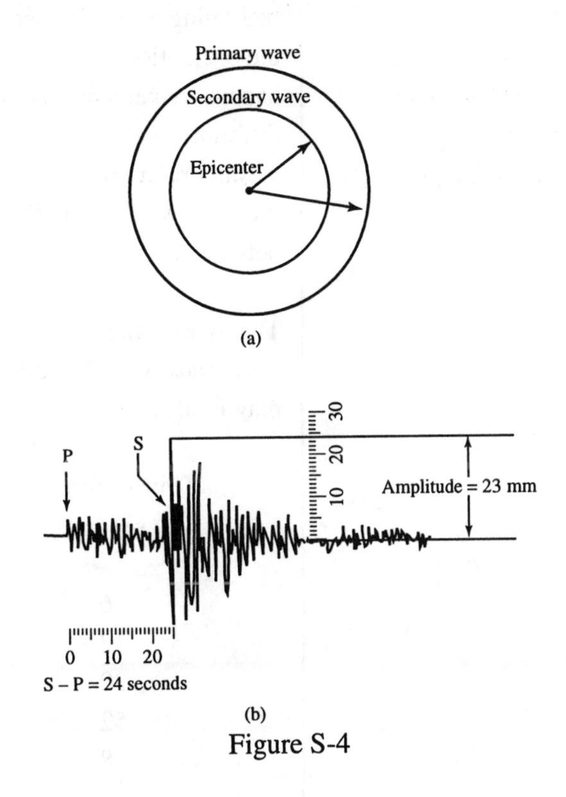

Figure S-4

Source: Bruce A. Bolt, *Earthquakes: A Primer*, W. H. Freeman and Co.

Tremors caused by an earthquake are recorded on an instrument called a **seismograph**. The simplest type of seismograph, now obsolete, has a pen attached to a mechanical arm which oscillates up and down as the earth moves. The oscillations produce marks on a long strip of paper which is fed slowly past the pen's position. The marks show the time difference between the arrival of the primary and secondary waves. They can thus be used to find the seismograph's distance from the epicenter.

In Figure S-4b, the primary wave caused the oscillations beginning at P. The secondary wave arrived later, causing the oscillations beginning at S. Knowing the speed of the paper's movement, scientists calculated that the time difference between P and S was 24 seconds. To find the seismograph's distance from the epicenter, the scientists must know how that distance relates to the time difference between P and S.

The hypothetical data in Table S3, based on a discussion in *Earthquakes: A Primer*, is typical of an earthquake having magnitude 5.2 on the Richter scale.

Time difference (t seconds)	Distance from epicenter (d kilometers)
6	52.2
12	104.4
30	261.0
52	452.4
98	852.6

Table S3

Exercises

1. How does Table S3 suggest that t and d are linearly related?

2. (Creating Models)

    a. Sketch a graph of d as a function of t.

    b. How does your graph suggest that t and d are linearly related?

3. (Creating Models)

    a. Write an equation to express d as a linear function of t.

    b. What are the domain and range of the function if d and t are considered as abstract variables?

    c. What are the domain and range of the function in the context of the physical situation?

4. (Interpreting Mathematics)

    a. If the primary wave arrives 24 seconds before the secondary wave, use your graph to estimate the distance of the seismograph from the epicenter.

    b. Extend your graph to predict the time difference for a seismograph located 2000 kilometers from the epicenter.

5. (Interpreting Mathematics) Use your equation from Exercise 3 to answer the questions in Exercises 4.

6. (Writing to Learn) Compare the two models you used in completing Exercises 4 and 5, and list some advantages and disadvantages of each.

A Mathematical Looking Glass

Tire Radius

(to follow Section 3-4)

Why is a pickup truck with oversized tires more likely to get speeding tickets than other vehicles? To answer this question, let's look at the relationship between tire size and speed. The speedometer of any vehicle measures the number of revolutions per minute (rpm) made by the wheels. For example, if your truck's tires have a 14-inch radius, then at 55 mph your wheels are turning about 660 rpm. If you put on tires with a bigger (or smaller) radius, they roll a greater (or lesser) distance with each revolution. In Exercise 1, you can verify that if your tire radius is r inches, a turning rate of 660 rpm corresponds to a speed of 3.93 mph. However, your speedometer will still read 55 mph.

Dealers recommend that you should adjust your speedometer when you change tire sizes, but most trucks made after 1989 have computerized speed sensors requiring specialized equipment to adjust. If you do not have the speedometer adjusted, then you need to know the size of the error in your speedometer reading. The error when your speedometer reads 55 mph is a function $E(r)$ of your tire radius. Let's look for an analytical expression for $E(r)$.

The error is equal to the size of the difference between your actual speed and your speedometer reading, without regard to its sign. For example, $E(r) = 2$ if your actual speed is either 53 mph or 57 mph. The size of a number without regard to its sign is its absolute value. Therefore, the error in your speedometer reading is the absolute value of the difference between $3.93r$ (your actual speed) and 55 (your speedometer reading). Symbolically, $E(r) = |3.93r - 55|$, or equivalently, $E(r) = 3.93|r - 14|$.

Exercises

1. (Creating Models) If the radius of your tires is r inches, and your wheels are turning at 660 rpm, your actual speed is about $3.93r$ mph. Verify that

$$\frac{660 \text{ revolutions}}{1 \text{ minute}} = \frac{3.93r \text{ miles}}{1 \text{ hour}}$$

(*Hint*: With each revolution, the truck travels a distance equal to the circumference of the tire.)

2. (Interpreting Mathematics) Chris Ford and Chris Baiker are friends of Lynn's daughter Carrie. When they were students at Fox Chapel High School, they wanted to update the look of Ford's Ford pickup truck by changing the tire size. They couldn't afford to have the speedometer adjusted, and Ford knew that he would usually forget to adjust his speed. If his actual speed on open highways is too much less than 55 mph, Baiker drives him nuts by tapping his fingernails on the dashboard. However, if it is more than 61 mph, he risks being ticketed for speeding. He decided that 6 mph is an acceptable error in the speedometer reading. What tire radii, in inches, meet this condition?

Chapter 3

Section 3-1

1. a. $\frac{\Delta S}{\Delta T} = \frac{6.1}{10} = 0.61$

   b. $\frac{\Delta S}{\Delta T} = \frac{12.2}{20} = 0.61$

   c. Choices will vary.

   d. 0.61

   e. Calculations will vary.

3. $\frac{\Delta y}{\Delta x}$ has a value of 4 between the first two data points, and a value of 2 between the last two, so the table does not fit a linear function.

5. $\frac{\Delta y}{\Delta x} = -20$ between each pair of data points, so the table fits a linear function.

7. The rate of change in S with respect to T is 0.61. This means that each increase of 1° in temperature results in an increase of 0.61 meters per second in the speed of sound.

9. The rate of change in the fan speed with respect to the motor pulley diameter is 340. This represents the increase in rpm corresponding to each 1-inch increase in the diameter of the pulley.

11. The rate of change in cost with respect to distance driven is 0.15. This represents the cost in dollars of each additional mile driven.

13. a.

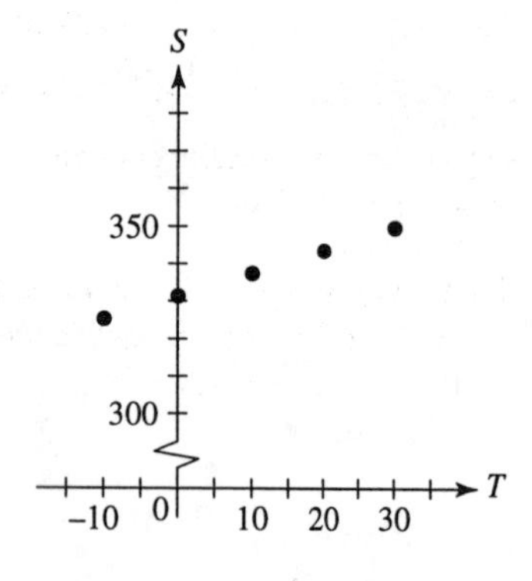

The points are collinear.

   b. The slope is 0.61.

15. a. The function is increasing.

    b. The rate of change in y with respect to x is 4.

17. a. The function is neither increasing nor decreasing.

    b. The rate of change in y with respect to x is 0.

19. The function f is linear; $y = 3x - 9$.

21. The function h is not linear, because it cannot be written $h(x) = mx + b$.

23. The function R is not linear, because it cannot be written $R(x) = mx + b$.

25. The graphs of the functions in Exercises 19, 20, 22, and 24 should appear to be lines.

27. a. The graph has slope 4 and y-intercept (0, 4). Its equation is $y = 4x + 4$.

    b. The graph has slope −1 and y-intercept (0, 2). Its equation is $y = -x + 2$.

    c. The graph has slope 0 and y-intercept (0, −2). Its equation is $y = -2$.

    d. The graph has slope $-\frac{2}{3}$ and y-intercept (0, −2). Its equation is $y = -\frac{2}{3}x - 2$.

29. a. The function can be rewritten as $h(x) = -x + 40$, so that $m = -1$.

    b. The function is decreasing.

31. a. $m = 0$

    b. The function is neither increasing nor decreasing.

33. a. The rate of change in $h(r)$ with respect to r is $\frac{1}{4}$.

b. The function is increasing.

c. The rate of change represents the rise in temperature corresponding to each additional chirp in a one-minute period.

35. a. The function is not linear, since it cannot be written $y = mx + b$.

37. a. The function is linear.

b. $f(x) = 0.5x - \sqrt{2}$

c. The function is increasing.

d. 0.5

39. a. The function is linear.

b. $\Sigma(x) = \pi x$ (Think $y = \pi x + 0$.)

c. The function is increasing.

d. π

41. a. The function is linear.

b. $a(x) = \frac{1}{3}x$ (Think $y = \frac{1}{3}x + 0$.)

c. The function is increasing.

d. $\frac{1}{3}$

43. a. The function is linear.

b. $z(x) = (x^2 + 6x + 9) - x^2$
$z(x) = 6x + 9$

c. The function is increasing.

d. 6

45. a. The table fits a linear function, because $\frac{\Delta y}{\Delta x} = 4$ for each pair of points.

b. The rate of change in y with respect to x is 4.

c. The function is increasing.

d. Using the point (1, 13), the equation in point-slope form is $y - 13 = 4(x - 1)$. Equivalently, $y = 4x + 9$.

47. a. The table fits a linear function, because $\frac{\Delta y}{\Delta x} = -1$ for each pair of points.

b. The rate of change in y with respect to x is -1.

c. The function is decreasing.

d. Using the point (32, 68), the equation in point-slope form is $y - 68 = -(x - 32)$. Equivalently, $y = 100 - x$.

49. a. The table does not fit any function, linear or otherwise, because there is more than one value of y corresponding to $x = 3$.

51. a. If v = the number of valentines purchased, and c = the cost of the valentines, in dollars, then $c = 2.50v$.

b. The rate of change in c with respect to v is 2.50. This agrees with the conclusion in Exercise 8.

53. a. $s = 100 - 2.5m$

b. The rate of change in s with respect to m is -2.5. This agrees with the conclusion in Exercise 10.

55. a. At the beginning of the descent the airliner's altitude is 36,000 feet, and 20 minutes later its altitude is 0 feet. Therefore, the graph of the function is a line through (0, 36,000) and (20, 0). Its slope is $-\frac{36,000}{20} = -1800$. In slope-intercept form, its equation is $A = -1800t + 36,000$.

b. The rate of change in A with respect to t is -1800. This says that the airliner descends 1800 feet each minute.

c. When the airliner is 100 miles from the airport its altitude is 36,000 feet, and when it reaches the airport its altitude is 0 feet. Therefore, the graph of the function is a line through (100, 36 000) and (0, 0). Its slope is $\frac{36\,000}{100} = 360$. In slope-intercept form, its equation is $A = 360d$.

d. The rate of change in A with respect to d is 360. This says that as the airliner gets one mile closer to the airport, it descends 360 feet.

57. $\frac{\Delta y}{\Delta x} = \frac{92-100}{2-1} = -8$, so that $\Delta y = -8\Delta x$.
Between the x-values of 2 and 3, $\Delta x = 1$, so $\Delta y = -8(1) = -8$, and $y = 92 - 8 = 84$.
Between the x values of 2 and 5 $\Delta x = 3$, so $\Delta y = -8(3) = -24$, and $y = 92 - 24 = 68$.

59. $\frac{\Delta y}{\Delta x} = \frac{22-2}{24-14} = 2$, so that $\Delta y = 2\Delta x$.
Between the x-values of 24 and 30, $\Delta x = 6$, so $\Delta y = 2(6) = 12$, and $y = 22 + 12 = 34$. Between the y-values of 2 and 7, $\Delta y = 5 = 2\Delta x$, so $\Delta x = 2.5$, and $x = 14 + 2.5 = 16.5$

61. No, because the calculation of the missing entries was made possible by the constancy of the ratio $\frac{\Delta y}{\Delta x}$.

63. a. $\frac{\Delta y}{\Delta x} = 1$ for each pair of points.

b. Choices of points will vary.

b. Choices will vary. One possibility is [–5, 5] by [–5, 5].

65.

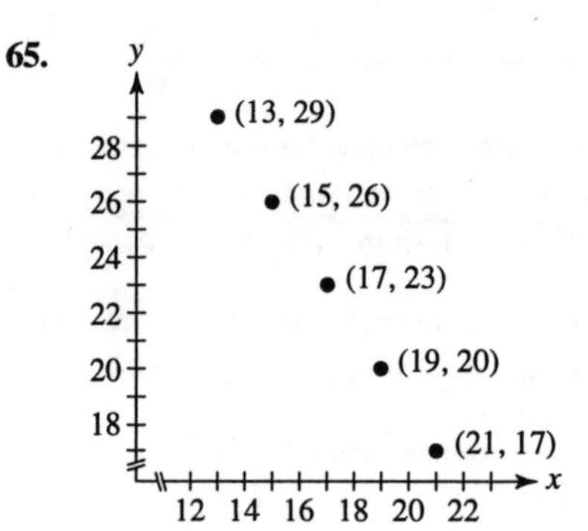

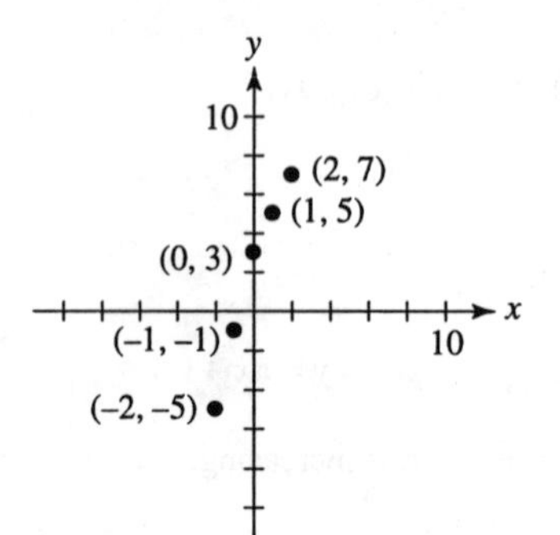

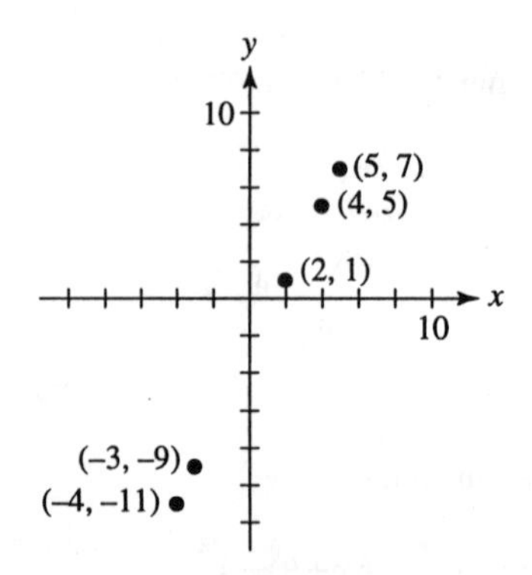

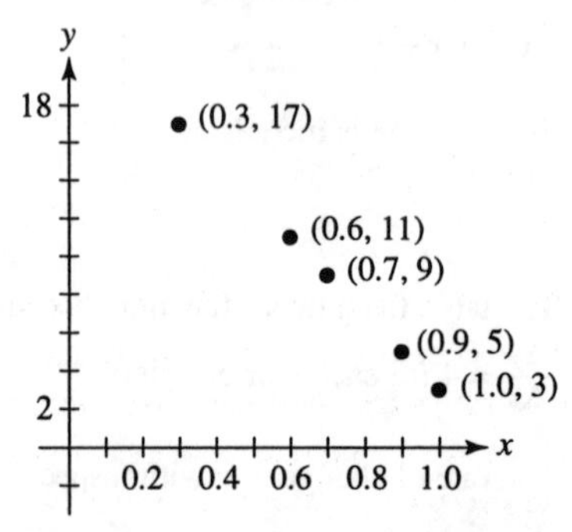

The points from Exercises 2, 4, and 5 are collinear.

Section 3-2

1. If $C = 10$, then $M = \frac{10-0.48}{0.67} \cong 14.21$.
However, if Carrie talks for 14.21 minutes, she must pay for 15 minutes. Therefore, no call will cost exactly $10.

3. a. If M and C are as in Exercise 1, then $M = \frac{C-0.63}{1.01}$. If $C = 10$, then $M = \frac{10-0.63}{1.01} \cong 9.28$ Since a call of 9.28 minutes will result in a charge for 10 minutes, no call will cost exactly $10.

   b. The inequality is $M \le \frac{C-0.63}{1.01}$, and the solution is $M \le 9.28$. In the context of the problem, the solution is $M \le 9$.

5. $y = mx$, and $y = 35$ when $x = 21$, so $35 = m(21)$
$m = \frac{35}{21} = \frac{5}{3}$
When $x = 39$, the value of y is $\frac{5}{3}(39) = 65$.

7. $p = mq$, and $p = 15$ when $q = 2$, so
$15 = m(2)$
$m = \frac{15}{2}$
When $q = 40$, the value of p is $\frac{15}{2}(40) = 300$.

9. The ratio $\frac{y}{x} = 4$ for all points. The equation $y = 4x$ fits the table.

11. The ratio $\frac{y}{x} = 0.2$ for all points. The equation $y = 0.2x$ fits the table.

13. a. The rate of change in M with respect to W is constant. Specifically, each time W increases by 1 week, M increases by 0.25 miles.

    b. The line relating W and M has a slope of 0.25, since that is the rate of change in M with respect to W. It also passes through (1, 1), since Tim runs 1 mile per day during the first week. The equation in point-slope form is therefore $M - 1 = 0.25(W - 1)$. In slope-intercept form, it is $M = 0.25W + 0.75$

15. The sequence is arithmetic, with a difference of 3 between consecutive terms. Since the first term is $a_1 = 4$, a formula for the nth term of the sequence is $a_n = 4 + 3(n-1)$. The value of the 425th term is $4 + 3(425 - 1) = 1276$.

17. The sequence is arithmetic, with a difference of –1.5 between consecutive terms. Since the first term is $a_1 = 7.5$, a formula for the nth term of the sequence is $a_n = 7.5 - 1.5(n-1)$. The value of the 425th term is $7.5 - 1.5(425 - 1) = -628.5$.

19. If a_n represents the amount of the deposit during the nth year, then the values of a_n form an arithmetic sequence with a first term of 1200, and a difference of 120 between consecutive terms. Thus, $a_n = 1200 + 120(n-1)$, and $a_{18} = 1200 + 120(17) = 3240$, so the amount of the deposit will be $3240.

21. The difference between consecutive terms in the series is 2, and the first term is 3, so $a_n = 3 + 2(n-1)$. To see which term is equal to 2001:
$2001 = 3 + 2(n - 1)$
$1998 = 2(n - 1)$
$999 = n - 1$
$n = 1000$
The following figure indicates that twice the required sum can be obtained by adding together 1000 pairs of numbers, each with a sum of 2004.

$$\begin{array}{cccccccc} 3 & 5 & 7 & \cdots & \cdots & 1997 & 1999 & 2001 \\ \uparrow & \uparrow & \uparrow & & & \uparrow & \uparrow & \uparrow \\ 2001 & 1999 & 1997 & \cdots & \cdots & 7 & 5 & 3 \end{array}$$

Therefore, the sum of the series is $\frac{(1000)(2004)}{2} = 1{,}002{,}000$.

23. The 73rd term is $6 + 4(73 - 1) = 294$. Pairing the terms as in Figure 3-3 results in 73 pairs, each with a sum of 300. The sum is $\frac{73(300)}{2} = 10{,}950$.

25. If the terms in the series are written down twice, as in Figure 3-3, there are N pairs, each with a sum of $(a_1 + a_N)$. The sum of all the pairs is $N(a_1 + a_N)$, and this is twice the required sum, so $S_N = \frac{N(a_1 + a_N)}{2}$.

27. The difference between consecutive terms is 5, and the first term is –15, so $a_n = -15 + 5(n-1)$. To see which term is equal to 235:
$235 = -15 + 5(N - 1)$
$250 = 5(N - 1)$
$50 = N - 1$
$N = 51$
The sum is $S_N = \dfrac{51(-15+235)}{2} = 5610.$

29. a. If d = horizontal distance from the low end, in feet, and h = elevation, in feet,

then d and h are related by a linear equation whose graph passes through the points (0, 10) and (4000, 170). Its slope is $\dfrac{160}{4000} = 0.04$, and its y-intercept is 10, so its equation is $h = 10 + 0.04d$.

b. If $h = 20$, then
$20 = 10 + 0.04d$
$0.04d = 10$
$d = 250$
The sea wall should extend to a point 250 feet from the low end of the roadway.

c. The graph of $h = 10 + 0.04d$ is above the line $h = 20$ when $d > 250$, and below it when $d < 250$. The sea wall should include all values of d less than 250. This means all points at most 250 feet from the low end.

d. Responses will vary.

31. If n = the number of people in the group, and p = the price per person, in dollars, then for $n \geq 50$, n and p are related by a linear equation whose graph passes through the point (50, 800), and has a slope of $-\frac{15}{2} = -7.5$ (indicating that p decreases by \$15 whenever n increases by 2). Its equation, in point-slope form, is
$p - 800 = -7.5(n - 50)$
If $p = 635$, then
$635 - 800 = -7.5(n - 50)$
$-165 = -7.5(n - 50)$
$n - 50 = \dfrac{165}{7.5} = 22$
$n = 72$ people

33. $t = ms$, and $t = 57$ when $s = 126$, so $57 = m(126)$, and $m = \dfrac{57}{126} = \dfrac{19}{42}$. When $s = 168$,
$t = \frac{19}{42}(168) = 76.$

35. $A = 0.7B$, so when $B = 62$, the value of A is $0.7(62) = 43.4$.

37. a. Yes, because doubling the burning time doubles the cost.

b. Yes, because doubling the number of people approximately doubles the water usage.

c. No, because doubling the number of people cuts each person's closet space in half.

d. Yes, because doubling the number of pages doubles the weight of the book.

39. a. The sequence is arithmetic, with a difference of 3 between consecutive terms.

b. Since the first term is 7, the nth term is $a_n = 7 + 3(n-1)$.

c. $a_{100} = 7 + 3(99) = 304$
$a_{425} = 7 + 3(424) = 1279$

d. The sum of the first 100 terms is
$\dfrac{100(7+304)}{2} = 15,550.$
The sum of the first 425 terms is
$\dfrac{425(7+1279)}{2} = 273,275.$

41. a. The sequence is arithmetic, with a difference of –0.1 between consecutive terms.

b. Since the first term is 3487.6, the nth term is $a_n = 3487.6 - 0.1(n-1)$.

c. $a_{100} = 3487.6 - 0.1(99) = 3477.7$
$a_{425} = 3487.6 - 0.1(424) = 3445.2$

d. The sum of the first 100 terms is
$\dfrac{100(3487.6+3477.7)}{2} = 348,265.$
The sum of the first 425 terms is
$\dfrac{425(3487.6+3445.2)}{2} = 1,473,220.$

43. The sequence is not arithmetic, since the difference between consecutive terms is not constant.

45. **a.** The sequence is arithmetic, with a difference of $\pi - 1$ between consecutive terms.

b. Since the first term is $\pi - 2$, the nth term is $a_n = \pi - 2 + (\pi - 1)(n - 1)$.

c. $a_{100} = \pi - 2 + (\pi - 1)(99) = 100\pi - 101$
$a_{425} = \pi - 2 + (\pi - 1)(424) = 425\pi - 426$

d. The sum of the first 100 terms is
$$\frac{100[(\pi-2)+(100\pi-101)]}{2} = 5050\pi - 5150$$
The sum of the first 425 terms is
$$\frac{425[(\pi-2)+(425\pi-426)]}{2}$$
$$= 90{,}525\pi - 90{,}950$$

47. **a.** The difference between consecutive terms is 32.

b. The first term is 16, so the nth term is $a_n = 16 + 32(n - 1)$. The 60th term is $a_{60} = 16 + 32(60 - 1) = 1904$ feet.

c. The sum of the first 60 terms is $\frac{60(16+1904)}{2} = 57{,}600$ feet (about 11 miles).

Section 3-3

1. The ratio $\frac{\Delta y}{\Delta x}$ is not the same for all pairs of points.

3.

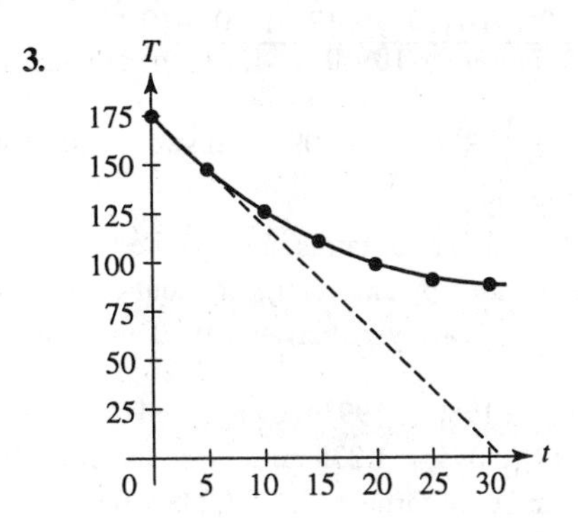

5. Entries in the second and fourth columns will vary. The missing y-values from the linear equation are listed in Exercise 4.

7. To cool 25° at a rate of 5.6° per minute requires $\frac{25}{5.6} \cong 4.46$ minutes. This agrees with the result obtained in Exercise 6.

9. Suppose t = time in years after 1980, and y = value of the house, in thousands of dollars. Then the line relating t and y passes through the points (0, 52) and (15, 97). Its slope is $\frac{97-52}{15-0} = 3$, indicating that the value of the house increased by $3000 per year. Using the rate of change to complete a table:

year	value
1980	$52,000
1981	$55,000
1982	$58,000
1983	$61,000
1984	$64,000
1985	$67,000
1986	$70,000
1987	$73,000
1988	$76,000
1989	$79,000

11. For values of n in [16, 36], the average rate of change in $\sqrt{n}$ with respect to n is $\frac{6-4}{16-36} = 0.1$, meaning that on the average, the value of $\sqrt{n}$ increases by 0.1 each time n increases by 1. Using the rate of change to complete a table:

n	$\sqrt{n}$
16	4.0
17	4.1
18	4.2
19	4.3
20	4.4
21	4.5
22	4.6
23	4.7
24	4.8
25	4.9

n	$\sqrt{n}$
26	5.0
27	5.1
28	5.2
29	5.3
30	5.4
31	5.5
32	5.6
33	5.7
34	5.8
35	5.9
36	6.0

You may wish to compare these estimates with the actual estimates with the actual values of $\sqrt{n}$. The error is no more than 0.1 in each case.

13. The average rate of change is $\frac{86-100}{30-20} = -1.4$, indicating that the coffee cooled at an average rate of 1.4° per minute during the time interval [20, 30].

15. The average rate of change in hours of daylight with respect to the number of days past June 21 is $\frac{12-21}{92-0} \cong -0.096$, indicating that on the average, each day between June 21 and September 21 Yellowknife receives about 0.096 hours less daylight than on the previous day.

17. a. $\frac{P(510)-P(500)}{510-500}$

$= \frac{4462.00-4387.50}{510-500} = 7.45$

b. \$7.45

c. From part (a), as your sales level increases from 500 to 510 shirts, each additional shirt sold produces about \$7.45 in additional profit, At a sales level of 505 shirts, your marginal profit is \$7.45.

19. Increasing your sales level from 505 to 506 shirts increases your profit by about \$7.45, while increasing from 525 to 526 shirts increases your profit by only about \$7.25.

21. a. The lines passes through $(2, f(2)) = (2, 1)$ and $(5, f(5)) = (5, 2)$. Its slope is $\frac{2-1}{5-2} = \frac{1}{3}$. Its equation in point-slope form is $y - 1 = \frac{1}{3}(x-2)$. Equivalently, $L(x) = \frac{1}{3}x + \frac{1}{3}$.

b. $L(2) = \frac{1}{3}(2) + \frac{1}{3} = 1 = f(2)$

$L(5) = \frac{1}{3}(5) + \frac{1}{3} = 2 = f(5)$

c. $L(3) - f(3) = \left[\frac{1}{3}(3) + \frac{1}{3}\right] - \sqrt{2} \cong -0.081$

$L(4) - f(4) = \left[\frac{1}{3}(4) + \frac{1}{3}\right] - \sqrt{3} \cong -0.065$

23. a. The line passes through $(2, f(2)) = (2, -8)$ and $(5, f(5)) = (5, -5)$. Its slope is $\frac{-5-(-8)}{5-2} = 1$. Its equation in point-slope form is $y + 8 = (1)(x - 2)$. Equivalently, $L(x) = x - 10$.

b. $L(2) = 2 - 10 = -8 = f(2)$

$L(5) = 5 - 10 = -5 = f(5)$

c. $L(3) - f(3) = [3 - 10] - (-9) = 2$

$L(4) - f(4) = [4 - 10] - (-8) = 2$

25. $\frac{(10^3 - 500) - (0^3 - 500)}{10 - 0} = 100$

27. $\frac{[7 + 10(10) - (10^2)] - [7 + 10(0) - (0)^2]}{10 - 0} = 0$

29. a. $\frac{2.2 - 6.0}{1980 - 1940} = -0.095$ (Answers will vary slightly.)

The number of farms in the United States decreased by an average of about 0.095 million each year between 1940 and 1980.

b. From 1980 to 1993 the predicted decrease is $(13)(0.095) = 1.235$ million, so the predicted number of farms is $2.2 - 1.235 = 0.965$ million. (Answers will vary slightly.)

31. a. $\frac{34.4-26.2}{1992-1983} \cong 0.91$
The percentage of persons between the ages of 18 and 24 enrolled in college increased by an average of 0.91 each year between 1983 and 1992.

b. Between 1983 and 1987 the predicted increase is (4)(0.91) = 3.64, so the predicted percentage is 26.2 + 3.64 = 29.8.

c. Between 1992 and 1993 the predicted increase is 0.91, so the predicted percentage is 34.4 + 0.91 = 35.3.

d. Between 1992 and 2063 the predicted increase is (71)(0.91) = 64, so the predicted percentage is 34.4 + 64 = 98.4.

e. The estimate in part (d) is probably not very accurate because it is a long-term prediction based on a short-term trend.

33. (Sample response) It is assumed that the airplane descends at a constant angle, and that it travels at a constant speed. The first assumption is probably reasonable, but the second is probably not, since the airplane must decrease its speed to land.

35. a. The average rate of change in milepost numbers with respect to time is $\frac{-226}{226} = -1$ mile per minute. At noon, 135 minutes had elapsed since 9:45 am, so the number of the milepost I was passing was 421 – 135 = 286.

b. The assumption of linearity means that the rate of change in milepost numbers with respect to time remains constant. In other words, it means that I am driving at a constant speed.

37. a. The equation of the line is $y = 175 - 5.6x$ (see Exercise 4). When $y = 90$, we have
$90 = 175 - 5.6x$
$5.6x = 85$
$x \cong 15.18$ minutes

b. The slope of the line is –1.6, so its equation in point-slope form is $y - 100 = -1.6(x-20)$. When $y = 90$, we have
$90 - 100 = -1.6(x-20)$
$-10 = -1.6(x-20)$
$x - 20 = 6.25$
$x = 26.25$ minutes

c. (Sample response) The interval (92, 100) is much closer to 90 than is (147, 175). We should therefore expect that the cooling of the coffee through the range of temperatures between 92° and 100° is a more accurate estimate of the time when the temperature will reach 90°.

Section 3-4

1. The first entry in the second column is (12)($73.30) = $879.60, representing the total cost of 12 monthly premiums. Under this plan Sarah must pay all of her medical expenses up to $1000. Therefore the other entries in the second column represent $879.60 in premiums plus the indicated medical expenses in the first column.

3. a. If Sarah's medical expenses are $1000 or less, the cost of the plan is the amount of her medical expenses plus the amount of her premiums. Symbolically, this is $x + 879.60$.

b. If Sarah's medical expenses are more than $1000, the cost of the plan is $1879.60 plus 20% of her medical expenses above $1000. Symbolically, this is
$1879.60 + 0.20(x - 1000) = 0.20x + 1679.60$.

5. a.

x	$f(x)$
2.5	4
2.6	4.2
2.7	4.4
2.8	4.6
2.9	4.8
3.0	5
3.1	4.9
3.2	4.8
3.3	4.7
3.4	4.6
3.5	4.5

b.

x	$f(x)$
2.95	4.9
2.96	4.92
2.97	4.94
2.98	4.96
2.99	4.98
3.00	5
3.01	4.99
3.02	4.98
3.03	4.97
3.04	4.96
3.05	4.95

c. Yes, because $f(3) = 5$ and $f(x)$ is near 5 when x is near 3.

7.

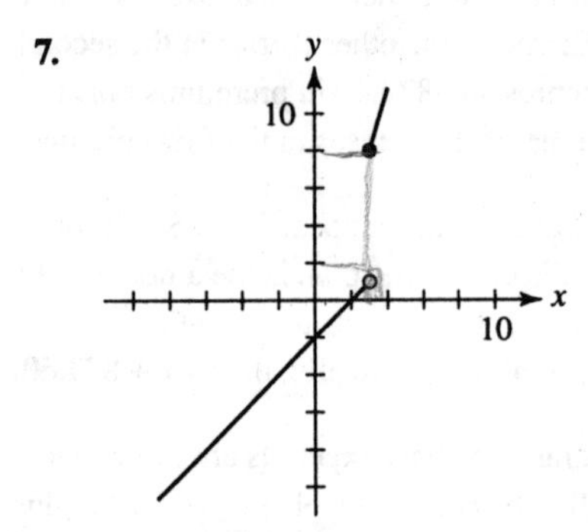

9.

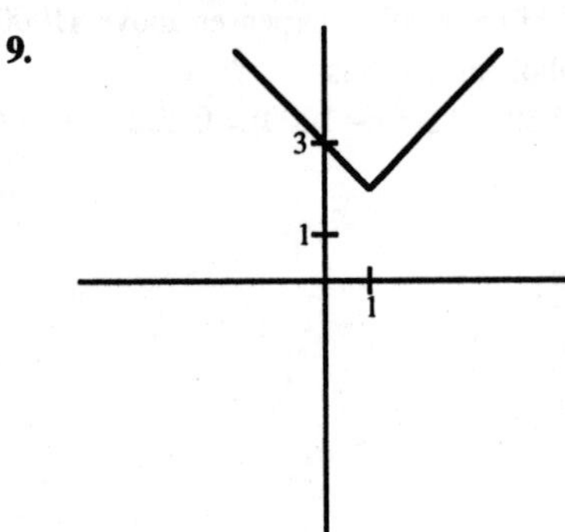

11.

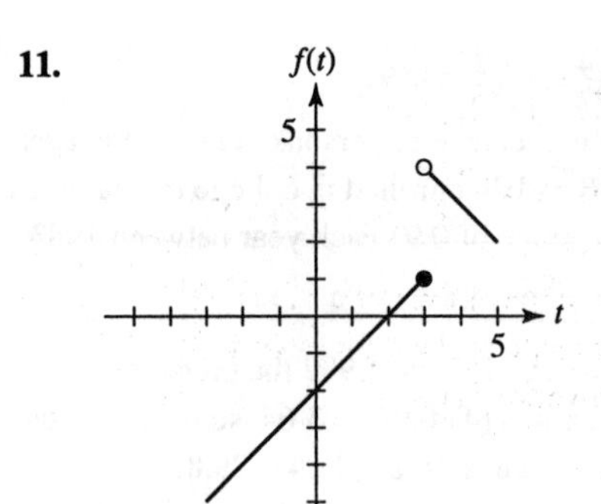

13. (Sample response) At an x-value where the rule defining the function changes, evaluate the two linear expressions that define the function on each side of that x-value. If the results are the same, the pieces join to form a continuous graph. Otherwise, they don't.

15. a. 0.1 cm; 0.05 cm; 0.05 cm; 0.1 cm

b. The deviation is the *size* of the difference $h - 13.90$ between actual height and designed height, without regard to the *sign* of the difference. Thus it is $|h - 13.90|$.

c. The difference between the actual capacity and designed capacity is $\pi(4.280)^2 h - \pi(4.280)^2(13.90)$ $= \pi(4.280)^2(h - 13.90)$. The deviation is the size of this difference, which is its absolute value $|\pi(4.280)^2(h - 13.90)|$ $= \pi(4.280)^2|h - 13.90| \cong 57.55|h - 13.90|$.

17. If $x \geq 0$, then $|x| = x$, so $g(x) = 3x + 4$. If $x < 0$, then $|x| = -x$, so $g(x) = -3x + 4$.

Thus, $g(x) = \begin{cases} -3x + 4 & \text{if } x < 0 \\ 3x + 4 & \text{if } x \geq 0 \end{cases}$

19. If $z \geq 0$, then $|z| = z$, so $H(z) = 5 - z$.
If $z < 0$, then $|z| = -z$, so $H(z) = 5 - (-z) = 5 + z$

Thus, $H(z) = \begin{cases} 5 + z & \text{if } z < 0 \\ 5 - z & \text{if } z \geq 0 \end{cases}$

21. a. $|x| = \begin{cases} -x & \text{if } x < 0 \\ x & \text{if } x \geq 0 \end{cases}$

Thus, $y = \begin{cases} -2x - 4 & \text{if } x < 0 \\ 2x - 4 & \text{if } x \geq 0 \end{cases}$

b. The slope of the left piece is –2.
The slope of the right piece is 2.
The vertex is (0, –4).

c.

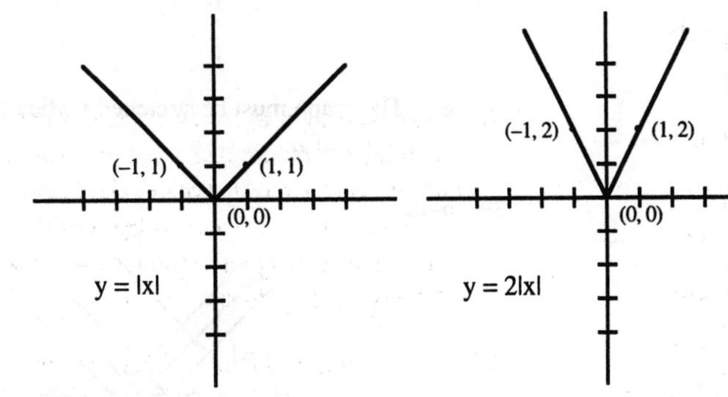

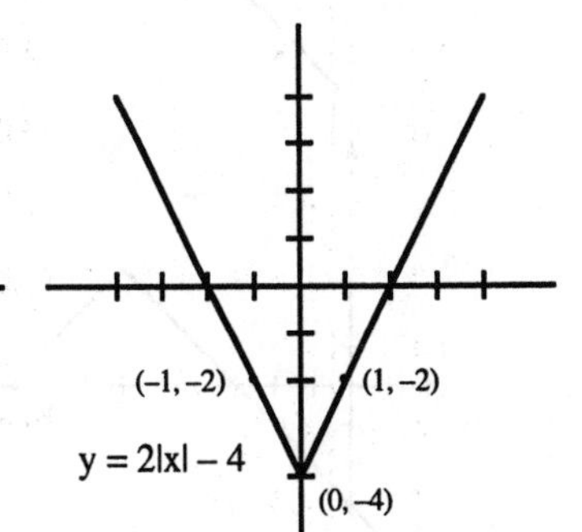

d. The axis of symmetry is the line $x = 0$ (the y-axis). The range is $[-4, \infty)$.

23. a. $|x - 6| = \begin{cases} -(x-6) & \text{if } x < 6 \\ x - 6 & \text{if } x \geq 6 \end{cases}$

Thus, $y = \begin{cases} (x-6)+3 & \text{if } x < 6 \\ -(x-6)+3 & \text{if } x \geq 6 \end{cases}$

b. The slope of the left piece is 1.
The slope of the right piece is –1.
The vertex is (6, 3).

c.

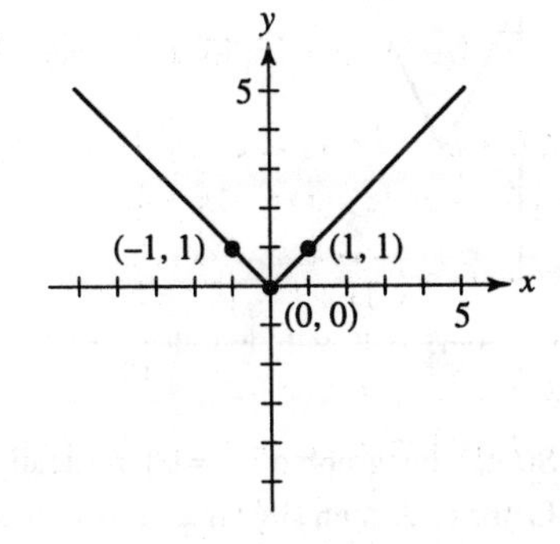

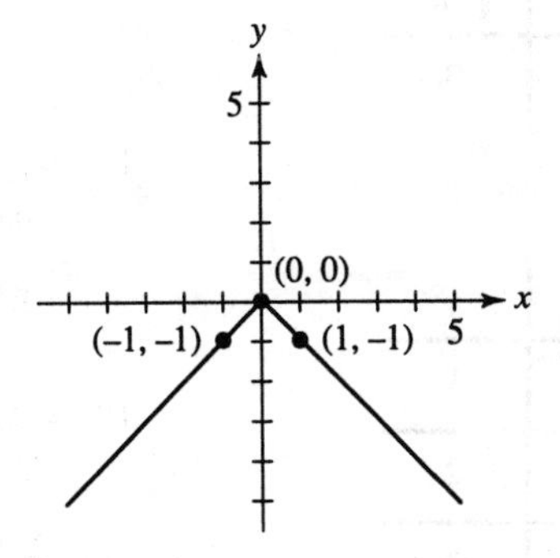

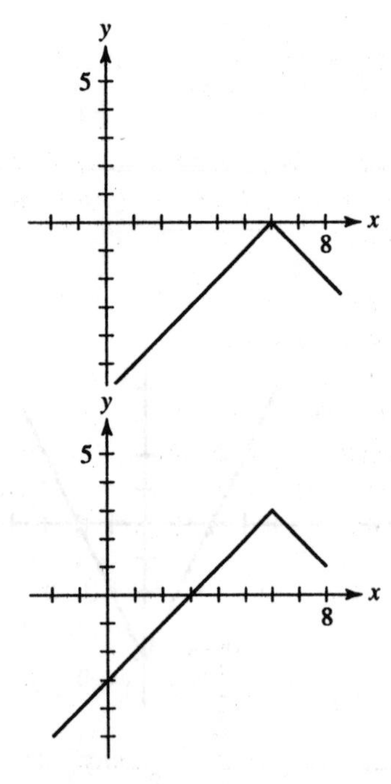

d. The axis of symmetry is the line $x = 6$.
The range is $(-\infty, 3]$.

25. a.

x	y
2	4
3	2
4	0
5	2
6	4

b.

x	y
–2	0
–1	–2
0	–4
1	–2
2	0

c.

x	y
–7	$\frac{5}{3}$
–6	$\frac{4}{3}$
–5	1
–4	$\frac{4}{3}$
–3	$\frac{5}{3}$

d.

x	y
4	1
5	2
6	3
7	2
8	1

e. The tables indicate the slope of each piece of the graph and the location of the vertex.

27. a–d.

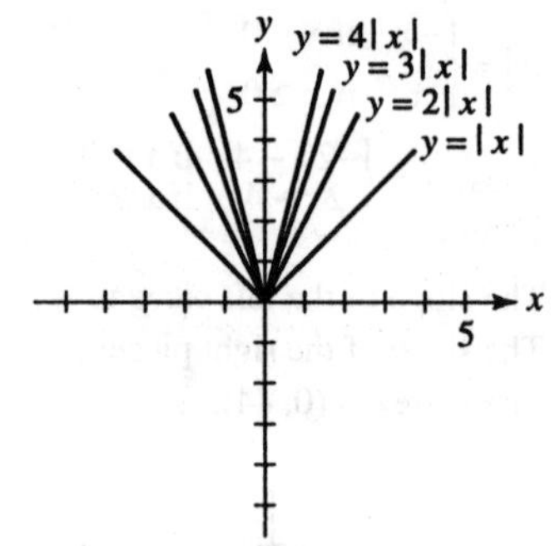

e. The graph must be stretched vertically by a factor of a.

29. a–d.

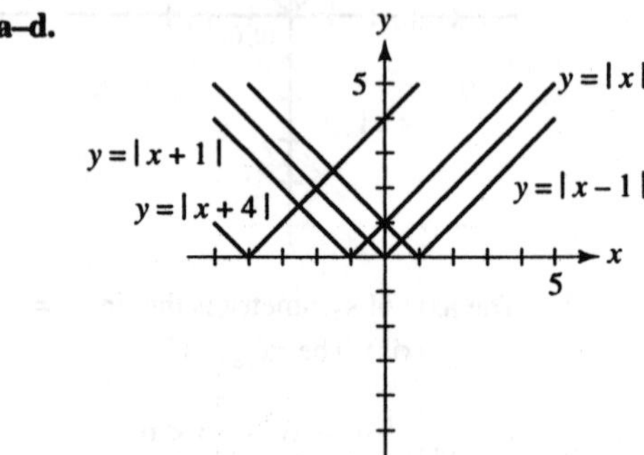

e. The graph must be shifted h units to the right. If $h < 0$, the actual shift is to the left.

31. a.

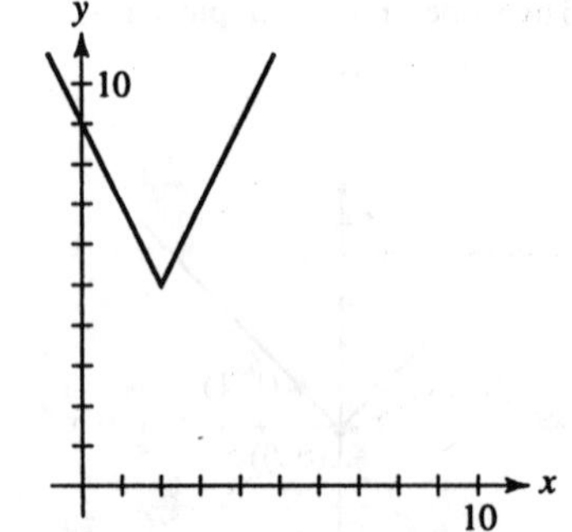

Stretch the graph of $y = |x|$ vertically by a factor of 2, then shift it 2 units to the right and 5 units up.

b.

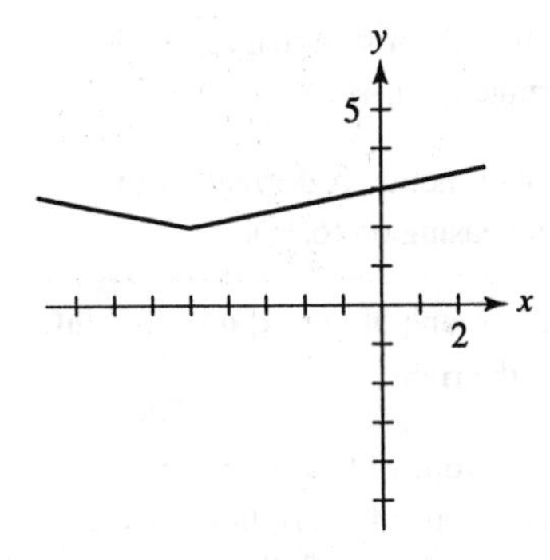

Compress the graph of $y = |x|$ vertically by a factor of 5, then shift it 5 units to the left and 2 units up.

c.

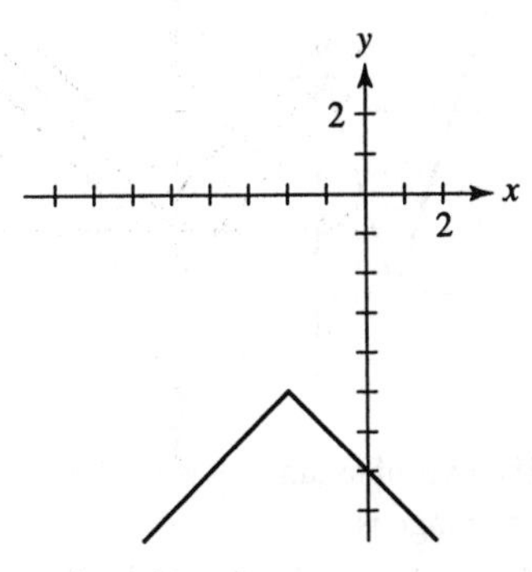

Reflect the graph of $y = |x|$ in the x-axis, then shift it 2 units to the left and 5 units down.

d.

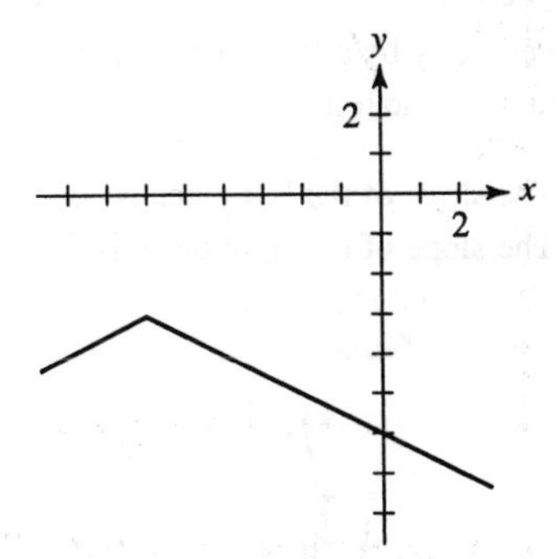

Compress the graph of $y = |x|$ vertically by a factor of 2, reflect it in the x-axis, then shift it 6 units to the left and 3 units down.

33. Analytically: $|2t - 1| = 5$
$2t - 1 = -5$ or $2t - 1 = 5$
$t = -2$ or $t = 3$
Graphically: The graph of $y = |2t - 1| - 5$ has t-intercepts at (–2, 0) and (3, 0), so the equation has solutions $t = -2, 3$.

35. Analytically: $|5z + 100| \geq 12$
$5z + 100 \leq -12$ or $5z + 100 \geq 12$
$5z \leq -112$ or $5z \geq -88$
$z \leq -22.4$ or $z \geq -17.6$
Graphically: The inequality is true when the graph of $y = |5z + 100| - 12$ is on or above the z-axis, that is, in $(-\infty, -22.4] \cup [-17.6, \infty)$.

37. a.

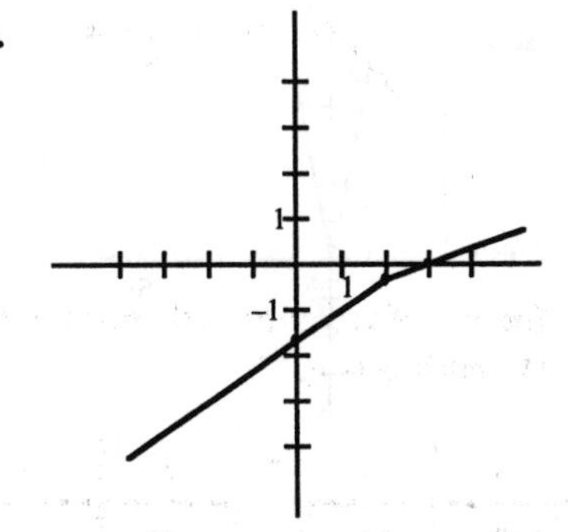

b. The domain is $(-\infty, \infty)$. The graph indicates that the range is also $(-\infty, \infty)$.

c. The graph indicates that the function is increasing on $(-\infty, \infty)$. (You can also draw this conclusion analytically, since the left piece of the graph has a slope of $\frac{2}{3}$, and the right piece has a slope of $\frac{1}{3}$.)

39. a.

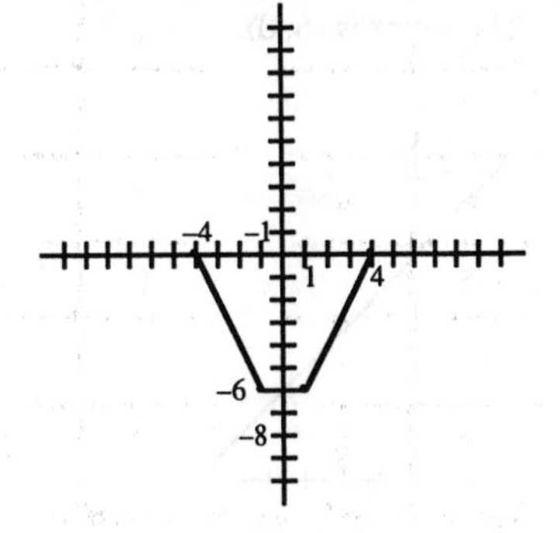

b. The domain is $[-4, 4]$. The graph indicates that the range is $[-6, 0]$.

c. The graph indicates that the function is decreasing on $[-4, -1]$, neither decreasing nor increasing on $[-1, 1]$, and increasing on $[1, 4]$. (You can also draw this conclusion analytically, since the left piece of the graph has a slope of -2, the middle piece has a slope of 0, and the right piece has a slope of 2.)

41. a. The slope of the left piece is -6. The slope of the right piece is 6. The vertex is $(0, 0)$.

b.

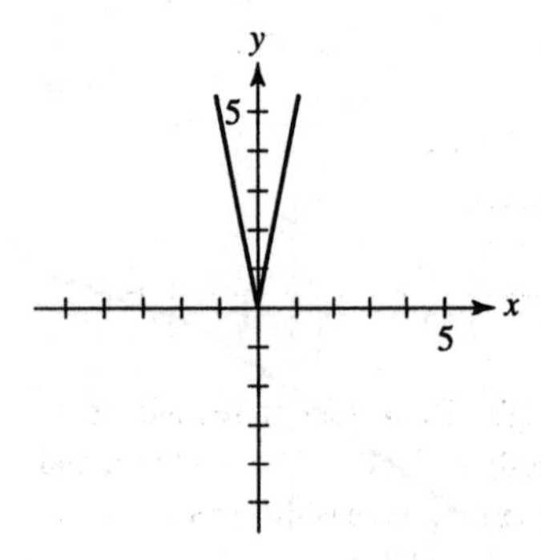

c. The axis of symmetry is the line $x = 0$ (the y-axis). The range is $[0, \infty)$.

d. The function is decreasing on $(-\infty, 0)$ and increasing on $(0, \infty)$.

e. The graph of $y = |x|$ must be stretched vertically by a factor of 6.

43. a. The slope of the left piece is -1.
The slope of the right piece is 1.
The vertex is $(6, 0)$.

b.

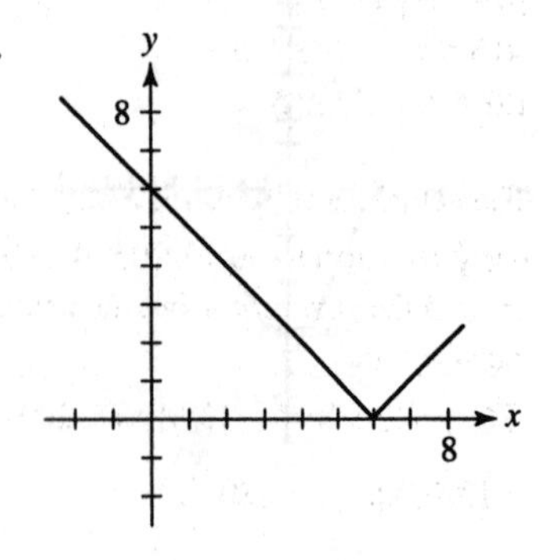

c. The axis of symmetry is the line $x = 6$. The range is $[0, \infty)$.

d. The function is decreasing on $(-\infty, 6)$ and increasing on $(6, \infty)$.

e. The graph of $y = |x|$ must be shifted 6 units to the right.

45. a. The slope of the left piece is -3.
The slope of the right piece is 3.
The vertex is $(-2, 0)$.

b.

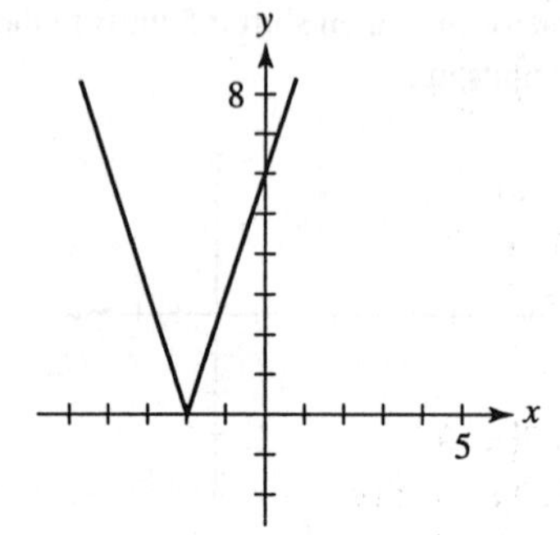

c. The axis of symmetry is the line $x = -2$. The range is $[0, \infty)$.

d. The function is decreasing on $(-\infty, -2)$ and increasing on $(-2, \infty)$.

e. The graph of $y = |x|$ must be stretched vertically by a factor of 3, then shifted 2 units to the left.

47. a. The slope of the left piece is -5.
The slope of the right piece is 5.
The vertex is $(-1, 7)$.

b.

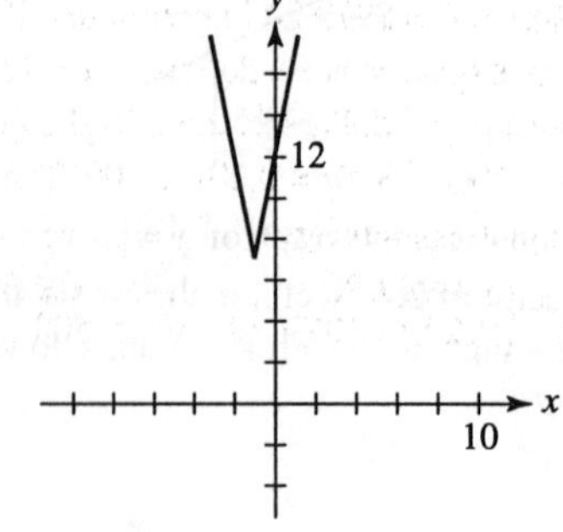

c. The axis of symmetry is the line $x = -1$. The range is $[7, \infty)$.

d. The function is decreasing on $(-\infty, -1)$ and increasing on $(-1, \infty)$.

e. The graph of $y = |x|$ must be stretched vertically by a factor of 5, then shifted 1 unit to the left and 7 units up.

49. $6|x| - 2 = 8$
$6|x| = 10$
$|x| = \frac{5}{3}$
$x = -\frac{5}{3}, \frac{5}{3}$

51. $|3 - 0.2x| = 0$
$3 - 0.2x = 0$
$0.2x = 3$
$x = 15$

53. $|3x + 7| < 19$
$-19 < 3x + 7 < 19$
$-26 < 3x < 12$
$-\frac{26}{3} < x < 4$

55. $|\pi x - 2\pi| > 2\pi$
$\pi x - 2\pi < -2\pi$ or $\pi x - 2\pi > 2\pi$
$\pi x < 0$ or $\pi x > 4\pi$
$x < 0$ or $x > 4$

57. $\left|\frac{x}{3} - 2\right| \geq -4$
Since $\left|\frac{x}{3} - 2\right|$ is always nonnegative, any real number is a solution.

59. Under Plan 2 Sarah must pay (12)($81.60) = $979.20 in premiums. If her medical expenses are x dollars and $x \leq 500$, then she must pay x dollars. If $x > 500$ she must pay $500 + 0.20(x - 500) = 0.20x + 400$. The total cost of the plan is therefore

$$g(x) = \begin{cases} x + 979.20 & \text{if } 0 \leq x \leq 500 \\ 0.20x + 1379.20 & \text{if } x > 500 \end{cases}$$

Under Plan 3, Sarah must pay (12)($88.80) = $1065.60 in premiums. If her medical expenses are x dollars and $x \leq 250$, then she must pay x dollars. If $x > 250$ she must pay $250 + 0.20(x - 250) = 0.20x + 200$. The total cost of the plan is therefore

$$h(x) = \begin{cases} x + 1065.60 & \text{if } 0 \leq x \leq 250 \\ 0.20x + 1265.60 & \text{if } x > 250 \end{cases}$$

61. Responses will vary.

63.

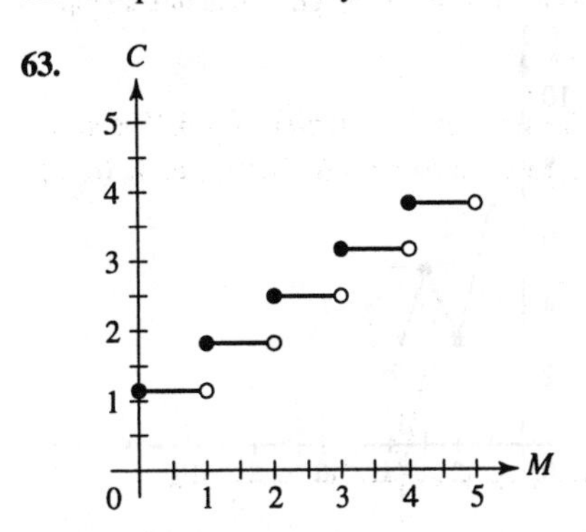

65. Pippig's official time in seconds is $2(3600) + 27(60) + 12 = 8832$ seconds. The error in recording time is therefore $E(t) = |t - 8832|$.

67. a. The measured length of one side of the room is 129 inches. If the actual length is L inches, then the error for that side is $|L - 129|$. The total error for the room will be $E(L) = 4|L - 129|$. (The doorways are irrelevant).

b. The error is $E(L) = 4|L - 129|$ inches, so the error will be no more than 2 inches as long as $4|L - 129| \leq 2$, or equivalently:
$|L - 129| \leq 0.5$
$-0.5 \leq L - 129 \leq 0.5$
$128.5 \leq L \leq 129.5$

69. a. The actual capacity is $13.90\pi r^2$, and designed capacity is $13.90\pi(4.280)^2$. The error is the deviation from the designed capacity, so

$$E(r) = \left|13.90\pi r^2 - 13.90\pi(4.280)^2\right| = 13.90\pi\left|r^2 - 4.280^2\right|.$$

b. The acceptable radii are the solutions to $13.90\pi\left|r^2 - 4.280^2\right| \le 10$. The graph of $y = 13.90\pi\left|r^2 - 4.280^2\right| - 10$ has r-intercepts at approximately $(\pm 4.25, 0)$ and $(\pm 4.31, 0)$, and is below the r-axis in $(-4.31, -4.25)$ and $(4.25, 4.31)$. The acceptable radii are in the interval $[4.25, 4.31]$.

71. a–b.

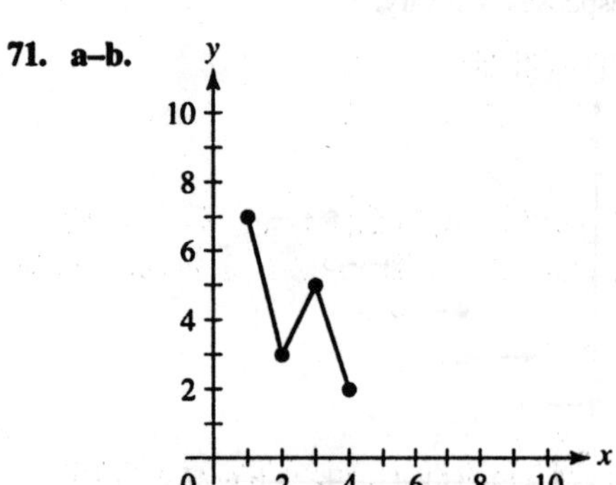

c. The slope of the first segment is $\frac{3-7}{2-1} = -4$. Its equation is $y - 7 = -4(x - 1)$, or $y = -4x + 11$. The slope of the second segment is $\frac{5-3}{3-2} = 2$. Its equation is $y - 3 = 2(x - 2)$, or $y = 2x - 1$. The slope of the third segment is $\frac{2-5}{4-3} = -3$. Its equation is $y - 5 = -3(x - 3)$, or $y = -3x + 14$.

d. $y = \begin{cases} -4x+11 & \text{if } 1 \le x < 2 \\ 2x-1 & \text{if } 2 \le x < 3 \\ -3x+14 & \text{if } 3 \le x \le 4 \end{cases}$

NUTSHELLS

Nutshell for Chapter 4 - LINEAR SYSTEMS

What's familiar?

- The primary focus of this chapter, systems of linear equations, should be familiar to you, especially if you have read Section A-10 in the Basic Algebra Review.
- Section 4-1 views systems of linear equations numerically, analytically and graphically.
- Section 4-1 explores the use of linear systems as models of physical problems.

What's new?

- Section 4-2 shows how certain linear systems can be solved more efficiently. This method only involves the coefficients of the linear equations which are put into a rectangular array called a *matrix*. This array can then be manipulated, by using certain rules, to solve the linear system.
- Section 4-3 looks at graphical solutions to systems of linear inequalities.
- In Section 4-3, you also learn a method for choosing a solution to a given system that is considered optimal ("best") in some respect.

List of hints for exercises in each chapter.

Chapter 4

- The terms *consistent*, *inconsistent*, *independent*, and *dependent* can be confusing at first. Write down the definitions several times before you begin the exercises.
- When you are going through the process of Gauss-Jordan elimination, be *very* careful. Mistakes are usually made through errors in arithmetic, not through lack of understanding.

Name:____________________

Chapter 4 Additional Exercises

Date:____________________

For Problems 1–2, solve the systems graphically.

1. $y = -2x + 5$
$y = 3x - 10$

1. ____________________

2. $2x + y = 10$
$3x - 2y = 15$

2. ____________________

For Problems 3–4, graph the systems. Decide whether each system is consistent and independent, inconsistent, or dependent.

3. $6x - 3y = 9$
$8x - 4y = 5$

3. ____________________

4. $8x - 2y = 10$
$3x - 15 = y$

4. ____________________

For Problems 5–10, solve the systems analytically.

5. $m - 3n = 7$
$4m + n = 2$

5. ____________________

6. $4S - 2R = 6$
$6S - R = -5$

6. ____________________

7. $2x - 2y = 4$
$3x + 5y = 14$

7. ____________________

8. $3x = 5y + 11$
$2x = 6y + 2$

8. ____________________

9. $a + b - 4c = 0$
$2a - 3b - 4c = 0$
$a - 2b + 3c = 0$

9. ____________________

10. $-x + 3y - 3z = -11$
$2x - y + 3z = 15$
$x + 2y + z = 8$

10. ____________________

11. The system
$4x + 2y = 8$
$16x + 8y = a$
is consistent for every value of a except one. What is the exceptional value of a?

11. ____________________

12. Write the augmented matrix of the system
$l + 4m - n = 4$
$-11m + 5n = -7$
$m - \frac{1}{5}n = \frac{7}{5}$

12. ____________________

13. Write a system of linear equations for the augmented matrix

$\begin{bmatrix} 2 & 3 & 1 & 9 \\ -1 & 4 & 3 & -7 \end{bmatrix}$ using any symbols you like as variables.

13. ____________________

Chapter 4 Additional Exercises *(cont.)*

Name:____________________

For Problems 14–15, decide whether the matrix is in reduced row-echelon form.

14. $\begin{bmatrix} 5 & 0 & 5 \\ 0 & 1 & 3 \\ 0 & 0 & 0 \end{bmatrix}$

14. ____________________

15. $\begin{bmatrix} 1 & 0 & 3 \\ 0 & 1 & -4 \end{bmatrix}$

15. ____________________

For Problems 16–17, use Gauss-Jordan elimination to obtain reduced row-echelon matrices.

16. $\begin{bmatrix} 2 & 1 & 12 \\ 3 & -1 & 8 \end{bmatrix}$

16. ____________________

17. $\begin{bmatrix} 1 & 1 & 1 & 6 \\ 3 & 1 & -1 & 6 \\ 2 & 1 & -1 & 4 \end{bmatrix}$

17. ____________________

For Problems 18–24, use Gauss-Jordan elimination to solve each linear system.

18. $\begin{aligned} 2a - b &= 8 \\ a + 3b &= 11 \end{aligned}$

18. ____________________

19. $\begin{aligned} l - 3m &= 7 \\ 5l + 2m &= 1 \end{aligned}$

19. ____________________

20. $\begin{aligned} x + y &= -1.5 \\ 0.75x - 0.5y &= -0.5 \end{aligned}$

20. ____________________

21. $\begin{aligned} x + y - z &= 7 \\ x + 2y + 2z &= 4 \\ 2x - y + z &= -1 \end{aligned}$

21. ____________________

22. $\begin{aligned} a + 2b - 3c &= 5 \\ 2a + b + c &= -4 \\ a + b - 2c &= 3 \end{aligned}$

22. ____________________

23. $\begin{aligned} 2x + 4y &= 10 \\ x + 2y &= 5 \end{aligned}$

23. ____________________

24. $\begin{aligned} 3m - 9n &= 9 \\ -m + 3n &= 6 \end{aligned}$

24. ____________________

For Problems 25–26, write the reduced linear system and write the solutions in standard form. Use any symbols you like as variables.

25. $\begin{bmatrix} 1 & 0 & -2 & -8 \\ 0 & 1 & 1 & 9 \end{bmatrix}$

25. ____________________

26. $\begin{bmatrix} 1 & 0 & 4 & 0 & 2 \\ 0 & 1 & -6 & 0 & 8 \\ 0 & 0 & 0 & 1 & -10 \end{bmatrix}$

26. ____________________

For Problems 27–28, solve the system and write the solutions in standard form.

27. $\begin{aligned} x + y + 4z &= -8 \\ y - z &= -4 \end{aligned}$

27. ____________________

Chapter 4 Additional Exercises *(cont.)*

Name:____________________

28.
$$\begin{aligned} A - 4B + C &= 18 \\ 2A - 7B - 2C &= 4 \\ 3A - 11B - C &= 22 \end{aligned}$$

28. ____________________

For Problems 29–31, graph the inequality.

29. $2x - y > 5$

29.

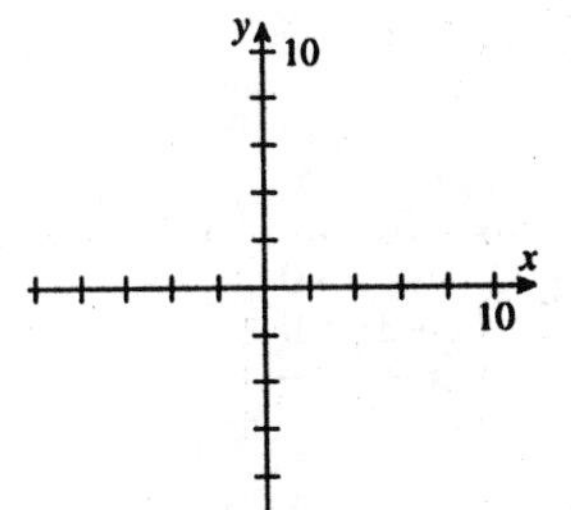

30. $3x + y \geq 4$

30.

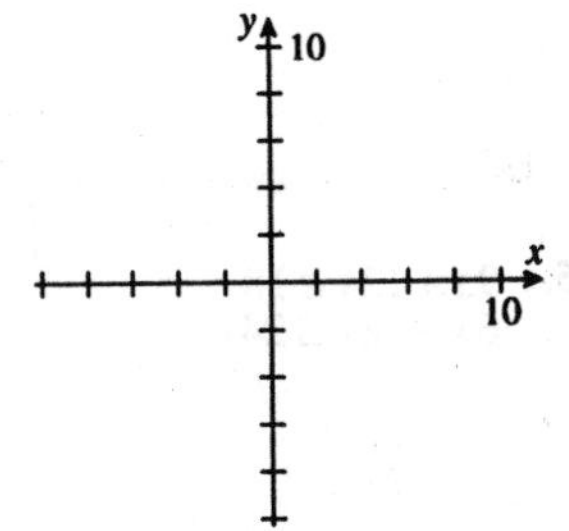

31. $5x - 2y < 9$

31.

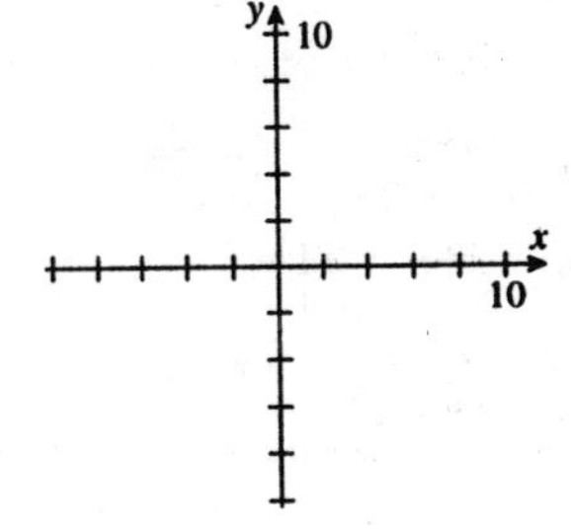

For Problems 32–37, find the feasible region.

32.
$$\begin{aligned} 4x + y &\leq 8 \\ 2x + 3y &\leq 14 \\ x &\geq 0 \\ y &\geq 0 \end{aligned}$$

32.

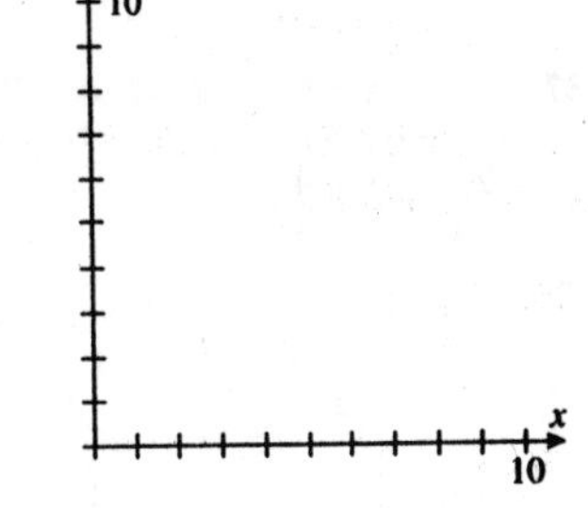

Chapter 4 Additional Exercises *(cont.)*

Name:________________________

33. $4x + y \le 8$
$x - y \le -2$
$x \ge 0$

33.

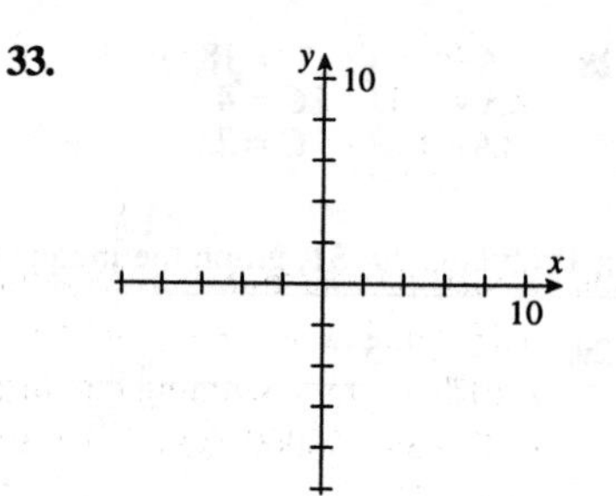

34. $y > 0$
$4x - 3y > 0$
$x - y < -1$

34.

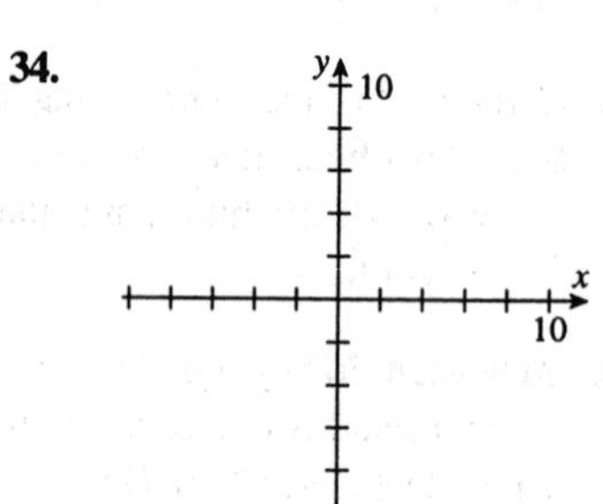

35. $9x + 12y \ge 18$
$6x + 8y \le 24$

35.

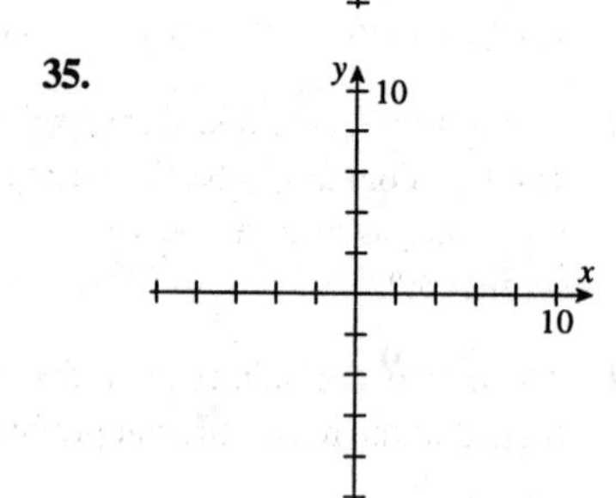

36. $x + y \ge 1$
$2x + 3y < 1$
$x > -3$

36.

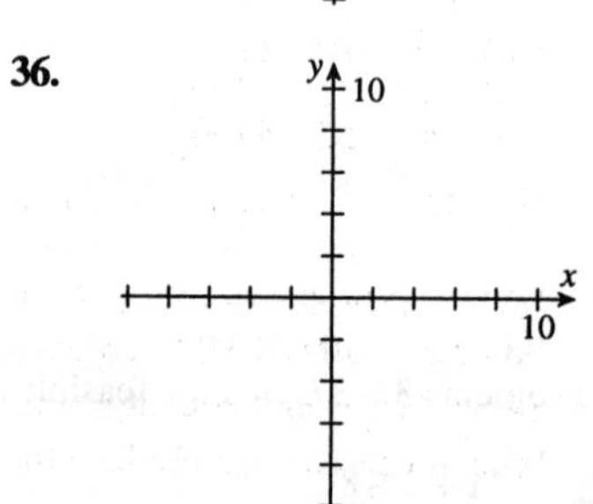

37. $x < 4$
$x - 5y = 2$
$2x - y > 4$
$x - y < 4$

37.

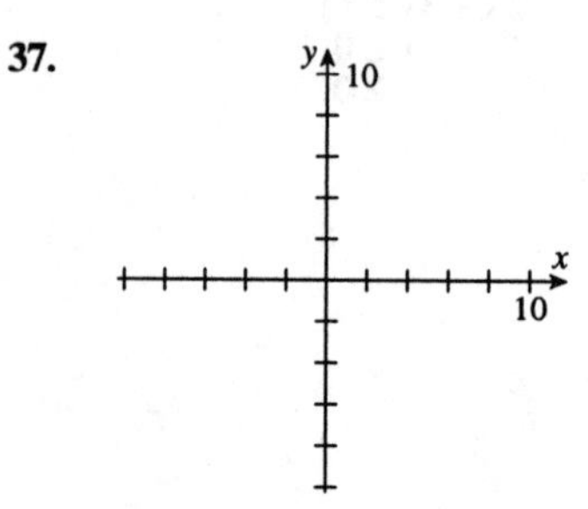

38. Maximize $2x - y$ subject to:
$4x + y \le 8$
$x - y \ge -2$
$x \ge 0$
$y \ge 0$

38. ________________________

Chapter 4 Additional Exercises *(cont.)*

Name:____________________

39. Minimize $5x + 2y$ subject to

$$\begin{aligned} x + y &\geq 4 \\ x &\geq 2 \\ y &\geq 0 \end{aligned}$$

39. ____________________

40. Robert invests money in two stocks. Stock *AB* yields a return of 5.5% and is a stable investment. Stock *CD* is more risky, but yields a 10% return. Assuming that these rates remain the same, how much of Robert's $6000 should he invest in *CD* if he wants a total return of 8.5%?

40. ____________________

41. Carin has two jobs, one paying $9.00 per hour and the other paying $12.50 per hour. If she worked a total of 43 hours last week and her pay was $453.50, how many hours did she work at the job paying $12.50 per hour?

41. ____________________

42. Kyle needs to buy food for lunch for the next two weeks. He decides to buy packages of rice soup which cost $0.59 and cans of lentil soup which cost $2.10. How many of each type of soup must he buy to get a total of 17 packages and cans at a cost of $20.60?

42. ____________________

43. A charity receives donations from three individuals, totaling $550. The first donation is half as large as the second, and the third is four times as large as the second. How much did each individual contribute?

43. ____________________

44. A company specializes in producing three products. The company finds that the most efficient production shedule is the solution of the matrix

$$\begin{bmatrix} 10 & 20 & 20 & 1900 \\ 30 & 40 & 30 & 3400 \\ 60 & 40 & 50 & 4700 \end{bmatrix}$$

How many of each product should be produced?

44. ____________________

45. Leon is printing up two types of brochures. The black and white brochure costs $1.90 to produce, while the color version costs $4.85. Write an inequality to express the combinations of brochures Leon can produce for no more than $375.

45. ____________________

Chapter 4 Additional Exercises *(cont.)*

Name:____________________

For Problems 46–48, use the objective function $5x + 4y$ subject to the constraints

$$\begin{aligned} x + y &\le 11 \\ 2x - y &\le -5 \\ 3x - y &\le 13 \\ x &\ge 0 \\ y &\ge 0 \end{aligned}$$

46. Find the feasible region of the system of inequalities. **46.**

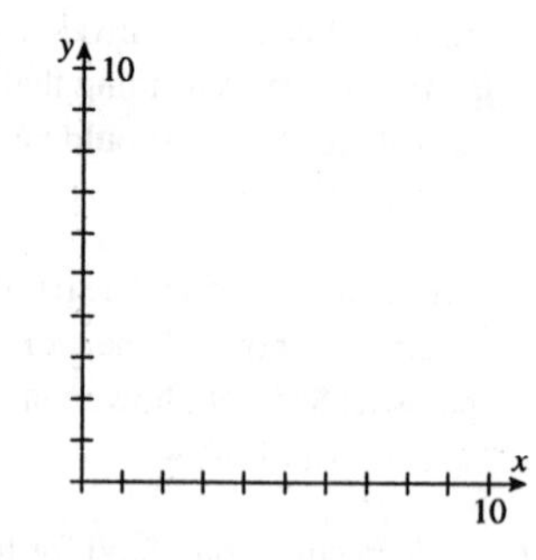

47. Find the coordinates of the corner points of the feasible region. **47.** ____________________

48. Evaluate the objective function at each corner point.
Which point gives the maximum value of the objective function? **48.** ____________________

Lisa wants to save money in two accounts. One account provides 5.4% interest and has a minimum required deposit of \$1500. The other account provides 8.6% interest, but has a limit of \$1000 on the amount deposited. Lisa has \$3000 to deposit.

49. Find Lisa's objective function and constraints. **49.** ____________________

50. How much should Lisa deposit in each account? **50.** ____________________

A Mathematical Looking Glass
Fertilizers
(to follow Section 4-1)

When you buy fertilizer at a garden store, it comes in bags which display three numbers. These numbers represent the percentages of total nitrogen, available phosphoric acid and soluble potash, respectively. Last fall, farmers Melvin and Jeannie Stauss of Mequon, WI, bought 500 pounds of 27-6-4 and 800 pounds of 5-10-5 on sale. Their corn requires a 10-20-10 mix, which they attempted to obtain by mixing the bought fertilizer with some combination of 35-0-0, 0-46-0, and/or 0-0-64.

Exercises

1. (Making Observations) If Jeannie writes a system of equations to find how much of each fertilizer they should use:

   a. How many solutions can she expect the system to have?

   b. How many variables will she use?

   c. How many equations should she write?

2. (Problem-Solving) Write and solve a system of linear equations to solve Melvin and Jeannie's problem.

3. (Writing to Learn) Since one of the purchased fertilizers was already more nitrogen-heavy than their desired mix, you might have expected that they could solve their problem without adding any 35-0-0 to the mix. Explain why this did not occur.

Chapter 4

Section 4-1

1.

Price per shirt	Demand	Supply
$3.50	2500	400
$5.50	2100	600
$7.50	1700	800
$9.50	1300	1000
$11.50	900	1200
$13.50	500	1400

3. If x represents the price per shirt, in dollars, and y represents the demand for shirts, then the graph of the demand function passes through (3.50, 2500) and (13.50, 500). Its equation is $y - 2500 = -200(x - 3.50)$, or equivalently, $y = 3200 - 200x$.

5. Subtracting,
$$\begin{aligned} y &= 3200 - 200x \\ y &= 50 + 100x \\ \hline 0 &= 3150 - 300x \end{aligned}$$
$300x = 3150$
$x = 10.50$
$y = 3200 - 200(10.50) = 1100$

7. The solution of the system is the point where the graphs intersect, at (1, 5).

9. Solve both equations for y, and graph $y = \frac{1}{2}(7 - x)$ and $y = 3x - 7$. The solution of the system is the point where the graphs intersect, at (3, 2).

11. For the system in Exercise 8:
Analytical solution:
$L = 1.5W - 0.2$
$L = 3.5W + 6.4$
Replacing L by $3.5W + 6.4$ in the first equation:
$3.5W + 6.4 = 1.5W - 0.2$
$2W = -6.6$
$W = -3.3$
$L = 1.5W - 0.2 = 1.5(-3.3) - 0.2 = -5.15$
Numerical solution:

W	$1.5W - 0.2$	$3.5W + 6.4$
0	−0.2	6.4
1	1.3	9.9
2	2.8	13.4

Stop! The difference between the values in the second and third columns is increasing as W becomes larger. Try going in the other direction.

−1	−1.7	2.9
−2	−3.2	−0.6
−3	−4.7	−4.1
−4	−6.2	−7.6

When $W = -3$, the value of $1.5W - 0.2$ is less than that of $3.5W + 6.4$. When $W = -4$, the value of $1.5W - 0.2$ is greater than that of $3.5W + 6.4$. Therefore, the solution has a W-value between −3 and −4. Generate more entries in the table.

−3.1	−4.85	−4.45
−3.2	−5.00	−4.80
−3.3	−5.15	−5.15

Since the L-values are both −5.15 when $W = -3.3$, the solution is $(W, L) = (-3.30, -5.15)$.

For the system in Exercise 9:
Analytical solution:
$x + 2y = 7$
$3x - y = 7$
Solving for y in the second equation, $y = 3x - 7$.
Replacing y by $3x - 7$ in the first equation:
$x + 2(3x - 7) = 7$
$x + 6x - 14 = 7$
$7x = 21$
$x = 3$
$y = 3x - 7 = 3(3) - 7 = 2$
Numerical solution:
Solve each equation for y:

$y = \frac{7-x}{2}$
$y = 3x - 7$

x	$\frac{7-x}{2}$	$3x-7$
1	3	–4
2	2.5	–1
3	2	2

Since the y-values are both 2 when $x = 3$, the solution is (3, 2).

13. The terms $0.65L$, $0.80M$, and $0.65C$ represent the number of days per month Paul and his helpers will spend on landscaping, masonry, and carpentry, respectively. The equation $0.65L + 0.80M + 0.65C = 72$ expresses the condition that the total amount of time Paul and his helpers spend working should equal 72 days per month.

15. Multiplying each equation in the system by 20 will eliminate decimals:
$13L + 16M + 13C = 1440$
$5L + 3M + 4C = 480$
$2L + M + 3C = 240$
Multiply the third equation by –16, and add it to the first equation to eliminate M:

$$\begin{aligned} 13L + 16M + 13C &= 1440 \\ -32L - 16M - 48C &= -3840 \\ \hline -19L \qquad\quad - 35C &= -2400 \end{aligned}$$

Multiply the third equation by –3, and add it to the second equation to eliminate M:

$$\begin{aligned} 5L + 3M + 4C &= 480 \\ -6L - 3M - 9C &= -720 \\ \hline -L \qquad\quad - 5C &= -240 \end{aligned}$$

In the resulting system in L and C, multiply the second equation by –19, and add it to the first equation to eliminate L:

$$\begin{aligned} -19L - 35C &= -2400 \\ 19L + 95C &= 4560 \\ \hline 60C &= 2160 \end{aligned}$$

$C = 36$
Replacing C by 36 in the equation $-L - 5C = -240$:
$-L - 5(36) = -240$
$L = 60$
Replacing L and C by 60 and 36, respectively, in the equation $2L + M + 3C = 240$:
$2(60) + M + 3(36) = 240$
$M = 12$
Everyone can be utilized to full capacity if the company schedules 60 days of landscaping work, 12 days of masonry, and 36 days of carpentry each month.

17.

x	y (1st equation)	y (2nd equation)
1	0	0
2	1	1
3	2	2
4	3	3

The system has infinitely many solutions. The table indicates that for every value of x, the corresponding values of y in the two equations are equal.

19. The graphs coincide, so the system is dependent.

21. The graphs coincide, so the system is dependent.

23. The system in Exercise 19 can be rewritten
$3S - 5T = -4$
$9S - 15T = -12$
Multiplying the first equation by 3 yields
$9S - 15T = -12$
$9S - 15T = -12$
Subtracting yields $0 = 0$, so the system is dependent.

25. Analytically:
$y = 5$
$4x - 3y = 2$
Replace y by 5 in the second equation:
$4x - 3(5) = 2$
$4x = 17$
$x = 4.25$
The solution is $(x, y) = (4.25, 5)$.
Graphically: Solve both equations for y, and graph $y = 5$ and $y = \frac{1}{3}(4x - 2)$. The solution of the system is the point where the graphs intersect, at (4.25, 5).

27. Analytically:
$2T + 5V = 80$
$3T - 2V = 82$

Multiply the first equation by 2, the second by 5, and add:

$$\begin{array}{l} 4T+10V=160 \\ \underline{15T-10V=410} \\ 19T \qquad\quad =570 \end{array}$$

$T=30$

Replace T by 30 in the first equation:

$2(30)+5V=80$

$5V=20$

$V=4$

The solution is $(T, V)=(30, 4)$.

Graphically: Solve both equations for V, and graph $V=\frac{1}{5}(80-2T)$ and $V=\frac{1}{2}(3T-82)$.

The solution of the system is the point where the graphs intersect, at (30, 4).

29. Analytically:

$y=2.5x-7$

$x=0.4y+1$

Replace y by $2.5x-7$ in the second equation:

$x=0.4(2.5x-7)+1$

$x=x-1.8$

$0=-1.8$

The system has no solution.

Graphically: Solve both equations for y, and graph $y=2.5x-7$ and $y=\frac{1}{0.4}(x-1)$. The graphs are parallel, so the system has no solution.

31. Multiply the second equation by −2 and the third equation by 3, and add.

$$\begin{array}{l} -4x-6y+6z=-16 \\ \underline{3x+24y-6z=\ \ 24} \\ -x+18y \qquad =\ \ 8 \end{array}$$

Add the resulting equation to the first equation in the system.

$$\begin{array}{l} x-2y=\ 8 \\ \underline{-x+18y=\ 8} \\ 16y=16 \end{array}$$

$y=1$

Substitute 1 for y in the equation $x-2y=8$ to obtain $x=10$. Substitute 10 for x and 1 for y in the equation $x+8y-2z=8$ to obtain $z=5$.

33. Add the first and third equations.

$$\begin{array}{l} 5p+2q+r=0.3 \\ \underline{3p-2q+6r=0.3} \\ 8p \qquad +7r=0.6 \end{array}$$

Multiply the second equation by 2 and add it to the first.

$$\begin{array}{l} 5p+2q+r=0.3 \\ \underline{4p-2q+6r=0.2} \\ 9p \qquad +7r=0.5 \end{array}$$

In the resulting 2×2 system, subtract the first equation from the second.

$$\begin{array}{l} 9p+7r=\ 0.5 \\ \underline{8p+7r=\ 0.6} \\ p \qquad =-0.1 \end{array}$$

Substitute −0.1 for p in the equation $8p+7r=0.6$ to obtain $r=0.2$. Substitute −0.1 for p and 0.2 for r in the equation $2p-q+3r=0.1$ to obtain $q=0.3$.

35. The graphs intersect at a unique point, so the system is consistent and independent.

37. The graphs coincide, so the system is dependent.

39. a.

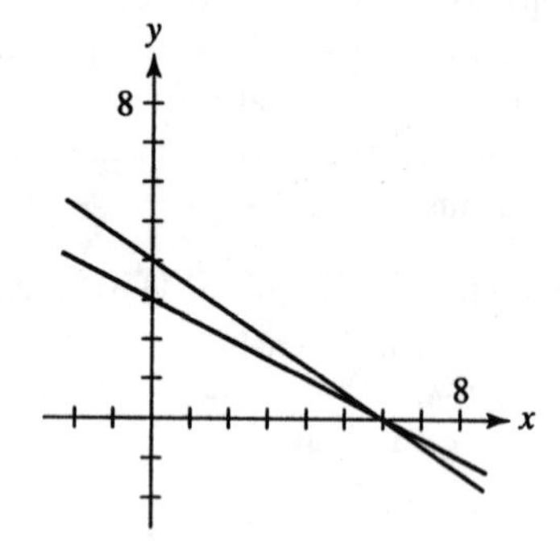

b.

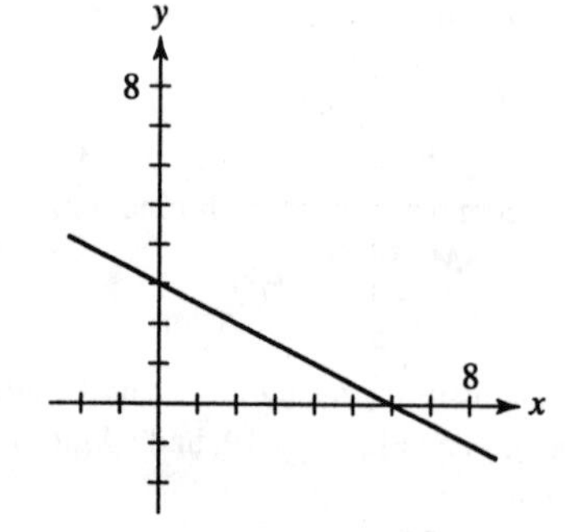

c.

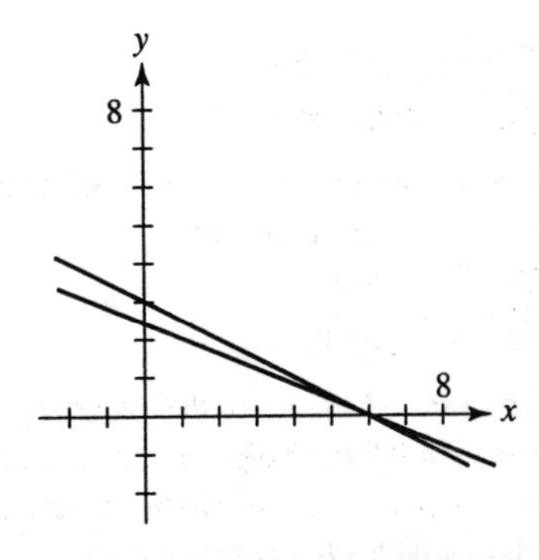

The system in part (b) is dependent because the graphs coincide.

41. a.

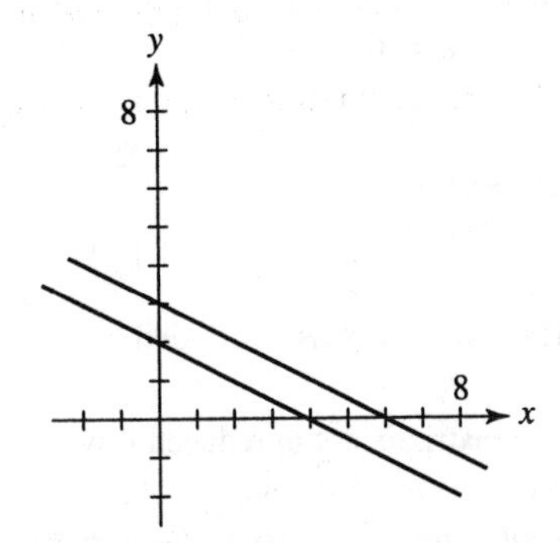

b.

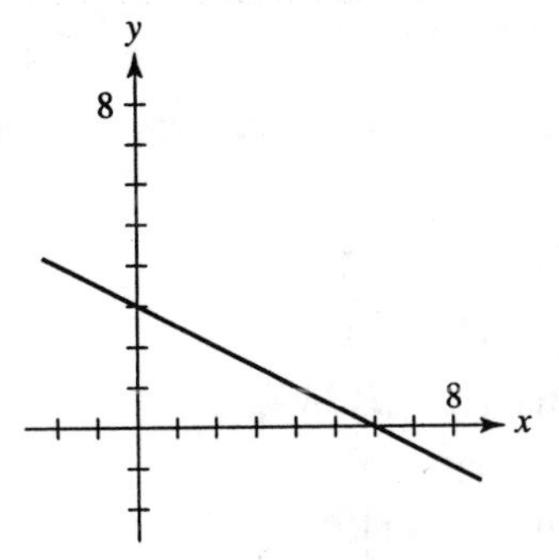

c.

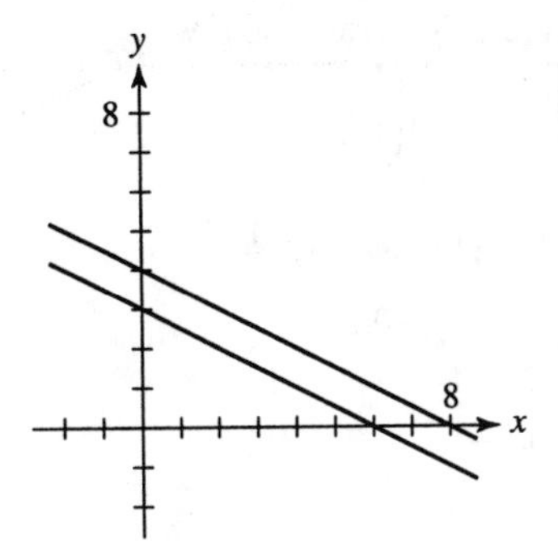

The system in part (b) is dependent because the graphs coincide.

43. If T = the amount of Dave's contribution which goes to TIAA, and C = the amount which goes to CREF, then $T + C = 4000$, since the sum of the two portions is the entire contribution.
Also, the interest from TIAA will be $0.087T$ and the interest from CREF will be $0.105C$. The sum of these two quantities must be equal to $0.09(4000) = 360$. Thus:
$T + C = 4000$
$0.087T + 0.105C = 0.09$
Solve for C in the first equation.
$C = 4000 - T$
Replace C by $4000 - T$ in the second equation.
$0.087T + 0.105(4000 - T) = 360$
$87T + 105(4000 - T) = 360{,}000$
$-18T = -60{,}000$
$T = \frac{10\ 000}{3} \cong 3333.33$
Thus, \$3333.33 of Dave's contribution should go to TIAA.

45. If L = the portion of McLean Deluxes in McDonald's hamburger sales, and F = the portion of other hamburgers, then $L + F = 1$, since the sum of the two portions is the entirety of hamburger sales.
Also, each pound of McLean Deluxe hamburger contains 90 grams. If each pound of hamburger sold by McDonald's consists of L pounds of McLean Deluxe and F pounds of other hamburger, then it will contain $40L + 90F$ grams of fat. Since the previous level of fat in McDonald's hamburgers was 90 grams per pound, a 10% reduction would result in a level of 81 grams per pound. Thus, $40L + 90F = 81$. The resulting system is:
$L + F = 1$
$40L + 90F = 81$
Solving for F in the first equation:
$F = 1 - L$
Replacing F by $1 - L$ in the second equation:
$40L + 90(1 - L) = 81$
$-50L = -9$
$L = 0.18$
Thus 18% of McDonald's total sales must be McLean Deluxes.

47. Let $x =$ the weight of the first alloy in pounds and $y =$ the weight of the second alloy in pounds. Then the weight of the entire mixture is $200{,}000 + x + y$. The amount of chromium in the mixture is $0.15(200{,}000) + 0.50x$. Since this must be 18% of the weight of the entire mixture, we have
$0.15(200{,}000) + 0.50x = 0.18(200{,}000 + x + y)$
Equivalently, $0.32x - 0.18y = 6000$
The amount of nickel in the mixture is $0.05(200{,}000) + 0.50y$. Since this must be 8% of the weight of the entire mixture, we have
$0.05(200{,}000) + 0.50y = 0.08(200{,}000 + x + y)$
Equivalently, $-0.08x + 0.42y = 6000$
Multiply the second equation by 4 and add it to the first.

$$\begin{aligned} 0.32x - 0.18y &= 6000 \\ -0.32x + 1.68y &= 24000 \\ \hline 1.50y &= 30000 \\ y &= 20{,}000 \end{aligned}$$

Substitute 20,000 for y in the equation $0.32x - 0.18y = 6000$ to obtain $x = 30{,}000$.

49. Responses will vary, but most students will probably mention *Writing an Equation* and *Examining a Related Problem.*

51. A system of three linear equations in two variables usually has no solutions.

53. A system of one linear equation in two variables has infinitely many solutions.

55. The first three planes usually have just one point in common (see Figure 4-2 in the text). The fourth plane usually misses that point, so there are usually no points common to all four planes.

57. The two planes usually intersect in a line, containing infinitely many points.

59. Exercises 55–58 suggest that an $m \times 3$ linear system usually has no solutions if $m > 3$, and infinitely many solutions if $m < 3$. We already know that there is usually exactly one solution if $m = 3$.

Section 4-2

1. $\begin{bmatrix} 3 & 2 & 6 \\ 1 & 7 & 5 \end{bmatrix}$

3. $\begin{bmatrix} 2 & 3 & 5 \\ -1 & -4 & 0 \\ 3 & 0 & 12 \\ 0 & 4 & -4 \end{bmatrix}$

5. If the variables are x, y, and z:
$x + 3y + 9z = 5$
$\frac{1}{3}x + 4y - 13z = 7$
$-2y + 3z = 16$

7. If the variables are s and t:
$23s + 6t = -2$
$7s + 31t = 12$

9. The third row was multiplied by $\frac{1}{60}$.

11. The matrix is not in reduced row-echelon form because the third column, which contains the leading nonzero entry in the third row, contains other nonzero entries.

13. The matrix is in reduced row-echelon form.

15. Add –4 times row 1 to row 2:
$\begin{bmatrix} 1 & -3 & 5 \\ 0 & 17 & -17 \end{bmatrix}$
Multiply row 2 by $\frac{1}{17}$:
$\begin{bmatrix} 1 & -3 & 5 \\ 0 & 1 & -1 \end{bmatrix}$
Add 3 times row 2 to row 1:
$\begin{bmatrix} 1 & 0 & 2 \\ 0 & 1 & -1 \end{bmatrix}$

17. Multiply row 1 by $\frac{1}{2}$:
$\begin{bmatrix} 1 & 0 & -0.5 & -1 \\ 0 & 1 & -2 & 1 \\ 4 & -1 & 1 & 0 \end{bmatrix}$

Add –4 times row 1 to row 3:

$$\begin{bmatrix} 1 & 0 & -0.5 & -1 \\ 0 & 1 & -2 & 1 \\ 0 & -1 & 3 & 4 \end{bmatrix}$$

Add row 2 to row 3:

$$\begin{bmatrix} 1 & 0 & -0.5 & -1 \\ 0 & 1 & -2 & 1 \\ 0 & 0 & 1 & 5 \end{bmatrix}$$

Add 0.5 times row 3 to row 1:

$$\begin{bmatrix} 1 & 0 & 0 & 1.5 \\ 0 & 1 & -2 & 1 \\ 0 & 0 & 1 & 5 \end{bmatrix}$$

Add 2 times row 3 to row 2:

$$\begin{bmatrix} 1 & 0 & 0 & 1.5 \\ 0 & 1 & 0 & 11 \\ 0 & 0 & 1 & 5 \end{bmatrix}$$

19. From the reduced row-echelon matrix in Exercise 15, the solution is $(x, y) = (2, -1)$.

21. From the reduced row-echelon matrix in Exercise 17, the solution is $(x, y, z) = (1.5, 11, 5)$.

23. The system can be constructed in a manner similar to that used in Exercises 13 and 14 of Section 4-1. The system is:

$$0.25L + 0.15M + 0.20C = 24$$
$$0.60L + 0.80M + 0.70C = 56$$
$$0.15L + 0.05M + 0.10C = 8$$

25. a. The augmented matrix of the system is

$$\begin{bmatrix} 2 & -3 & 10 \\ 4 & -6 & 15 \end{bmatrix}$$

Adding –2 times row 1 to row 2:

$$\begin{bmatrix} 2 & -3 & 10 \\ 0 & 0 & -5 \end{bmatrix}$$

Since row 2 now corresponds to the equation $0 = -5$, the system has no solution.

b–c. The graph shows that the system has no solution because the graphs of the two equations in the system have no point in common. The matrix shows the same thing because the system is equivalent to one containing the self-contradictory equation $0 = -5$.

27. The augmented matrix of the system is:

$$\begin{bmatrix} 0.25 & 0.15 & 0.20 & 16 \\ 0.60 & 0.80 & 0.70 & 56 \\ 0.15 & 0.05 & 0.10 & 8 \end{bmatrix}$$

Multiply row 1 by 4:

$$\begin{bmatrix} 1 & 0.60 & 0.80 & 64 \\ 0.60 & 0.80 & 0.70 & 56 \\ 0.15 & 0.05 & 0.10 & 8 \end{bmatrix}$$

Add –0.60 times row 1 to row 2:

$$\begin{bmatrix} 1 & 0.60 & 0.80 & 64 \\ 0 & 0.44 & 0.22 & 17.6 \\ 0.15 & 0.05 & 0.10 & 8 \end{bmatrix}$$

Add –0.15 times row 1 to row 3:

$$\begin{bmatrix} 1 & 0.60 & 0.80 & 64 \\ 0 & 0.44 & 0.22 & 17.6 \\ 0 & -0.04 & -0.02 & -1.6 \end{bmatrix}$$

Interchange rows 2 and 3:

$$\begin{bmatrix} 1 & 0.60 & 0.80 & 64 \\ 0 & -0.04 & -0.02 & -1.6 \\ 0 & 0.44 & 0.22 & 17.6 \end{bmatrix}$$

Multiply row 2 by –25:

$$\begin{bmatrix} 1 & 0.60 & 0.80 & 64 \\ 0 & 1 & 0.50 & 40 \\ 0 & 0.44 & 0.22 & 17.6 \end{bmatrix}$$

Add –0.60 times row 2 to row 1:

$$\begin{bmatrix} 1 & 0 & 0.50 & 40 \\ 0 & 1 & 0.50 & 40 \\ 0 & 0.44 & 0.22 & 17.6 \end{bmatrix}$$

Add –0.44 times row 2 to row 3:

$$\begin{bmatrix} 1 & 0 & 0.50 & 40 \\ 0 & 1 & 0.50 & 40 \\ 0 & 0 & 0 & 0 \end{bmatrix}$$

29. If the variables are x, y, and z, then the reduced linear system is

$$x + 2z = 8$$
$$y + 4z = 7$$

Assign a value z_0 to the nonleading variable z. Then

$$x = 8 - 2z_0$$
$$y = 7 - 4z_0$$
$$z = z_0$$

31. If the variables are $w, x, y,$ and z, then the reduced linear system is

$$\begin{aligned} w+10y \quad &= 0.5 \\ x+20y \quad &= 2.5 \\ z &= 4 \end{aligned}$$

Assign a value y_0 to the nonleading variable y. Then

$$\begin{aligned} w &= 0.5-10y_0 \\ x &= 2.5-20y_0 \\ y &= y_0 \\ z &= 4 \end{aligned}$$

33. The augmented matrix of the system is

$$\begin{bmatrix} 1 & 1 & 1 & 1 & 1 \\ 0 & 1 & 1 & 1 & 2 \\ 0 & 0 & 1 & 1 & 3 \end{bmatrix}$$

Add –1 times row 2 to row 1:

$$\begin{bmatrix} 1 & 0 & 0 & 0 & -1 \\ 0 & 1 & 1 & 1 & 2 \\ 0 & 0 & 1 & 1 & 3 \end{bmatrix}$$

Add –1 times row 3 to row 2:

$$\begin{bmatrix} 1 & 0 & 0 & 0 & -1 \\ 0 & 1 & 0 & 0 & -1 \\ 0 & 0 & 1 & 1 & 3 \end{bmatrix}$$

Rewrite the reduced row-echelon matrix as a system of equations.

$$\begin{aligned} A &= -1 \\ B &= -1 \\ C+D &= 3 \end{aligned}$$

The solution is

$A=-1$

$B=-1$

$C=3-d$

$D=d$ for any real value of d.

35. The augmented matrix of the system is:

$$\begin{bmatrix} 1 & -1 & 1 & 1 & 0 \\ 1 & 1 & 1 & -1 & 0 \\ 0 & 1 & -1 & -1 & 0 \end{bmatrix}$$

Add –1 times row 1 to row 2:

$$\begin{bmatrix} 1 & -1 & 1 & 1 & 0 \\ 0 & 2 & 0 & -2 & 0 \\ 0 & 1 & -1 & -1 & 0 \end{bmatrix}$$

Multiply row 2 by 0.5:

$$\begin{bmatrix} 1 & -1 & 1 & 1 & 0 \\ 0 & 1 & 0 & -1 & 0 \\ 0 & 1 & -1 & -1 & 0 \end{bmatrix}$$

Add row 2 to row 1:

$$\begin{bmatrix} 1 & 0 & 1 & 0 & 0 \\ 0 & 1 & 0 & -1 & 0 \\ 0 & 1 & -1 & -1 & 0 \end{bmatrix}$$

Add –1 times row 2 to row 3:

$$\begin{bmatrix} 1 & 0 & 1 & 0 & 0 \\ 0 & 1 & 0 & -1 & 0 \\ 0 & 0 & -1 & 0 & 0 \end{bmatrix}$$

Multiply row 3 by –1:

$$\begin{bmatrix} 1 & 0 & 1 & 0 & 0 \\ 0 & 1 & 0 & -1 & 0 \\ 0 & 0 & 1 & 0 & 0 \end{bmatrix}$$

Add –1 times row 3 to row 1:

$$\begin{bmatrix} 1 & 0 & 0 & 0 & 0 \\ 0 & 1 & 0 & -1 & 0 \\ 0 & 0 & 1 & 0 & 0 \end{bmatrix}$$

Rewrite the reduced row-echelon matrix as a system of equations.

$$\begin{aligned} p &= 0 \\ q-s &= 0 \\ r &= 0 \end{aligned}$$

The solution is

$p=0$

$q=s_0$

$r=0$

$s=s_0$ for any real value of s_0.

37. $$\begin{bmatrix} 1 & 1 & 1 & 1 & 0 \\ 3 & -1 & 19 & 6 & \frac{1}{7} \\ 2 & 0 & 0 & 0 & 8 \end{bmatrix}$$

39. $$\begin{bmatrix} 2 & -3 & -4 & -5 & 0 & 0.001 \\ 0 & 2 & -3 & -4 & -5 & -0.001 \end{bmatrix}$$

41. The matrix is not in reduced row-echelon form, since the column that contains the leading nonzero entry in the second row also contains a nonzero entry above it.

43. The matrix is in reduced row-echelon form.

45. Multiply row 1 by $\frac{1}{2}$:

$$\begin{bmatrix} 1 & -\frac{7}{2} & 3 \\ 5 & -1 & 4 \end{bmatrix}$$

Add –5 times row 1 to row 2:

$$\begin{bmatrix} 1 & -\frac{7}{2} & 3 \\ 0 & 16.5 & -11 \end{bmatrix}$$

Multiply row 2 by $\frac{1}{16.5}$:

$$\begin{bmatrix} 1 & -\frac{7}{2} & 3 \\ 0 & 1 & -\frac{2}{3} \end{bmatrix}$$

Add $\frac{7}{2}$ times row 2 to row 1:

$$\begin{bmatrix} 1 & 0 & \frac{2}{3} \\ 0 & 1 & -\frac{2}{3} \end{bmatrix}$$

47. Add –1 times row 1 to row 2:

$$\begin{bmatrix} 1 & 2 & 3 & 0 \\ 0 & 0 & -2 & 0 \\ 2 & 1 & 1 & 0 \end{bmatrix}$$

Add –2 times row 1 to row 3:

$$\begin{bmatrix} 1 & 2 & 3 & 0 \\ 0 & 0 & -2 & 0 \\ 0 & -3 & -5 & 0 \end{bmatrix}$$

Interchange rows 2 and 3:

$$\begin{bmatrix} 1 & 2 & 3 & 0 \\ 0 & -3 & -5 & 0 \\ 0 & 0 & -2 & 0 \end{bmatrix}$$

Multiply row 2 by $-\frac{1}{3}$:

$$\begin{bmatrix} 1 & 2 & 3 & 0 \\ 0 & 1 & \frac{5}{3} & 0 \\ 0 & 0 & -2 & 0 \end{bmatrix}$$

Add –2 times row 2 to row 1:

$$\begin{bmatrix} 1 & 0 & -\frac{1}{3} & 0 \\ 0 & 1 & \frac{5}{3} & 0 \\ 0 & 0 & -2 & 0 \end{bmatrix}$$

Multiply row 3 by $-\frac{1}{2}$:

$$\begin{bmatrix} 1 & 0 & -\frac{1}{3} & 0 \\ 0 & 1 & \frac{5}{3} & 0 \\ 0 & 0 & 1 & 0 \end{bmatrix}$$

Add $\frac{1}{3}$ times row 3 to row 1:

$$\begin{bmatrix} 1 & 0 & 0 & 0 \\ 0 & 1 & \frac{5}{3} & 0 \\ 0 & 0 & 1 & 0 \end{bmatrix}$$

Add $-\frac{5}{3}$ times row 3 to row 2:

$$\begin{bmatrix} 1 & 0 & 0 & 0 \\ 0 & 1 & 0 & 0 \\ 0 & 0 & 1 & 0 \end{bmatrix}$$

49. The augmented matrix of the system is

$$\begin{bmatrix} 1 & -4 & 6 \\ 4 & -16 & 18 \end{bmatrix}$$

Add –4 times row 1 to row 2:

$$\begin{bmatrix} 1 & -4 & 6 \\ 0 & 0 & -6 \end{bmatrix}$$

Since row 2 now represents the equation $0 = -6$, the system is inconsistent.

51. The augmented matrix of the system is

$$\begin{bmatrix} 2 & 1 & 1 & 5 \\ 4 & -1 & 3 & 4 \end{bmatrix}$$

Multiply row 1 by $\frac{1}{2}$:

$$\begin{bmatrix} 1 & \frac{1}{2} & \frac{1}{2} & \frac{5}{2} \\ 4 & -1 & 3 & 4 \end{bmatrix}$$

Add –4 times row 1 to row 2:

$$\begin{bmatrix} 1 & \frac{1}{2} & \frac{1}{2} & \frac{5}{2} \\ 0 & -3 & 1 & -6 \end{bmatrix}$$

Multiply row 2 by $-\frac{1}{3}$:

$$\begin{bmatrix} 1 & \frac{1}{2} & \frac{1}{2} & \frac{5}{2} \\ 0 & 1 & -\frac{1}{3} & \frac{1}{2} \end{bmatrix}$$

Add $-\frac{1}{2}$ times row 2 to row 1:

$$\begin{bmatrix} 1 & 0 & \frac{2}{3} & \frac{9}{4} \\ 0 & 1 & -\frac{1}{3} & \frac{1}{2} \end{bmatrix}$$

Rewrite the matrix as a system of equations.

$$x + \frac{2}{3}z = \frac{9}{4}$$
$$y - \frac{1}{3}z = \frac{1}{2}$$

The solution is

$$x = -\frac{2}{3}z_0 + \frac{3}{2}$$
$$y = \frac{1}{3}z_0 + 2$$
$z = z_0$ for any real value of z_0.

53. The augmented matrix of the system is

$$\begin{bmatrix} 1 & -2 & 1 & 9 \\ 1 & -2 & -1 & 5 \end{bmatrix}$$

Add –1 times row 1 to row 2:

$$\begin{bmatrix} 1 & -2 & 1 & 9 \\ 0 & 0 & -2 & -4 \end{bmatrix}$$

Multiply row 2 by –0.5:

$$\begin{bmatrix} 1 & -2 & 1 & 9 \\ 0 & 0 & 1 & 2 \end{bmatrix}$$

Add –1 times row 2 to row 1:

$$\begin{bmatrix} 1 & -2 & 0 & 7 \\ 0 & 0 & 1 & 2 \end{bmatrix}$$

Rewrite the matrix as a system of equations.

$R-2S+T=9$

$-2T=-4$

The solution is

$R=2s+7$

$S=s$

$T=2$ for any real value of s.

55. The augmented matrix of the system is

$$\begin{bmatrix} 1 & 1 & 1 & 0 \\ 2 & 1 & 1 & 1 \\ 1 & -2 & -2 & 0 \\ 1 & 1 & -1 & 1 \end{bmatrix}$$

Add –2 times row 1 to row 2:

$$\begin{bmatrix} 1 & 1 & 1 & 0 \\ 0 & -1 & -1 & 1 \\ 1 & -2 & -2 & 0 \\ 1 & 1 & -1 & 1 \end{bmatrix}$$

Add –1 times row 1 to row 3:

$$\begin{bmatrix} 1 & 1 & 1 & 0 \\ 0 & -1 & -1 & 1 \\ 0 & -3 & -3 & 0 \\ 1 & 1 & -1 & 1 \end{bmatrix}$$

Add –1 times row 1 to row 4:

$$\begin{bmatrix} 1 & 1 & 1 & 0 \\ 0 & -1 & -1 & 1 \\ 0 & -3 & -3 & 0 \\ 0 & 0 & -2 & 1 \end{bmatrix}$$

Multiply row 2 by –1:

$$\begin{bmatrix} 1 & 1 & 1 & 0 \\ 0 & 1 & 1 & -1 \\ 0 & -3 & -3 & 0 \\ 0 & 0 & -2 & 1 \end{bmatrix}$$

Add 3 times row 2 to row 3:

$$\begin{bmatrix} 1 & 1 & 1 & 0 \\ 0 & 1 & 1 & -1 \\ 0 & 0 & 0 & -3 \\ 0 & 0 & -2 & 1 \end{bmatrix}$$

Since row 3 now corresponds to the equation $0=-3$, the system is inconsistent.

57. If the variables are x, y, and z, the reduced linear system is

$x=-2$

$y+0.1z=4$

The solution is

$x=-2$

$y=4-0.1z_0$

$z=z_0$ for any real value of z_0.

59. If the variables are v, w, x, y, and z, the reduced linear system is

$v+3x=25$

$w+x=0$

$y+5z=15$

The solution is

$v=25-3x_0$

$w=-x_0$

$x=x_0$

$y=15-5z_0$

$z=z_0$ for any real values of x_0 and z_0.

61. a. Since 250 vehicles entered at A and proceeded to either B or D, we have $x_1+x_2=250$.

Since 300 vehicles entered at C and proceeded to either B or D, we have $y_1+y_2=300$.

Since 200 vehicles exited at B and came from either A or C, we have $x_1+y_1=200$.

Since 350 vehicles exited at D and came from either A or C, we have $x_2+y_2=350$.

b. The augmented matrix of the system is

$$\begin{bmatrix} 1 & 1 & 0 & 0 & 250 \\ 0 & 0 & 1 & 1 & 300 \\ 1 & 0 & 1 & 0 & 200 \\ 0 & 1 & 0 & 1 & 350 \end{bmatrix}$$

Add –1 times row 1 to row 3:

$$\begin{bmatrix} 1 & 1 & 0 & 0 & 250 \\ 0 & 0 & 1 & 1 & 300 \\ 0 & -1 & 1 & 0 & -50 \\ 0 & 1 & 0 & 1 & 350 \end{bmatrix}$$

Interchange rows 2 and 4:

$$\begin{bmatrix} 1 & 1 & 0 & 0 & 250 \\ 0 & 1 & 0 & 1 & 350 \\ 0 & -1 & 1 & 0 & -50 \\ 0 & 0 & 1 & 1 & 300 \end{bmatrix}$$

Add –1 times row 2 to row 1:

$$\begin{bmatrix} 1 & 0 & 0 & -1 & -100 \\ 0 & 1 & 0 & 1 & 350 \\ 0 & -1 & 1 & 0 & -50 \\ 0 & 0 & 1 & 1 & 300 \end{bmatrix}$$

Add row 2 to row 3:

$$\begin{bmatrix} 1 & 0 & 0 & -1 & -100 \\ 0 & 1 & 0 & 1 & 350 \\ 0 & 0 & 1 & 1 & 300 \\ 0 & 0 & 1 & 1 & 300 \end{bmatrix}$$

Add –1 times row 3 to row 4:

$$\begin{bmatrix} 1 & 0 & 0 & -1 & -100 \\ 0 & 1 & 0 & 1 & 350 \\ 0 & 0 & 1 & 1 & 300 \\ 0 & 0 & 0 & 0 & 0 \end{bmatrix}$$

Rewrite the matrix as a system of equations.

$x_1 - y_2 = -100$
$x_2 + y_2 = 350$
$y_1 + y_2 = 300$

The solution is

$x_1 = -100 + y_2$
$x_2 = 350 - y_2$
$y_1 = 300 - y_2$ for any real value of y_2.

In the context of the problem,

x_1, x_2, y_1, and y_2 must all be nonnegative.

c. Some choices for (x_1, x_2, y_1, y_2) are (100, 150, 100, 200), (150, 100, 50, 250), (50, 200, 150, 150), and (0, 250, 200, 100).

d. One possibility would be to set up a counter on the straight-away between C and D, positioned between the exit to B and the entrance from A as shown.

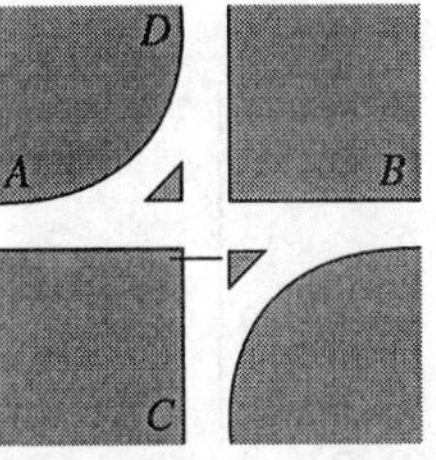

This counter would provide the value of y_2 directly, and the values of the other variables could be calculated from the expressions in part (b).

e. One possibility would be to move the counter at C to the position shown in part (d). This would eliminate the second equation from the system in part (a), but would provide the value of y_2 directly. It would still be possible to calculate the values of the other variables from the remaining equations in the system.

63. **a.** Rewrite the system of equations as

$0.20C + 0.25L + 0.15M = 16$
$0.70C + 0.60L + 0.80M = 56$
$0.10C + 0.15L + 0.05M = 8$

b. A system in which all the variables are leading cannot have infinitely many solutions. Therefore, the system in ***Sunsilk 3*** must have at least one nonleading variable.

65. The conditions lead to a system of equations

$0.2P + 0.5B + 6.4S + 5.5A + 2.6M = 15$
$32.8P + 8.1B + 22.0A + 11.7M = 100$
$4.0P + 5.6B + 30.8S + 1.1A + 8.0M = 50$
$40S + 10M = 50$
$145P + 47B + 192S + 139A + 102M = 600$

The augmented matrix of the system is

$$\begin{bmatrix} 0.2 & 0.5 & 6.4 & 5.5 & 2.6 & 15 \\ 32.8 & 8.1 & 0 & 22.0 & 11.7 & 100 \\ 4.0 & 5.6 & 30.8 & 1.1 & 8.0 & 50 \\ 0 & 0 & 40 & 0 & 10 & 50 \\ 145 & 47 & 192 & 139 & 102 & 600 \end{bmatrix}$$

The reduced row-echelon form is

$$\begin{bmatrix} 1 & 0 & 0 & 0 & 0 & -8.11 \\ 0 & 1 & 0 & 0 & 0 & 6.99 \\ 0 & 0 & 1 & 0 & 0 & -8.13 \\ 0 & 0 & 0 & 1 & 0 & -5.89 \\ 0 & 0 & 0 & 0 & 1 & 37.50 \end{bmatrix}$$

The solution is (P, B, S, A, M)
$= (-8.11, 6.99, -8.13, -5.89, 37.50)$. Since some variables have negative values, the solution has no physical meaning.

Section 4-3

1. The cost of x bales of timothy at \$1.50 per bale is $1.50x$ dollars, and the cost of y bales of alfalfa at \$2.50 per bale is $2.50y$ dollars. The combined cost is $1.50x + 2.50y$ dollars, which must be equal to her expenditure of \$165.

3. (Sample response) Every linear inequality can be written in one of the forms $y > f(x)$, $y < f(x)$, $y \geq f(x)$, or $y \leq f(x)$. The inequality $y > f(x)$ is true at all points above the graph of $y = f(x)$. The inequality $y < f(x)$ is true at all points below the graph of $y = f(x)$. The inequality $y \geq f(x)$ is true at all points on or above the graph of $y = f(x)$. The inequality $y \leq f(x)$ is true at all points on or below the graph of $y = f(x)$.

5.

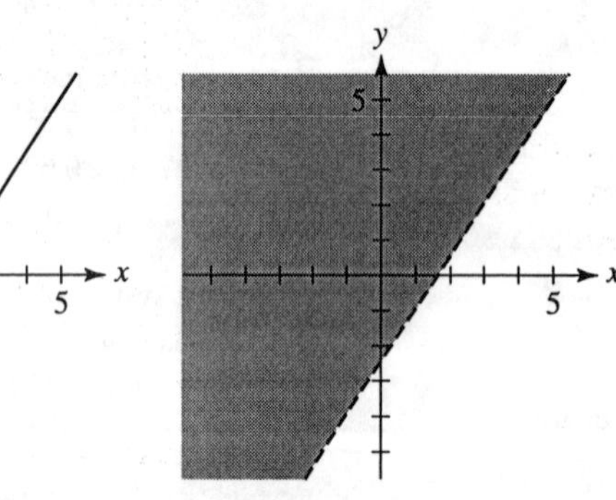

7.

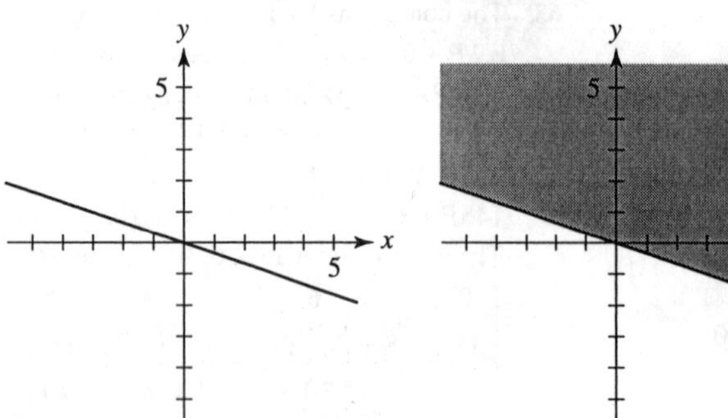

9.

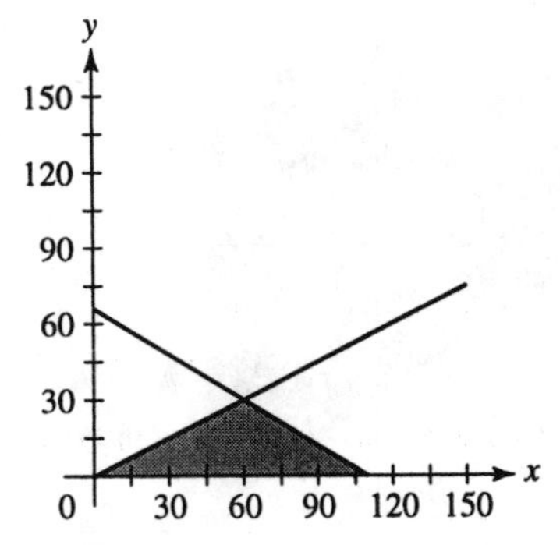

11.

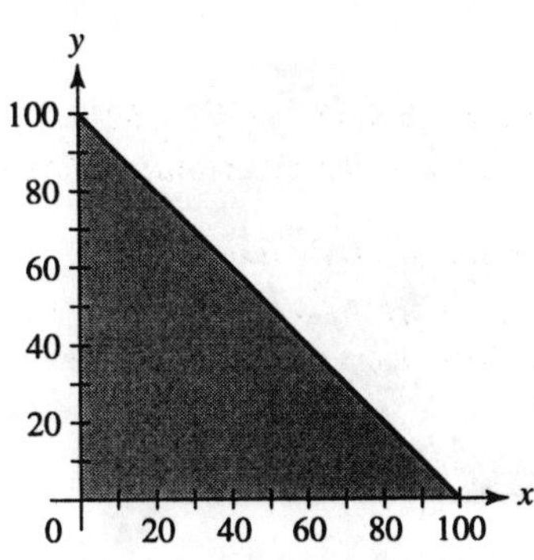

13. The feasible region represents the combinations of at most 100 bales of timothy and alfalfa Carrie can buy for no more than \$165, containing at least twice as much timothy as alfalfa.

15.

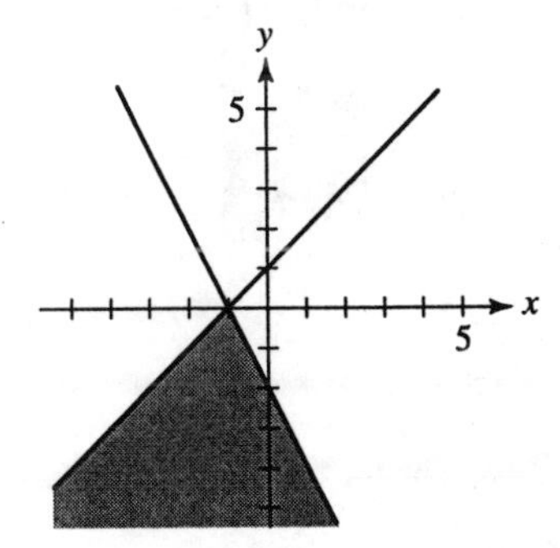

17.

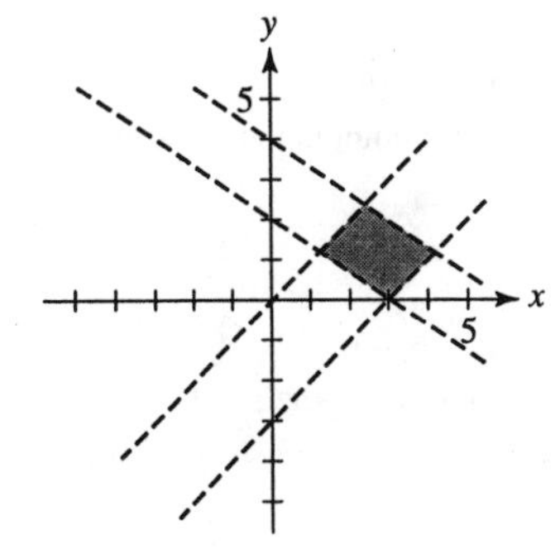

19.

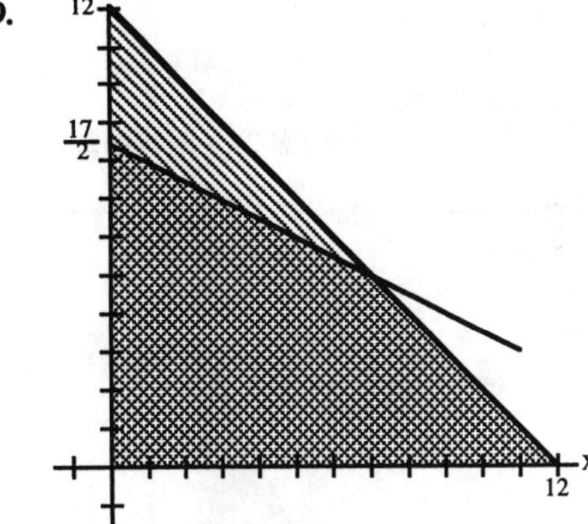

21. At (0, 0), the value of $50 - 2x - 3y$ is 50.
At (17, 0), the value of $50 - 2x - 3y$ is 16.
At (0, 12), the value of $50 - 2x - 3y$ is 14.
At (7, 5), the value of $50 - 2x - 3y$ is 21.

The minimum value is 14, and occurs at (0, 12).

The maximum value is 34, and occurs at (4, 10).

23. The graph of the feasible region is shown below.

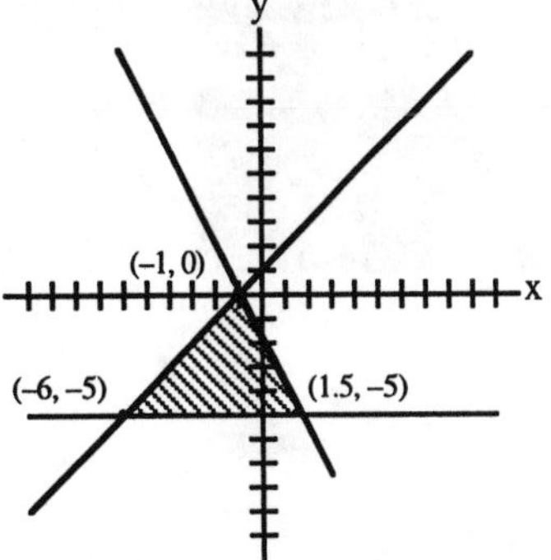

The feasible region has three extreme points.
One is the solution to the system
$y = x + 1$

$y = -2x - 2$
The solution is (–1, 0).
A second extreme point is the solution to the system
$y = x + 1$
$y = -5$
The solution is (–6, –5).
A third extreme point is the solution to the system
$y = -2x - 2$
$y = -5$
The solution is (1.5, –5).
At (–1, 0), the value of $2x + 3y$ is –2.
At (–6, –5), the value of $2x + 3y$ is –27.
At (1.5, –5), the value of $2x + 3y$ is –12.

The minimum value is –27, and occurs at (–6, –5).

25.

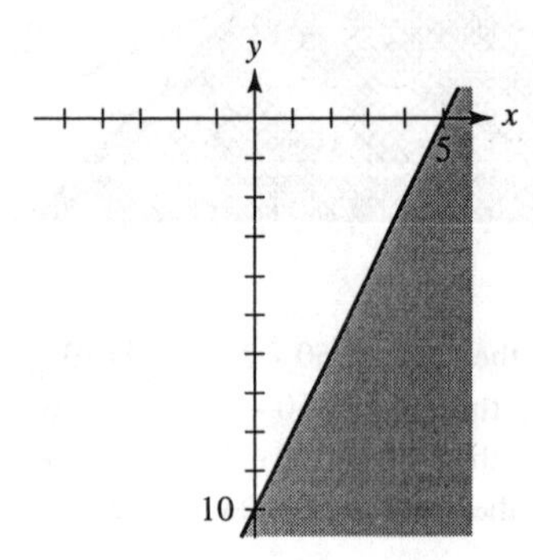

27.

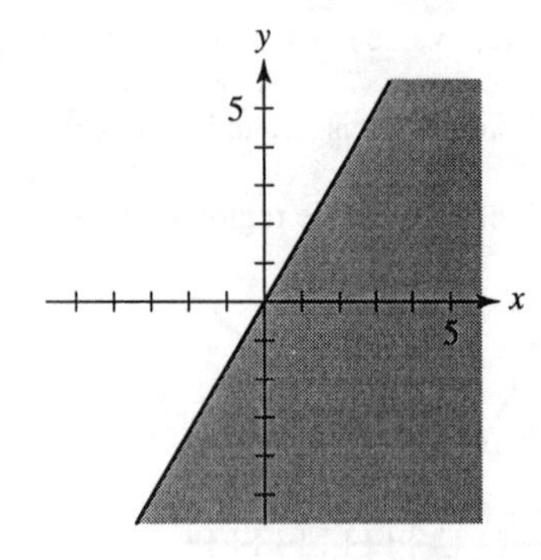

29.

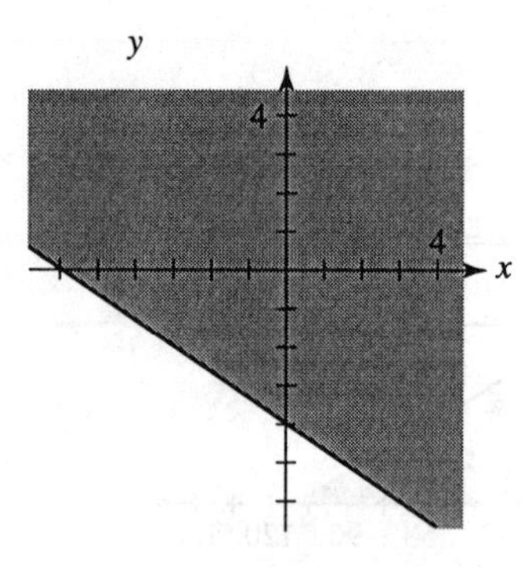

31.

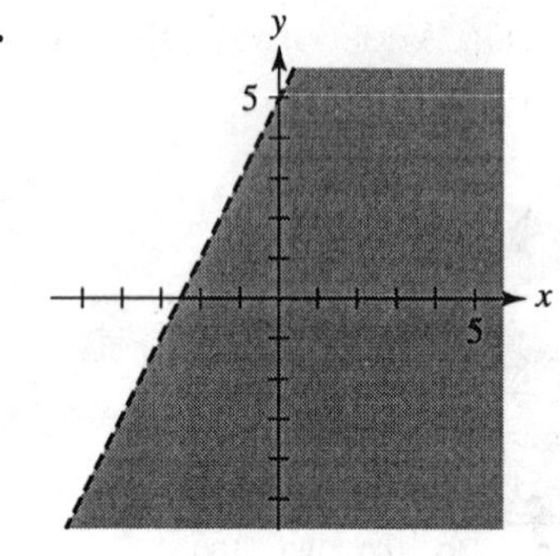

33.

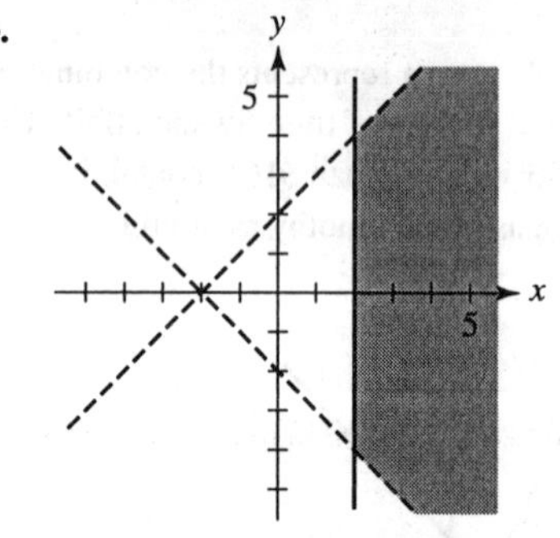

35.

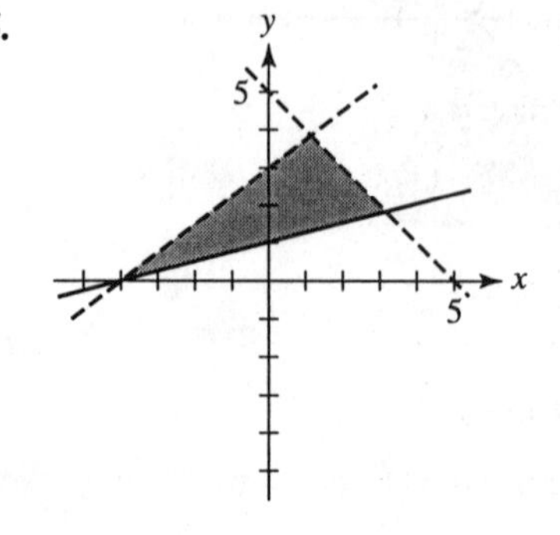

37.

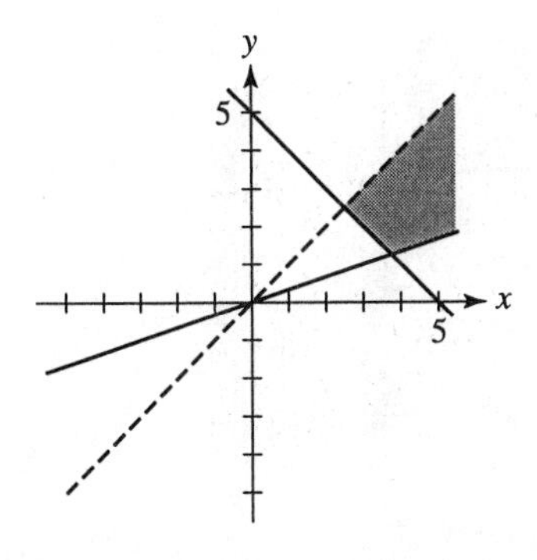

39. The graph of the feasible region, shown below, has four extreme points.

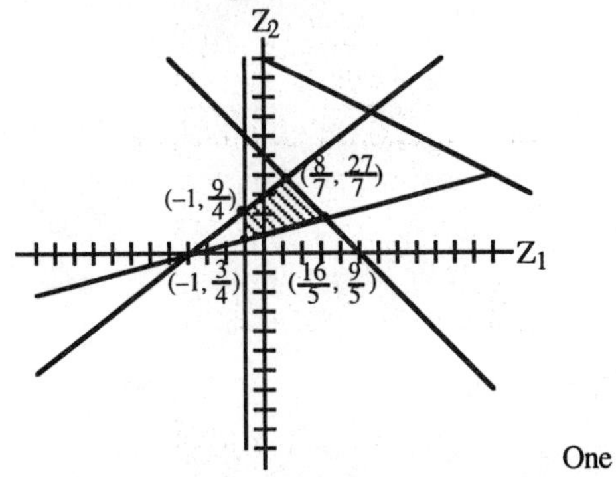

One is the solution to the system

$Z_2 = \frac{3}{4}Z_1 + 3$

$Z_2 = 5 - Z_1$

The solution is $\left(\frac{8}{7}, \frac{27}{7}\right)$.

A second extreme point is the solution to the system

$Z_2 = 5 - Z_1$

$Z_2 = \frac{1}{4}Z_1 + 1$

The solution is $\left(\frac{16}{5}, \frac{9}{5}\right)$.

A third extreme point is the solution to the system

$Z_2 = \frac{3}{4}Z_1 + 3$

$Z_1 = -1$

The solution is $\left(-1, \frac{9}{4}\right)$.

The fourth extreme point is the solution to the system

$Z_2 = \frac{1}{4}Z_1 + 1$

$Z_1 = -1$

The solution is $\left(-1, \frac{3}{4}\right)$.

At $\left(\frac{8}{7}, \frac{27}{7}\right)$, the value of $3Z_1 - Z_2$ is $-\frac{3}{7}$.

At $\left(\frac{16}{5}, \frac{9}{5}\right)$, the value of $3Z_1 - Z_2$ is $\frac{39}{5}$.

At $\left(-1, \frac{9}{4}\right)$, the value of $3Z_1 - Z_2$ is $-\frac{21}{4}$.

At $\left(-1, \frac{3}{4}\right)$, the value of $3Z_1 - Z_2$ is $-\frac{15}{4}$.

The minimum value is $-\frac{21}{4}$, and occurs at $\left(-1, \frac{9}{4}\right)$.

41. The graph of the feasible region, shown below, has five extreme points.

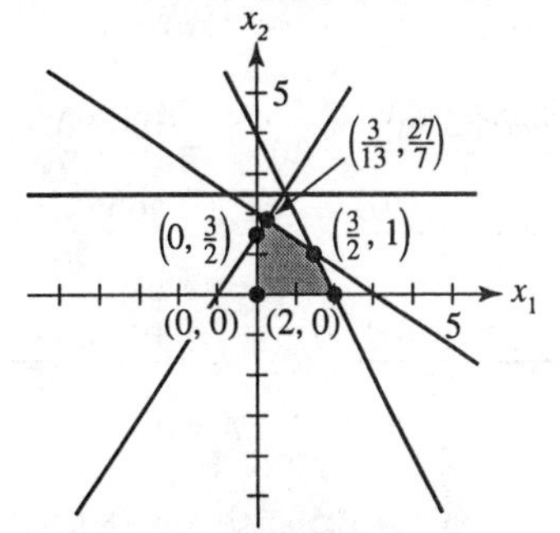

They are solutions to the systems

(a) $x_2 = \frac{6 - 2x_1}{3}$
$x_2 = \frac{3x_1 + 3}{2}$

(b) $x_2 = \frac{6 - 2x_1}{3}$
$x_2 = 4 - 2x_1$

(c) $x_2 = \frac{3x_1 + 3}{2}$
$x_1 = 0$

(d) $x_2 = 4 - 2x_1$
$x_2 = 0$

(e) $x_1 = 0$
$x_2 = 0$

The solutions are (a) $\left(\frac{3}{13}, \frac{24}{13}\right)$ (b) $\left(\frac{3}{2}, 1\right)$ (c) $\left(0, \frac{3}{2}\right)$ (d) (2, 0), (e) (0, 0)

At $\left(\frac{3}{13}, \frac{24}{13}\right)$, the value of $4x_1 + 3x_2$ is $\frac{84}{13}$.

At $\left(\frac{3}{2}, 1\right)$, the value of $4x_1 + 3x_2$ is 9.

At $\left(0, \frac{3}{2}\right)$, the value of $4x_1 + 3x_2$ is $\frac{9}{2}$.
At (2, 0), the value of $4x_1 + 3x_2$ is 8.
At (0, 0), the value of $4x_1 + 3x_2$ is 0.
The maximum value is 9, and occurs at $\left(\frac{3}{2}, 1\right)$.

43. If newspaper ads and TV spots cost $\$n$ and $\$10n$ each, respectively, then the cost of N newspaper ads and T TV spots is $nN + 10nT = n(N + 10T)$ dollars. The cost is minimized if the objective function $N + 10T$ is minimized.

The feasible region in Exercise 32 has four extreme points, which are the solutions to the systems

(a) $T = \frac{9000 - 20N}{400}$, $T = \frac{3000 - 10N}{50}$ (b) $T = \frac{10\,000 - 70N}{100}$, $T = \frac{3000 - 10N}{50}$

(c) $T = \frac{9000 - 20N}{400}$, $T = 0$ (d) $T = \frac{10\,000 - 70N}{100}$, $N = 0$

The solutions are (a) (250, 10) (b) (80, 44) (c) (450, 0) (d) (0, 100)

At (250, 10), the value of $N + 10T$ is 350.
At (80, 44), the value of $N + 10T$ is 520.
At (450, 0), the value of $N + 10T$ is 450.
At (0, 100), the value of $N + 10T$ is 1000.

The minimum value of $N + 10T$ occurs at (250, 10). You should purchase 250 newspapers ads and 10 TV spots.

NUTSHELLS

Nutshell for Chapter 5 - QUADRATIC FUNCTIONS

What's familiar?

- Section 5-1 involves the process of solving quadratic equations. You have done this before, but it still might be a good idea to review Sections A-11 and A-12.
- Section 5-1 looks at quadratic functions from the three views: numerical, analytical and graphical.
- You will discover how to obtain the graph of every quadratic function of the form $y = a(x-h)^2 + k$ by transforming the basic graph $y = x^2$. This continues a theme that originated with linear absolute value functions in Chapter 3.
- Chapter 3 used a method to discover whether or not a table with equally spaced x-values fits a linear function. In Section 5-2 you use a similar method to make a similar decision about tables and quadratic functions.
- Section 5-2 also applies the methods of Chapter 2 to solving quadratic inequalities.

What's new?

- Section 5-2 shows you how to find the *maximum* or *minimum* (largest or smallest) value of a given quadratic function.
- Section 5-2 discusses an analytical method for solving quadratic inequalities.
- Section 5-2 also shows you a method for fitting a quadratic function to a table by constructing and solving a linear system.
- The above three methods will be generalized so that we can apply them to the functions in Chapter 7.

List of hints for exercises in each chapter.

Chapter 5

- Notice the similarities between the graphs of quadratic functions and those of absolute value functions in Section 3-4. The transformation principles are the same for both, so you may want to review Section 3-4.
- Solve several inequalities both graphically and analytically and compare the processes. This will help you understand both methods of solution.

Chapter 5 Additional Exercises

Name:____________________

Date:____________________

For Problems 1–8, sketch the graph.

1. $y = 2x^2$

1.

2. $y = -3x^2$

2.

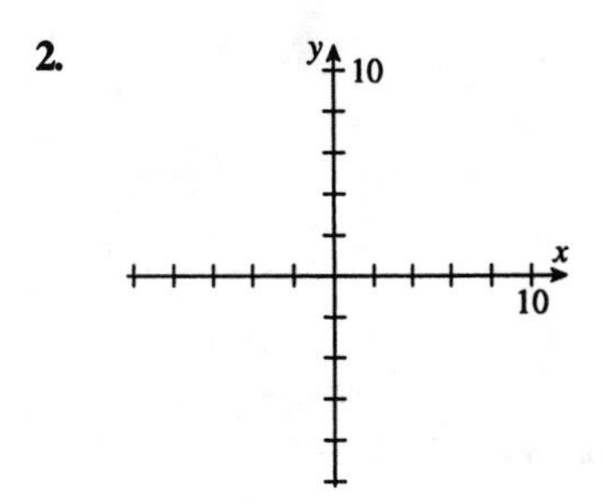

3. $y = x^2 + 3$

3.

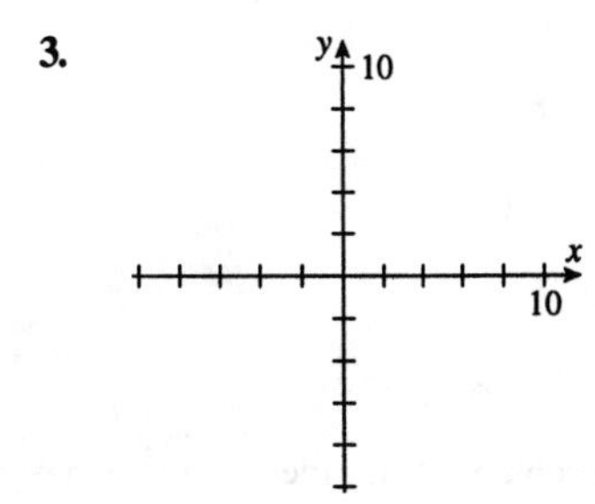

4. $y = (x+4)^2$

4.

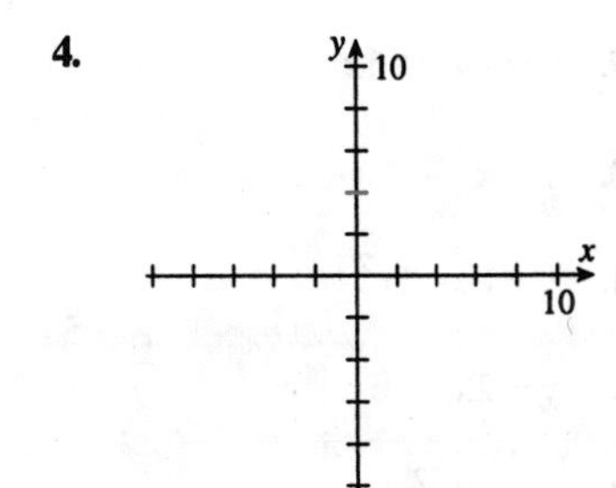

5. $y = (x-3)^2 + 1$

5.

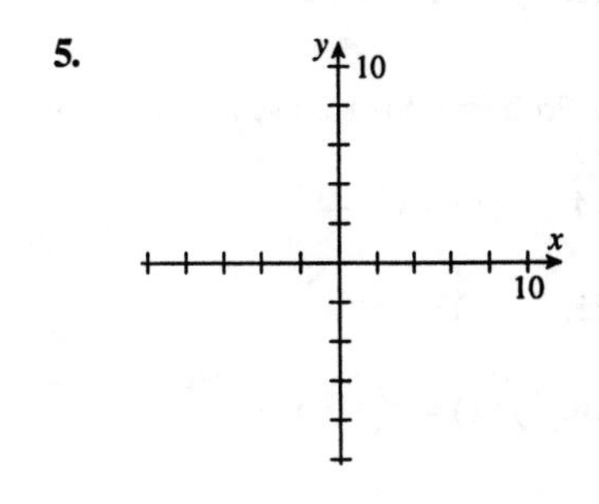

Chapter 5 Additional Exercises *(cont.)*

Name:____________________

6. $y = -3(x-5)^2 + 4$

6.

7. $y = 2(x-4)^2 + 3$

7.

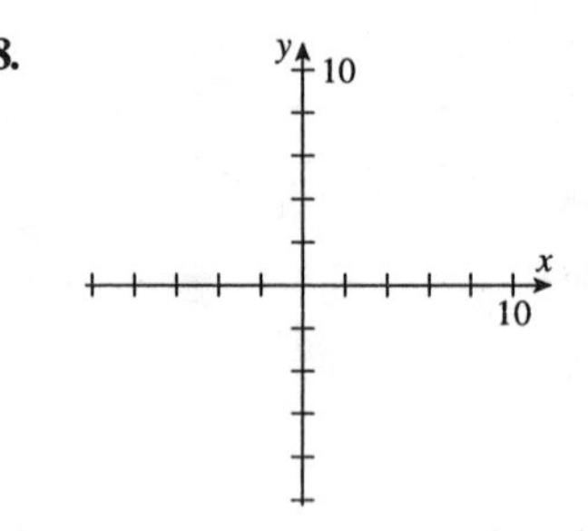

8. $y = x^2 - 2x - 8$

8.

For Problems 9–13, identify the vertex and the axis of symmetry.

9. $y = -x^2 + 6x - 7$ **9.** ____________________

10. $y = x^2 - 10x + 21$ **10.** ____________________

11. $y = 10 - 2x^2$ **11.** ____________________

12. $y = 2x^2 - 6x + 7$ **12.** ____________________

13. $y = -2x^2 + 6x - 4$ **13.** ____________________

For Problems 14–16, use the discriminant to determine the number of x-intercepts on the graph of $y = f(x)$.

14. $f(x) = x^2 - 4$ **14.** ____________________

15. $f(x) = -x^2 + 2x - 1$ **15.** ____________________

16. $f(x) = x^2 - 6x + 11$ **16.** ____________________

For Problems 17–19, write the function in factored form.

17. $f(y) = 3y^2 + 6y$ **17.** ____________________

Chapter 5 Additional Exercises *(cont.)*

Name:________________

18. $g(x) = 14x^2 - x - 3$ **18.** ________________

19. $H(t) = 4t^2 - 25$ **19.** ________________

For Problems 20–22, obtain the factored form of $H(x)$ by solving the equation $H(x) = 0$ first.

20. $H(x) = x^2 - 4x - 1$ **20.** ________________

21. $H(x) = x^2 + 2x - 4$ **21.** ________________

22. $H(x) = x^2 - 5x - 13$ **22.** ________________

For Problems 23–25, find the vertex and axis of symmetry of the graph of $y = f(x)$.

23. $f(x) = x(x-4) - 12$ **23.** ________________

24. $-2\left[x - (1+\sqrt{13})\right]\left[x - (1-\sqrt{13})\right]$ **24.** ________________

25. $f(x) = 3[x - (2-i)][x - (2+i)]$ **25.** ________________

For Problems 26–27, find the x-intercepts on the graph of $y = G(x)$.

26. $G(x) = -4x^2 + 12x - 7$ **26.** ________________

27. $G(x) = \frac{1}{8}x^2 - \frac{1}{2}x - 1$ **27.** ________________

For Problems 28–32, find the x-intercepts and identify the vertex.

28. $y = 2x^2 - x - 3$ **28.** ________________

29. $g(x) = 4x^2 + 4x - 3$ **29.** ________________

30. $y = (x+6)(x-3)$ **30.** ________________

31. $m = 4\left[n - (3+\sqrt{2})\right]\left[n - (3-\sqrt{2})\right]$ **31.** ________________

32. $f(x) = -5[x - (1+2i)][x - (1-2i)]$ **32.** ________________

For Problems 33–35, find an equation to fit the graph.

33. **33.** ________________

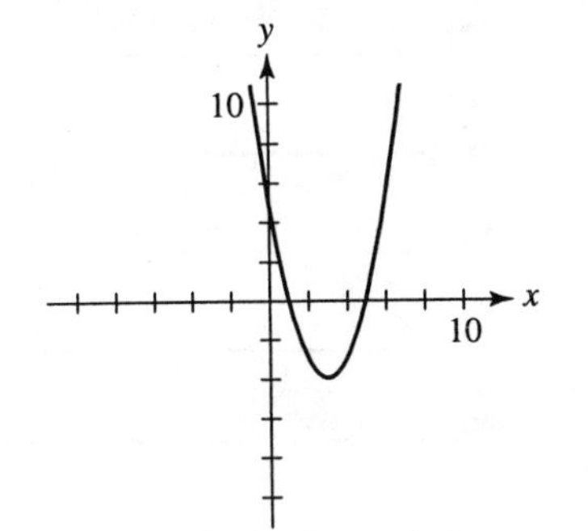

Name:______________

34.

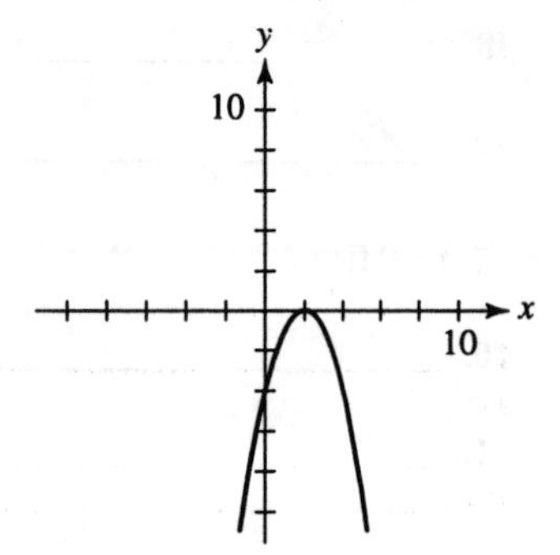

34. ______________

35.

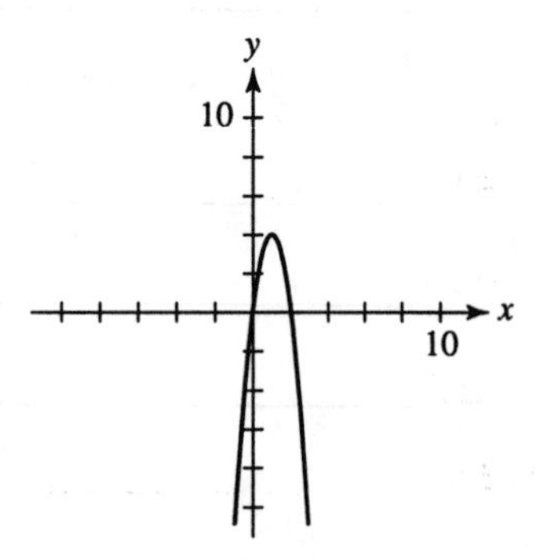

35. ______________

For Problems 36–37, obtain the factored form of $g(x)$ by solving $g(x) = 0$.

36. $g(x) = -4x^2 + 3x - 6$ **36.** ______________

37. $g(x) = 3x^2 + 6x + 5$ **37.** ______________

For Problems 38–41, solve the inequality analytically.

38. $3x^2 - 7x + 2 \le 0$ **38.** ______________

39. $x^2 + 2x - 24 > 0$ **39.** ______________

40. $9x^2 - 6x + 1 \ge 0$ **40.** ______________

41. $2x^2 - 3x \ge 5$ **41.** ______________

For Problems 42–45, find the extreme value of the function and state whether it is a maximum or minimum.

42. $f(x) = x^2 + 4x - 5$ **42.** ______________

43. $g(t) = -2x^2 - 3x + 35$ **43.** ______________

44. $y = -(x-2)^2 - 3$ **44.** ______________

45. $y = 3x^2 - 12x$ **45.** ______________

46. Find the equation of a quadratic function that fits the table **46.** ______________

x	y

 Name:________________

1	0
3	12
5	32
7	60

47. A company's weekly profit function is $P(x) = -x^2 + 160x - 3375$ Solve $P(x) \geq 0$ graphically and analytically. **47.** ________________

48. Leeanne wants to fence a rectangular dog pen against her garage. Since the garage is to serve as one of the sides, she needs to fence in only three sides. If she uses 100 linear feet of fencing, what are the dimensions of the pen that would give the maximum area possible? **48.** ________________

49. The daily cost for a window manufacturer is

$$F(w) = 0.02w^2 - 0.48w + 428.8$$

where $F(w)$ is the daily production cost in dollars, and w is the number of windows produced daily. How many windows should the manufacturer produce in order to minimize the daily production cost? **49.** ________________

50. What two numbers whose sum is 104 will yield a maximum product? **50.** ________________

A Mathematical Looking Glass

Just Do It

(to follow Section 5-2)

The Pittsburgh Racquet Club has an indoor running track consisting of two straightaways and two semicircular segments, as in Figure S-5.

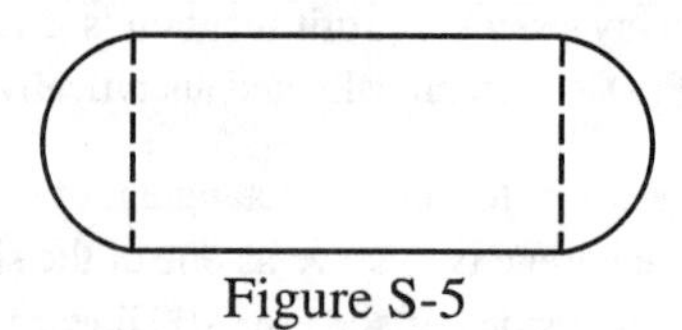

Figure S-5

When the track was put in, the owners of the club wanted its design to meet three requirements.

- One lap around the inside edge of the track should be 440 feet.
- There should be adequate exercise space in the two semicircular regions inside the track.
- There should be space for one or more basketball courts in the rectangular region. The edges of the court should be 6 feet from the track.

Exercises

1. (Creating Models) Let L represent the length of one straightaway. Express each of the following quantities in terms of L.

   a. the combined length of the two straightaways

   b. the combined length of the two semicircles

   c. the radius of each semicircle

   d. the total exercise area inside the track

   e. the total rectangular area inside the track

   f. the total area of the basketball courts

2. (Problem Solving)

   a. What value of L maximizes the area of the basketball courts? What is the maximum area?

   b. What value of L maximizes the exercise areas? What is the maximum exercise area?

   c. What values of L allow for at least one regulation (40 by 60 feet) basketball court?

   d. What values of L allow at least 6000 square feet of exercise area?

   e. What values of L meet both of the requirements in parts (c) and (d)?

3. (Writing to Learn) Imagine that you are responsible for installing the track. Write a letter to the club owners, recommending the dimensions you believe will best suit their needs. Use your results from Exercise 2 to justify your recommendation.

A Mathematical Looking Glass

Traffic Light

(to follow Section 5-2)

If a traffic light turns yellow as your car approaches it, your reaction will depend on both your speed and your distance from the intersection . To be safe, you must either stop before reaching the near side of the intersection, or reach the far side before the light turns red. You should therefore drive at a speed that allows you to do one or the other, no matter how far you are from the light when it turns.

Suppose you regularly drive across a 40-foot wide intersection whose traffic light stays yellow for 3 seconds before turning red. You might want to know what speeds are safe as you approach the light. One way to proceed is by *Breaking the Problem into Parts.*

- Find your safe stopping distance S as a function of your speed.

- Find the distance T you can drive before the light turns red (in 3 seconds) as a function of your speed.

- To be safe, you must either stop before reaching the intersection, or reach the far side before the light turns red without speeding. This means that your distance from the intersection must be either at least S, or at most T - 40. Your speed is safe as long as $S \leq T - 40$. See Figure S-6.

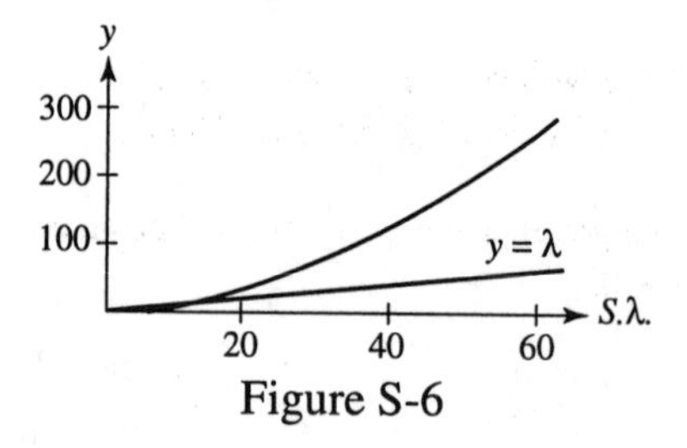

Figure S-6

Exercises

1. (Writing to Learn) Let D be your distance from the near side of the intersection when the light turns yellow. Explain why your speed is safe if either $S \leq D$ or $T \geq D + 40$, then explain why it is safe if $S \leq T - 40$.

2. (Creating Models) In **Stopping Distance** (page 160), you constructed the function $S(v) = .055v^2 + 1.25v - 10$ to represent stopping distance as a function of speed. Now construct a function $T(v)$ for the distance, in feet, that you can drive in 3 seconds at v miles per hour. (*Hint*: First convert miles per hour to feet per second.)

3. (Interpreting Mathematics) Solve the inequality $S(v) \leq T(v) - 40$. What speeds are safe as you approach the intersection?

Chapter 5

Section 5-1

1. a. At each sales level x, the profit is the y-coordinate on the graph of $y = P(x)$. The point V has the largest y-coordinate on the graph, and therefore represents the sales level with the maximum profit.

   b. Estimates will vary slightly, but should be around 1250 shirts.

3. From the graph it appears that the maximum profit occurs at a sales level halfway between the x-intercepts.

5. a. The vertex is at (0, 0).

   b. The graph is steeper than the graph of $y = x^2$ if $|a| > 1$ and flatter if $|a| < 1$.

   c. The graph opens up if $a > 0$ and down if $a < 0$.

7. a–d.

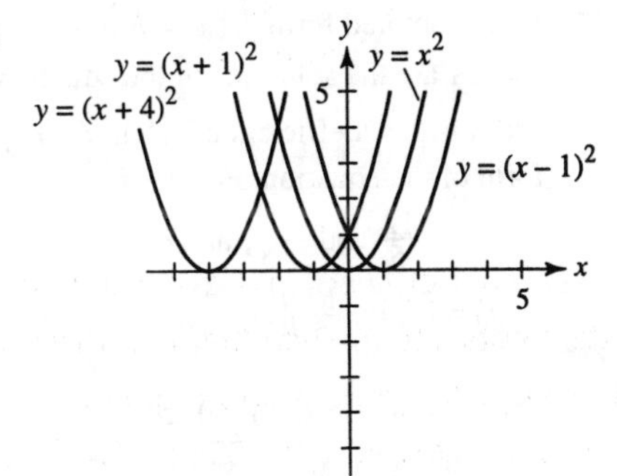

9. a–d.

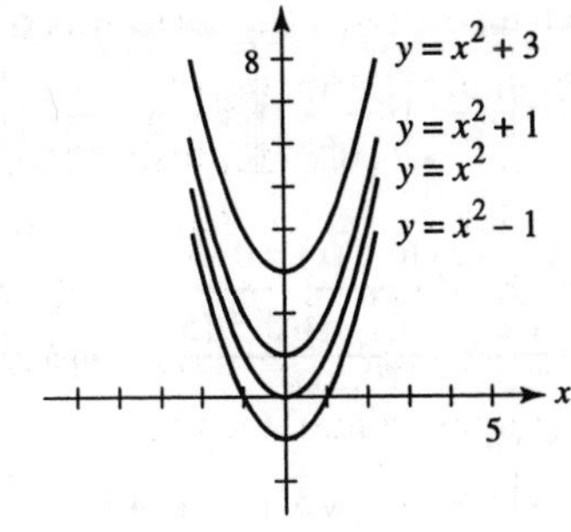

11. a.

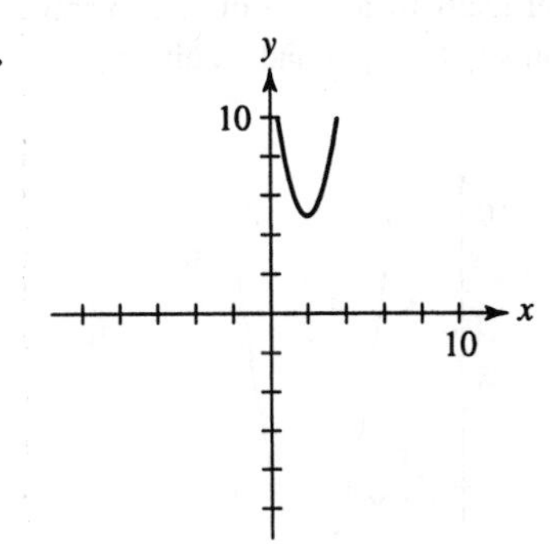

   b.

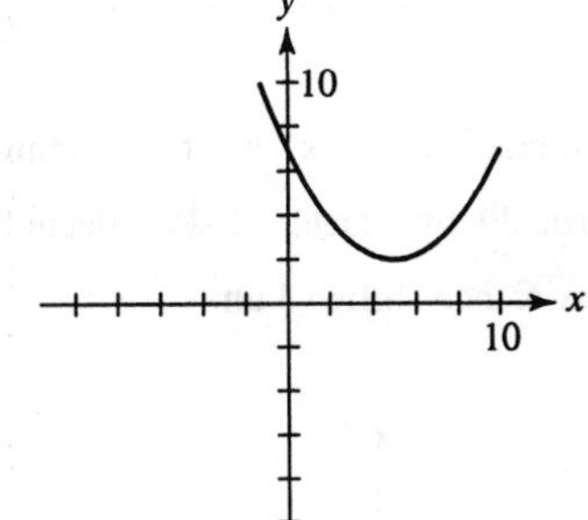

   c.

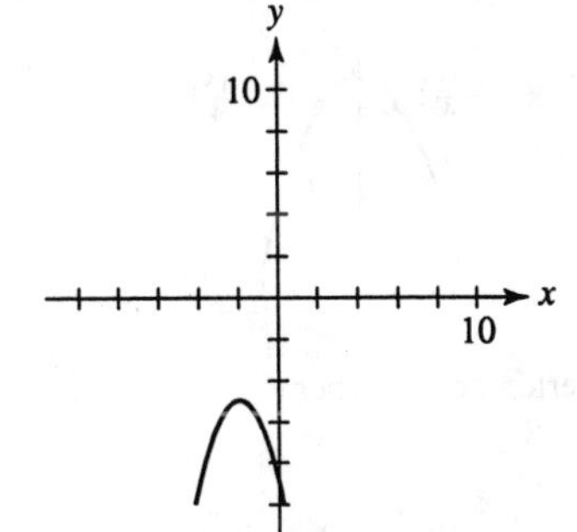

   d.

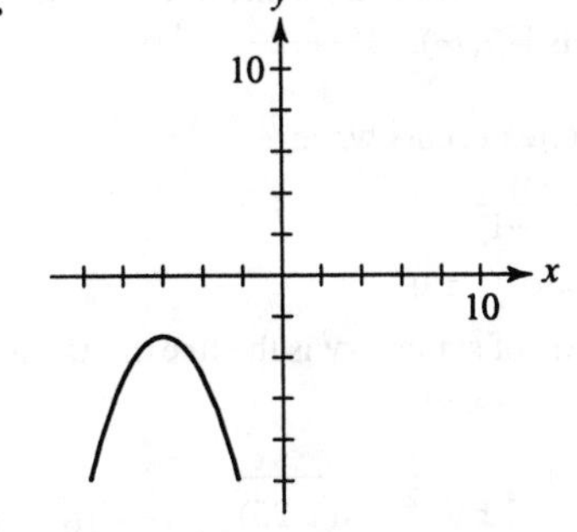

13. a. The graph of $y = x^2$ must be stretched vertically by a factor of 2, then shifted 4 units to the right and 1 unit up.

b.

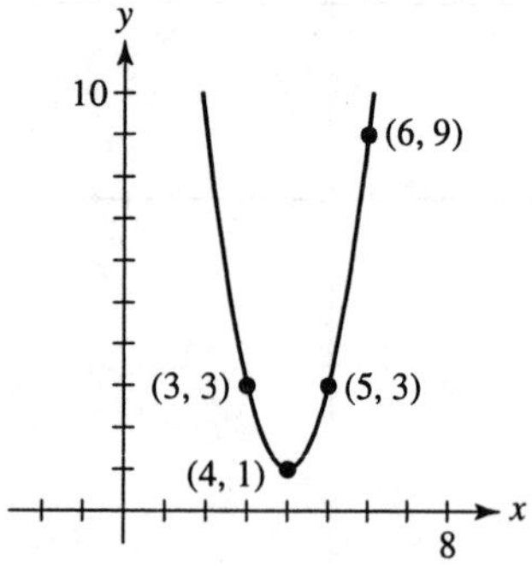

15. a. The graph of $y = x^2$ must be compressed vertically by a factor of $\frac{4}{3}$, reflected in the x-axis, then shifted 3 units down.

b.

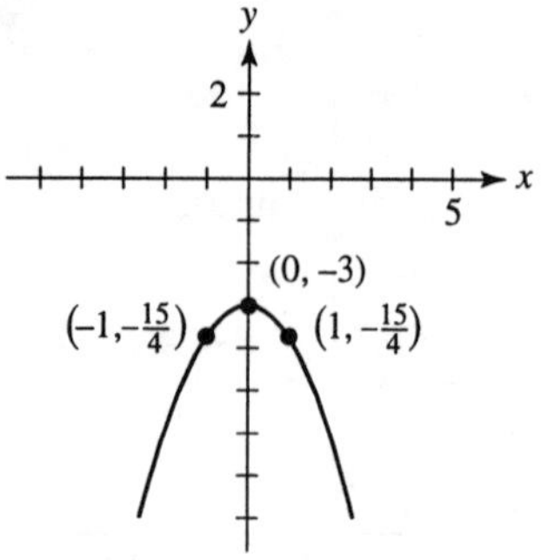

17. The vertex occurs when

$x = -\frac{8}{2(1)} = -4$

$y = (-4)^2 + 8(-4) + 9 = -7$

The axis of symmetry is the line $x = -4$, and the range is $[-7, \infty)$.

19. The vertex occurs when

$x = -\frac{0}{2(-1)} = 0$

$y = 9 - (0)^2 = 9$

The axis of symmetry is the line $x = 0$, and the range is $(-\infty, 9]$.

21. a. $x = \frac{5 \pm \sqrt{5^2 - 4(2.25)}}{2(1)} = \frac{5 \pm \sqrt{16}}{2} = \frac{1}{2}, \frac{9}{2}$

b. $x = \frac{5 \pm \sqrt{5^2 - 4(6.25)}}{2(1)} = \frac{5 \pm \sqrt{0}}{2} = \frac{5}{2}$

c. $x = \frac{5 \pm \sqrt{5^2 - 4(10.25)}}{2(1)}$

$= \frac{5 \pm \sqrt{-16}}{2} = \frac{5 \pm 4i}{2} = \frac{5}{2} \pm 2i$

23. A quadratic equation can have 0, 1, or 2 real solutions. The number of real solutions to a quadratic equation $f(x) = 0$ is equal to the number of x-intercepts on the graph of $y = f(x)$.

25. The discriminant is $0^2 - 4(1)(4) = -16$. There are no x-intercepts on the graph.

27. The discriminant is $4^2 - 4(1)(4) = 0$. There is one x-intercept on the graph.

29. $f(x) = x(3x + 8) = 3x\left(x + \frac{8}{3}\right)$

31. $P(t) = (3t + 5)(2t - 1) = 6\left(t + \frac{5}{3}\right)\left(t - \frac{1}{2}\right)$

33. In the factored form $f(x) = a(x - r_1)(x - r_2)$, r_1 and r_2 are the solutions of the equation $f(x) = 0$, and a is the coefficient of x^2 in the expanded form of the equation. For this function, r_1 and r_2 are $-\frac{12}{5}$ and $-\frac{1}{3}$, and $a = 15$. Therefore $f(x) = 15\left(x + \frac{12}{5}\right)\left(x + \frac{1}{3}\right)$.

35. The solutions to $Q(x) = 0$ are

$x = \frac{4 \pm \sqrt{4^2 - 4(2)(-1)}}{2(2)} = 1 \pm \frac{\sqrt{6}}{2}$.

Therefore,

$Q(x) = \left[x - \left(1 + \frac{\sqrt{6}}{2}\right)\right]\left[x - \left(1 - \frac{\sqrt{6}}{2}\right)\right]$.

37. The solutions to $Q(x) = 0$ are

$x = \frac{-4 \pm \sqrt{(-4)^2 - 4(0.5)(5)}}{2(0.5)} = -4 \pm \sqrt{6}$

Therefore,

$Q(x) = \left[x - (-4 + \sqrt{6})\right]\left[x - (-4 - \sqrt{6})\right]$.

39. In ***Breakeven Analysis*** (page 88) you saw that the solutions of $P(x) = 0$ are $x = 50$ and $x = 2450$.

Therefore, in factored form,
$P(x) = -0.005(x-50)(x-2450)$.

41. The x-coordinate at the vertex is the average of -2 and 8, which is $\frac{-2+8}{2} = 3$.

43. The vertex occurs when

$$x = \frac{\sqrt{7}+(-\sqrt{7})}{2} = 0$$
$$y = -(0+\sqrt{7})(0-\sqrt{7}) = 7$$

The axis of symmetry is the line $x = 0$, and the range is $(-\infty, 7]$.

45. The vertex occurs when

$$x = \frac{(1+3i)+(1-3i)}{2} = 1$$
$$y = 2[1-(1+3i)][1-(1-3i)] = 18$$

The axis of symmetry is the line $x = 1$, and the range is $[18, \infty)$.

47. a.

$$\begin{aligned} y &= 0.5x^2+4x+9 \\ &= 0.5(x^2+8x)+9 \\ &= 0.5(x^2+8x+16-16)+9 \\ &= 0.5(x^2+8x+16)-8+9 \\ &= 0.5(x+4)^2+1 \end{aligned}$$

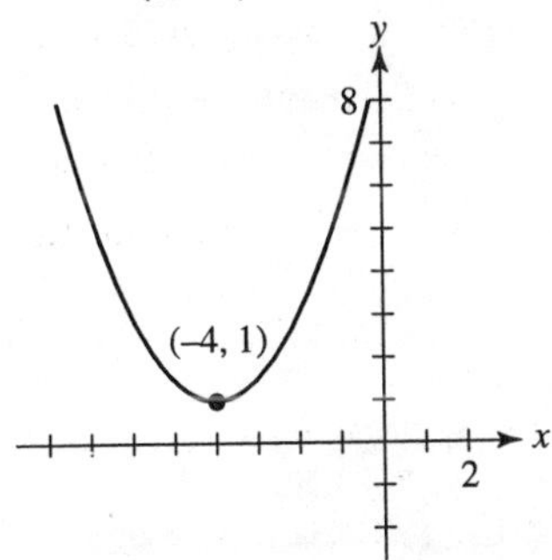

b. The axis of symmetry is the line $x = -4$. The range is $[1, \infty)$. The vertex is $(-4, 1)$.

49. a.

$$\begin{aligned} w &= -3z^2+6z \\ &= -3(z^2-2z) \\ &= -3(z^2-2z+1-1) \\ &= -3(z^2-2z+1)+3 \\ &= -3(z-1)^2+3 \end{aligned}$$

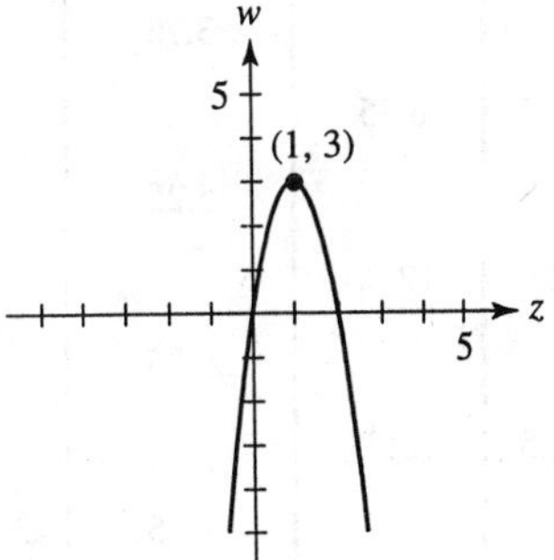

b. The axis of symmetry is the line $z = 1$. The range is $(-\infty, 3]$. The vertex is $(1, 3)$.

51. The t-intercepts occur when

$$\begin{aligned} (t+6)^2-4 &= 0 \\ (t+6)^2 &= 4 \\ t+6 &= \pm 2 \\ t &= -6\pm 2 \\ t &= -4,\ -8 \end{aligned}$$

53. The t-intercepts occur when

$$\begin{aligned} -2(t+3)^2+1 &= 0 \\ -2(t+3)^2 &= -1 \\ (t+3)^2 &= \frac{1}{2} \\ t+3 &= \pm\frac{1}{\sqrt{2}} \\ t &= -3\pm\frac{1}{\sqrt{2}} \end{aligned}$$

55. If a and k are both positive, then $a(x-h)^2 \geq 0$ for all x, so $a(x-h)^2+k > 0$ for all x. Similarly, if a and k are both negative, then $a(x-h)^2+k < 0$ for all x. In each case, $a(x-h)^2+k$ is never 0, so the graph has no x-intercepts.

57. a.

x	$y = 3x^2$	first differences	second differences
1.0	3.00		
		3.75	
1.5	6.75		1.5
		5.25	
2.0	12.00		1.5
		6.75	
2.5	18.75		1.5
		8.25	
3.0	27.00		

b.

x	$y = ax^2$	first differences	second differences
−1	a		
		$-a$	
0	0		$2a$
		a	
1	a		$2a$
		$3a$	
2	$4a$		$2a$
		$5a$	
3	$9a$		

c.

x	$y = ax^2$	first differences	second differences
x_0	ax_0^2		
		$2ahx_0 + ah^2$	
$x_0 + h$	$ax_0^2 + 2ahx_0 + ah^2$		$2ah^2$
		$2ahx_0 + 3ah^2$	
$x_0 + 2h$	$ax_0^2 + 4ahx_0 + 4ah^2$		$2ah^2$
		$2ahx_0 + 5ah^2$	
$x_0 + 3h$	$ax_0^2 + 6ahx_0 + 9ah^2$		$2ah^2$
		$2ahx_0 + 7ah^2$	
$x_0 + 4h$	$ax_0^2 + 8ahx_0 + 16ah^2$		

59.

x	y	first differences	second differences
3	23		
		−7	
4	16		2
		−5	
5	11		2
		−3	
6	8		2
		−1	
7	7		

The first differences are not constant, but the second differences are. The table fits a quadratic function.

61.

x	y	first differences	second differences
−1	−4		
		−2	
1	−6		2
		0	
3	−6		2
		2	
5	−4		2
		4	
7	0		2
		6	
9	6		

The first differences are not constant, but the second differences are. The table fits a quadratic function.

63. a. $\dfrac{S(40)-S(30)}{40-30}=\dfrac{128-77}{10}=5.1$

For speeds between 30 and 40 mph, every increase of 1 mph in speed results in an increase of about 5.1 feet in the required stopping distance.

b. $\dfrac{S(50)-S(40)}{50-40}=\dfrac{190-128}{10}=6.2>5.1$

As your speed increases, increases of 1 mph in speed produce increasingly larger increases in the required stopping distance.

65. a. The graph of $y=x^2$ must be shifted 2.5 units to the right and 6.8 units down.

b–c.

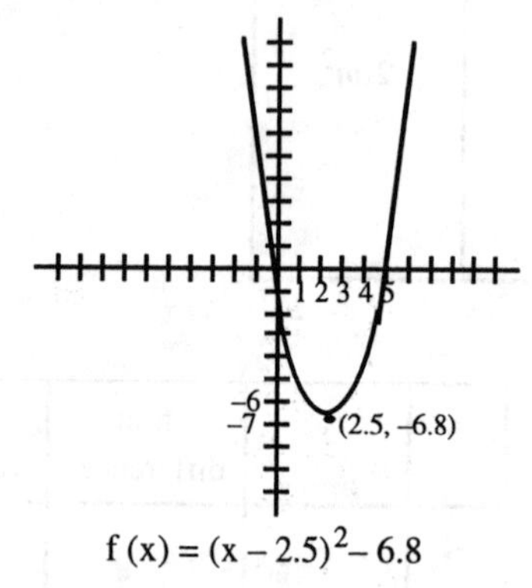

f (x) = (x – 2.5)^2 – 6.8

d. The axis of symmetry is the line $x = 2.5$ The range is [–6.8, ∞).

e. The function decreases on (–∞, 2.5) and increases on (2.5, ∞).

67. a. The graph of $y=r^2$ must be stretched vertically by a factor of 2π, then shifted 1 unit to the left and 2π units down.

b–c.

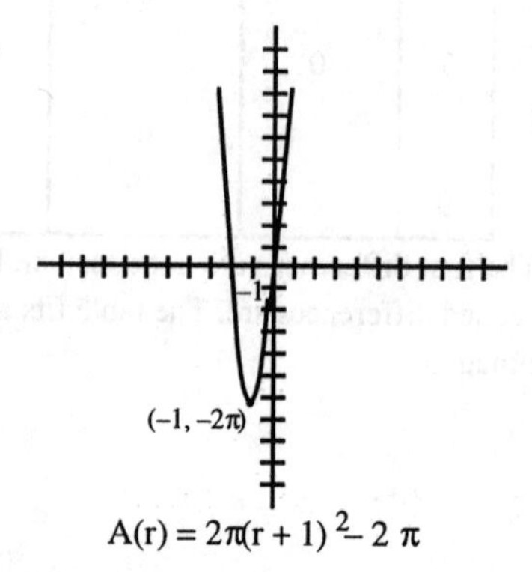

A(r) = 2π(r + 1)^2 – 2 π

d. The axis of symmetry is the line $r = -1$. The range is [–2π, ∞).

e. The function decreases on (–∞, –1) and increases on (–1, ∞).

69. a. The graph of $q=p^2$ must be stretched vertically by a factor of 100, then shifted 3 units to the left and 5 units down.

b–c.

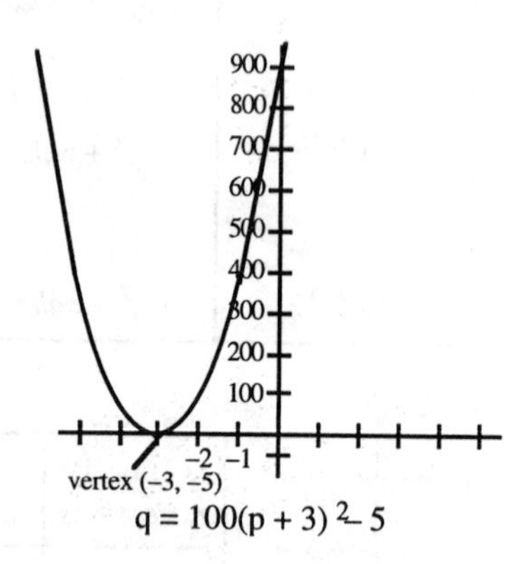

q = 100(p + 3)^2 – 5

d. The axis of symmetry is the line $p = -3$. The range is [–5, ∞).

e. The function decreases on (–∞, –3) and increases on (–3, ∞).

71. a. The x-intercepts occur when

$6+x-x^2=0$

$(2+x)(3-x)=0$

$x=-2, 3$

The vertex occurs when

$x=-\dfrac{1}{2(-1)}=\dfrac{1}{2}$

$y=6+\frac{1}{2}-\left(\frac{1}{2}\right)^2=\frac{25}{4}$

b.

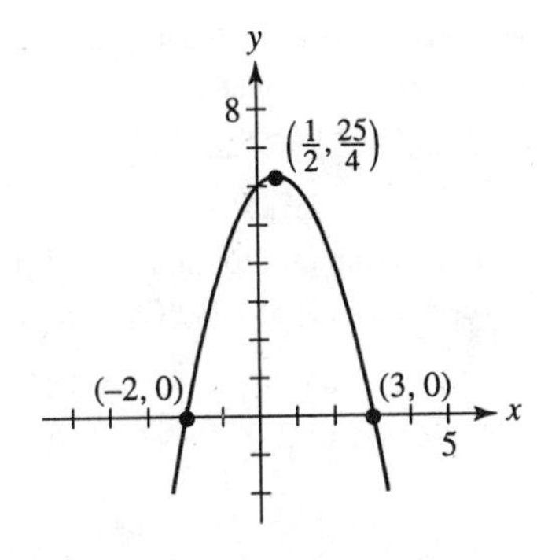

$y = 6 + x - x^2$

c. The axis of symmetry is the line $x = \frac{1}{2}$. The range is $\left(-\infty, \frac{25}{4}\right]$.

d. The function increases on $\left(-\infty, \frac{1}{2}\right)$ and decreases on $\left(\frac{1}{2}, \infty\right)$.

73. a. The t-intercepts occur when

$400 - 16t^2 = 0$
$25 - t^2 = 0$
$t^2 = 25$
$t = \pm 5$

The vertex occurs when

$t = -\frac{0}{2(-16)} = 0$
$y = 400 - 0^2 = 400$

b.

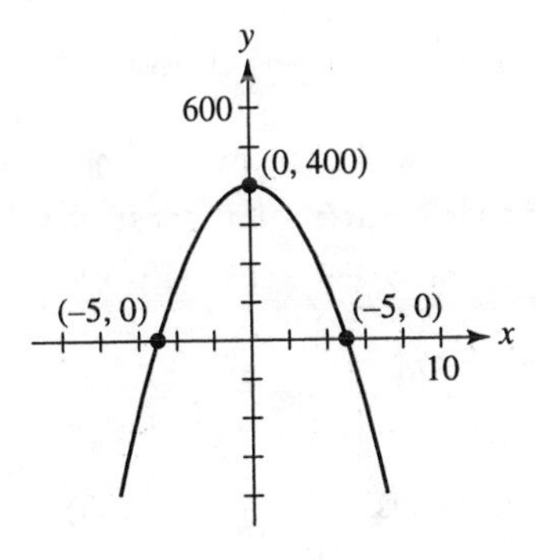

$s(t) = 400 - 16t^2$

c. The axis of symmetry is the line $t = 0$. The range is $(-\infty, 400]$.

d. The function increases on $(-\infty, 0)$ and decreases on $(0, \infty)$.

75. a. The x-intercepts occur when

$x^2 - 3x + 1 = 0$

$x = \frac{3 \pm \sqrt{(-3)^2 - 4(1)(1)}}{2(1)} = \frac{3 \pm \sqrt{5}}{2}$

The vertex occurs when

$x = -\frac{-3}{2(1)} = \frac{3}{2}$

$y = \left(\frac{3}{2}\right)^2 - 3\left(\frac{3}{2}\right) + 1 = -\frac{5}{4}$

b.

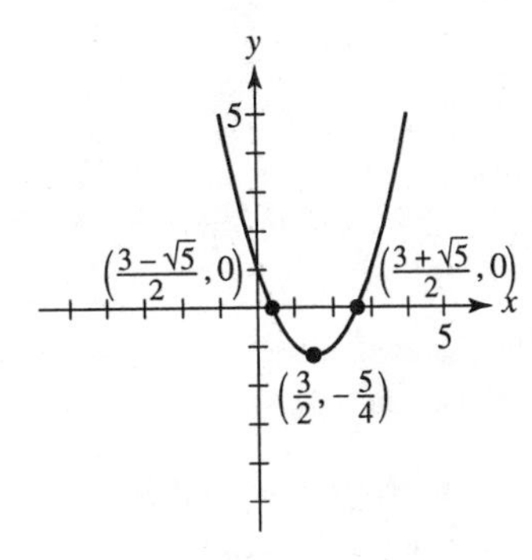

$y = x^2 - 3x + 1$

c. The axis of symmetry is the line $x = \frac{3}{2}$. The range is $\left[-\frac{5}{4}, \infty\right)$.

d. The function decreases on $\left(-\infty, \frac{3}{2}\right)$ and increases on $\left(\frac{3}{2}, \infty\right)$.

77. a. The u-intercepts occur when

$2u(u + 6) = 0$
$u = 0, -6$

The vertex occurs when

$u = \frac{0 + (-6)}{2} = -3$
$y = 2(-3)(-3 + 6) = -18$

b.

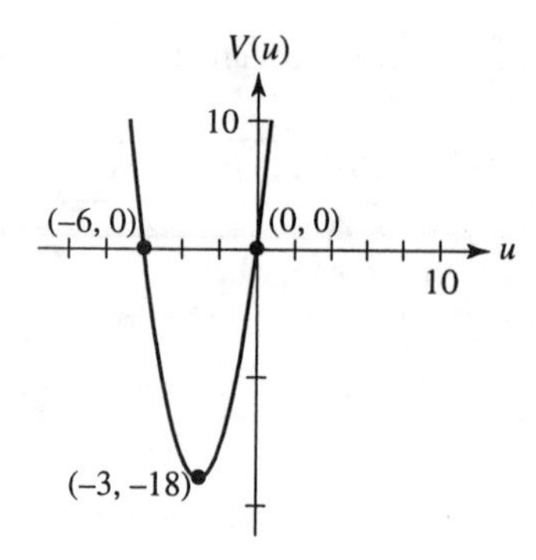

$V(u) = 2u(u+6)$

c. The axis of symmetry is the line $u = -3$. The range is $[-18, \infty)$.

d. The function decreases on $(-\infty, -3)$ and increases on $(-3, \infty)$.

79. a. The x-intercepts occur when
$(x+\sqrt{5})(x-\sqrt{5}) = 0$
$x = \pm\sqrt{5}$
The vertex occurs when
$x = \dfrac{\sqrt{5}+(-\sqrt{5})}{2} = 0$
$z = (0+\sqrt{5})(0-\sqrt{5}) = -5$

b.

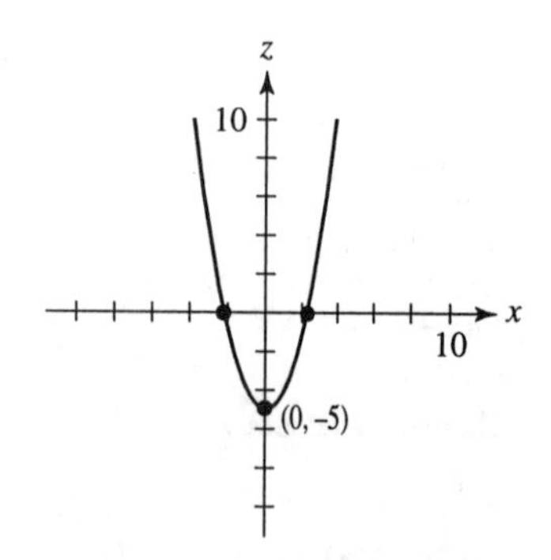

$x = (x+\sqrt{5})(x-\sqrt{5})$

c. The axis of symmetry is the line $x = 0$. The range is $[-5, \infty)$.

d. The function decreases on $(-\infty, 0)$ and increases on $(0, \infty)$.

81. a. There are no α-intercepts.
The vertex occurs when
$\alpha = \dfrac{(2+4i)+(2-4i)}{2} = 2$
$\beta = 5[2-(2+4i)][2-(2-4i)] = 80$
The points one unit to the left and right of the vertex are (1, 85) and (3, 85).

b.

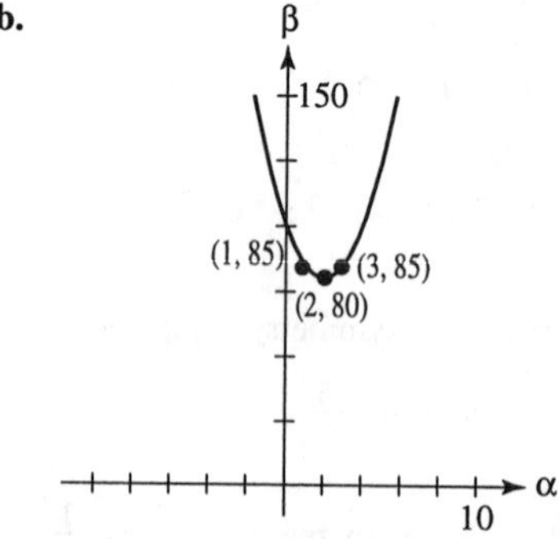

c. The axis of symmetry is the line $\alpha = 2$. The range is $[80, \infty)$.

d. The function decreases on $(-\infty, 2)$ and increases on $(2, \infty)$.

83. The x-intercepts are (–3, 0) and (2, 0), so the equation is $y = a(x+3)(x-2)$ for some value of a. Since the graph passes through (0, 4),
$4 = a(0+3)(0-2)$
$4 = -6a$
$a = -\frac{2}{3}$
The equation is $y = -\frac{2}{3}(x+3)(x-2)$.

85. The x-intercepts are (0, 0) and (100, 0), so the equation is $y = ax(x-100)$ for some value of a. Since the graph passes through (80, 500),
$500 = a(80)(80-100)$
$500 = -1600a$
$a = -\frac{5}{16}$
The equation is $y = -\frac{5}{16}x(x-100)$.

87. The vertex is (–4, 0), so the equation is $y = a(x+4)^2$ for some value of a. Since the graph passes through (1, 3),
$3 = a(1+4)^2$
$3 = 25a$

$a = 0.12$
The equation is $y = 0.12(x+4)^2$.

89. The vertex is (–10, –5), so the equation is $y = a(x+10)^2 - 5$ for some value of a.
Since the graph passes through (–5, –30),
$-30 = a(-5+10)^2 - 5$
$-30 = 25a - 5$
$25a = -25$
$a = -1$
The equation is $y = -(x+10)^2 - 5$.

91. It is the vertex.

93. a.
$$\begin{aligned} f(x) &= ax^2 + bx + c \\ &= a\left(x^2 + \frac{b}{a}x\right) + c \\ &= a\left(x^2 + \frac{b}{a}x + \frac{b^2}{4a^2} - \frac{b^2}{4a^2}\right) + c \\ &= a\left(x^2 + \frac{b}{a}x + \frac{b^2}{4a^2}\right) - \frac{b^2}{4a} + c \\ &= a\left(x + \frac{b}{2a}\right)^2 - \frac{b^2 - 4ac}{4a} \end{aligned}$$

b. From the standard graphing form in part (a) the vertex is $\left(-\frac{b}{2a}, \frac{b^2 - 4ac}{4a}\right)$.

95. a. The graph of $y = g(x)$ is steepest, because the coefficient of x^2 has the largest absolute value. Similarly, the graph of $y = h(x)$ is the least steep.

b. The graphs do not appear to support the conclusions in part (a). This is because the windows distort the steepness of each graph.

c. In the same window, the steepness of the graphs *relative to each other* is shown accurately. Thus, the graph of $y = g(x)$ appears steepest, and the graph of $y = h(x)$ appears the least steep.

97. The solutions to $F(x) = 0$ are
$$x = \frac{4 \pm \sqrt{(-4)^2 - 4(1)(13)}}{2(1)} = \frac{4 \pm \sqrt{-36}}{2} = 2 \pm 3i$$
In factored form,
$F(x) = [x - (2 + 3i)][x - (2 - 3i)]$.

99. The solutions to $F(x) = 0$ are
$$x = \frac{-8 \pm \sqrt{8^2 - 4(2)(9)}}{2(2)} = \frac{-8 \pm \sqrt{-8}}{4} = -2 \pm \frac{\sqrt{2}}{2}i.$$
In factored form,
$$F(x) = \left[x - \left(-2 + \frac{\sqrt{2}}{2}i\right)\right]\left[x - \left(-2 - \frac{\sqrt{2}}{2}i\right)\right]$$

Section 5-2

1. a. $s(t) = 0$ when the height of the rocket is 0, that is, when the rocket is at ground level. Since the flight lasts 10 seconds, $s(10) = 0$.

b. Since $s(0) = 0$ and $s(10) = 0$, the equation is $s(t) = at(t - 10)$ in factored form. The value of a is the coefficient of t^2 in the expanded form of the equation, so $a = -16$. Thus $s(t) = -16t(t - 10)$.

c. The vertex occurs when
$t = \frac{0+10}{2} = 5$
$y = -16(5)(10 - 5) = 400$
The vertex is (5, 400), and the parabola opens downward, so the vertex is the highest point on the graph. Therefore, the rocket reaches a height of 400 feet.

3. The vertex occurs when
$x = -\frac{4}{2(2)} = -1$
$y = 2(-1)^2 + 4(-1) + 5 = 3$
The extreme value is 3. Since $a = 2 > 0$, it is a minimum.

5. The vertex occurs when
$x = \frac{-20+8}{2} = -6$
$y = -6(-6+20)(-6-8) = 1176$
The extreme value is 1176. Since $a = -6 < 0$, it is a maximum.

7. The entire graph is above the x-axis.

9. First solve $2x^2 + x - 1 = 0$ to obtain $x = -1, \frac{1}{2}$.
These x-values divide the number line into the intervals $(-\infty, -1)$, $\left(-1, \frac{1}{2}\right)$ and $\left(\frac{1}{2}, \infty\right)$.

Choose a test value in $(-\infty, -1)$:
$2(-2)^2 + (-2) - 1 = 5 > 0$
The inequality is true.

Choose a test value in $\left(-1, \frac{1}{2}\right)$:
$2(0)^2 + (0) - 1 = -1 < 0$
The inequality is false.

Choose a test value in $\left(\frac{1}{2}, \infty\right)$:
$2(1)^2 + (1) - 1 = 2 > 0$
The inequality is true.

The solution is $(-\infty, -1) \cup \left(\frac{1}{2}, \infty\right)$.

11. First solve $100 + 15x - x^2 = 0$ to obtain $x = -5, 20$. These x-values divide the number line into the intervals $(-\infty, -5)$, $(-5, 20)$, and $(20, \infty)$.

Choose a test value in $(-\infty, -5)$:
$100 + 15(-10) - (-10)^2 = -150 < 0$
The inequality is false.

Choose a test value in $(-5, 20)$:
$100 + 15(0) - (0)^2 = 100 > 0$
The inequality is true.

Choose a test value in $(20, \infty)$:
$100 + 15(30) - (30)^2 = -350 < 0$
The inequality is false.

The solution is $[-5, 20]$.

13. $900a + 30b + c = 77$
$1600a + 40b + c = 128$
$2500a + 50b + c = 190$

Subtract the first equation from the second to obtain $700a + 10b = 51$. Subtract the second equation from the third to obtain $900a + 10b = 62$. In the resulting 2×2 system, subtract the first equation from the second to obtain $200a = 11$, so $a = 0.055$. Substitute 0.055 for a in the equation $700a + 10b = 51$ to obtain $b = 1.25$. Substitute 0.055 for a and 1.25 for b in the equation $900a + 30b + c = 77$ to obtain $c = -10$. The equation is $S(v) = 0.055v^2 + 1.25v - 10$.

This solution fits all rows of the table:

$$0.055(10)^2 + 1.25(10) - 10 = 77$$
$$0.055(20)^2 + 1.25(20) - 10 = 128$$
$$0.055(30)^2 + 1.25(30) - 10 = 190$$
$$0.055(40)^2 + 1.25(40) - 10 = 263$$

15. If $y = ax^2 + bx + c$ then

$$a(2)^2 + b(2) + c = -6$$
$$a(4)^2 + b(4) + c = 6$$
$$a(6)^2 + b(6) + c = 14$$

$$4a + 2b + c = -6$$
$$16a + 4b + c = 6$$
$$36a + 6b + c = 14$$

Subtract the first equation from the second to obtain $12a + 2b = 12$. Subtract the second equation from the third to obtain $20a + 2b = 8$. In the resulting 2×2 system, subtract the first equation from the second to obtain $a = -0.5$. Substitute -0.5 for a in the equation $12a + 2b = 12$ to obtain $b = 9$. Substitute -0.5 for a and 9 for b in the equation $4a + 2b + c = -6$ to obtain $c = -22$. The equation is $y = -0.5x^2 + 9x - 22$.

17. If $y = ax^2 + bx + c$, then

$$a(1.0)^2 + b(1.0) + c = 1$$
$$a(1.5)^2 + b(1.5) + c = 0$$
$$a(2.0)^2 + b(2.0) + c = 1$$

$$a + b + c = 1$$
$$2.25a + 1.5b + c = 0$$
$$4a + 2b + c = 1$$

Subtract the first equation from the second to obtain $1.25a + 0.5b = -1$. Subtract the second equation from the third to obtain $1.75a + 0.5b = 1$. In the resulting 2×2 system, subtract the first equation from the second to obtain $a = 4$.
Substitute 4 for a in the equation $1.25a + 0.5b = -1$ to obtain $b = -12$. Substitute 4 for a and -12 for b in the equation $a + b + c = 1$ to obtain $c = 9$. The equation is $y = 4x^2 - 12x + 9$.

19. If $h = at^2 + bt + c$, then

$$a(10)^2 + b(10) + c = 634$$
$$a(20)^2 + b(20) + c = 736$$
$$a(30)^2 + b(30) + c = 306$$

$$100a + 10b + c = 634$$
$$400a + 20b + c = 736$$
$$900a + 30b + c = 306$$

Subtract the first equation from the second to obtain $300a + 10b = 102$. Subtract the second equation from the third to obtain $500a + 10b = -430$. In the resulting 2×2 system, subtract the first equation from the second to obtain $a = -2.66$. Substitute -2.66 for a in the equation $300a + 10b = 102$ to obtain $b = 90$. Substitute -2.66 for a and 90 for b in the equation $100a + 10b + c = 634$ to obtain $c = 0$. The equation is $h = -2.66t^2 + 90t$

21. The vertex is (1, 0), so the extreme value of the function $y = 0$. Since $a = 3 > 0$, it is a minimum.

23. The vertex occurs at

$$x = -\frac{4}{2(1)} = -2$$
$$y = (-2)^2 + 4(-2) + 5 = 1$$

The extreme value of the function is $y = 1$. Since $a = 1 > 0$, it is a minimum.

25. The vertex occurs at

$$x = -\frac{-12}{2(-1)} = -6$$
$$y = -(-6)^2 - 12(-6) + 2 = 38$$

The extreme value of the function is $y = 38$. Since $a = -6 < 0$, it is a maximum.

27. The vertex occurs at

$$x = \frac{-3+9}{2} = 3$$
$$y = -3(3+3)(3-9) = 108$$

The extreme value of the function is $y = 108$. Since $a = -3 < 0$, it is a maximum.

29. First solve $(x + 5)(x - 13) = 0$ to obtain $x = -5, 13$. These x-values divide the number line into the intervals $(-\infty, -5)$, $(-5, 13)$, and $(13, \infty)$.

Choose a test value in $(-\infty, -5)$:
$(-6 + 5)(-6 - 13) = 19 > 0$
The inequality is false.

Choose a test value in $(-5, 13)$:
$(0 + 5)(0 - 13) = -65 < 0$
The inequality is true.

Choose a test value in $(13, \infty)$:
$(14 + 5)(14 - 13) = 19 > 0$
The inequality is false.

The solution is $[-5, -13]$.

31. First solve $2x^2 + 7x - 9 = 0$ to obtain $x = -4.5, 1$. These x-values divide the number line into the intervals $(-\infty, -4.5)$, $(-4.5, 1)$, and $(1, \infty)$.

Choose a test value in $(-\infty, -4.5)$:
$2(-5)^2 + 7(-5) - 9 = 6 > 0$
The inequality is true.

Choose a test value in $(-4.5, 1)$:
$2(0)^2 + 7(0) - 9 = -9 < 0$
The inequality is false.

Choose a test value in $(1, \infty)$:
$2(2)^2 + 7(2) - 9 = 13 > 0$
The inequality is true.

The solution is $(-\infty, -4.5) \cup (1, \infty)$.

33. First solve $x^2 - 8x + 16 = 0$ to obtain $x = 4$. This x-value divides the number line into the intervals $(-\infty, 4)$ and $(4, \infty)$.

Choose a test value in $(-\infty, 4)$:
$3^2 - 8(3) + 16 = 1 > 0$
The inequality is true.

Choose a test value in $(4, \infty)$:
$5^2 - 8(5) + 16 = 1 > 0$
The inequality is true.

The solution is $(-\infty, \infty)$.

35. First solve $x^2+5x+8=0$ to obtain

$$x=\frac{-5\pm\sqrt{5^2-4(1)(8)}}{2(1)}=\frac{-5\pm\sqrt{7}i}{2}$$

There are no real solutions, so any real number can serve as a test value.

$0^2+5(0)+8=8>0$
The inequality is false.

There are no solutions.

37.

x	y	first differences	second differences
–3	11		
		–5	
–2	6		2
		–3	
–1	3		2
		–1	
0	2		

The first differences are not constant, but the second differences are. The table fits a quadratic function.

If $y=ax^2+bx+c$, then

$$\begin{aligned}a(-2)^2+b(-2)+c&=6\\ a(-1)^2+b(-1)+c&=3\\ a(0)^2+b(0)+c&=2\end{aligned}$$

$$\begin{aligned}4a-2b+c&=6\\ a-b+c&=3\\ c&=2\end{aligned}$$

Subtract the second equation from the first to obtain $3a-b=3$. Subtract the third equation from the second to obtain $a-b=1$. In the resulting 2×2 system, subtract the second equation from the first to obtain $a=1$. Substitute 1 for a in the equation $a-b=1$ to obtain $b=0$. From the third equation in the original system, $c=2$. The equation is $y=x^2+2$.

39.

x	y	first differences	second differences
4	18		
		3	
6	21		–1
		2	
8	23		1
		3	
10	26		

Neither the first nor the second differences are constant. The table does not fit a quadratic function.

41.

x	y	first differences	second differences
–6	100		
		–6	
–2	94		4
		–2	
2	92		4
		2	
6	94		

The first differences are not constant, but the second differences are. The table fits a quadratic function.

If $y=ax^2+bx+c$, then

$$\begin{aligned}a(-2)^2+b(-2)+c&=94\\ a(2)^2+b(2)+c&=92\\ a(6)^2+b(6)+c&=94\end{aligned}$$

$$\begin{aligned}4a-2b+c&=94\\ 4a+2b+c&=92\\ 36a+6b+c&=94\end{aligned}$$

Subtract the first equation from the second to obtain $4b = -2$. Subtract the second equation from the third to obtain $32a + 4b = 2$.
From the first equation in the resulting 2×2 system, $b = -\frac{1}{2}$. Substitute $-\frac{1}{2}$ for b in the equation $32a + 4b = 2$ to obtain $a = \frac{1}{8}$.
Substitute $\frac{1}{8}$ for a and $-\frac{1}{2}$ for b in the equation $4a - 2b + c = 94$ to obtain $c = \frac{185}{2}$. The equation is $y = \frac{1}{8}x^2 - \frac{1}{2}x + \frac{185}{2}$.

43. From Exercise 6, if the width of the rented space is W feet, then the amount of floor space is $f(W) = 105W - W^2$ square feet. Since the graph of $f(W)$ opens downward, the maximum amount of floor space corresponds to the vertex. The vertex occurs when $W = -\frac{105}{2(-1)} = 52.5$. The rented space should be 52.5 feet on each side.

45. From Exercise 44, the revenue from the sale of x shirts is $R(x) = 16x - 0.005x^2$. The company will obtain more than \$10,000 in revenue if $16x - 0.005x^2 > 10{,}000$. To solve this inequality, first solve the equation

$16x - 0.005x^2 = 10{,}000$
$-0.005x^2 + 16x - 10{,}000 = 0$
$$x = \frac{-16 \pm \sqrt{16^2 - 4(-0.005)(-10000)}}{2(-0.005)}$$
$x \cong 851.67,\ 2348.33$

These x-values divide the number line into the intervals $(-\infty, 851.67)$, $(851.67, 2348.33)$, and $(2348.33, \infty)$.

Choose a test value in $(-\infty, 851.67)$:
$-0.005(0)^2 + 16(0) - 10{,}000 = -10{,}000 < 0$
The inequality is false.

Choose a test value in $(851,67, 2348.33)$:
$-0.005(1000)^2 + 16(1000) - 10{,}000 = 1000 > 0$
The inequality is true.

Choose a test value in $(2348.33, \infty)$:
$-0.005(10{,}000)^2 + 16(10{,}000) - 10{,}000$
$= -350{,}000 < 0$
The inequality is false.

The solution is approximately $(851.67, 2348.33)$. In the context of the problem the solution is $[852, 2348]$.

47. From Exercise 46, the revenue generated by n people is $1175n - 7.5n^2$ dollars. The revenue is at least \$45,000 if $1175n - 7.5n^2 \ge 45{,}000$. That is, $7.5n^2 - 1175n + 45{,}000 \le 0$.

First solve $7.5n^2 - 1175n + 45{,}000 = 0$ to obtain
$$n = \frac{1175 \pm \sqrt{1175^2 - 4(7.5)(45{,}000)}}{2(7.5)} = \frac{200}{3},\ 90$$
These values of n divide the number line into the intervals $\left(-\infty, \frac{200}{3}\right)$, $\left(\frac{200}{3}, 90\right)$, and $(90, \infty)$.

Choose a test value in $\left(-\infty, \frac{200}{3}\right)$:
$7.5(0)^2 - 1175(0) + 45{,}000 = 45{,}000 > 0$
The inequality is false.

Choose a test value in $\left(\frac{200}{3}, 90\right)$:
$7.5(80)^2 - 1175(80) + 45{,}000 = -1000 < 0$
The inequality is true.

Choose a test value in $\left(-\infty, \frac{200}{3}\right)$:
$7.5(100)^2 - 1175(100) + 45{,}000 = 2500 > 0$
The inequality is false.

The solution is $\left[\frac{200}{3}, 90\right]$. In the context of the problem the solution is $[67, 90]$.

49. In the coordinate system of Exercise 48, a height of 6 inches (0.5 feet) above the drain is represented by a y-coordinate of $1234.375 + 0.5 = 1234.875$. A point (x, y) on the road will be under water if and only if $0.00004x^2 - 0.05x + 1250 < 1234.875$.
Equivalently, $0.00004x^2 - 0.05x + 15.125 < 0$.

First, solve $0.00004x^2 - 0.05x + 15.125 = 0$ to obtain
$$x = \frac{0.05 \pm \sqrt{(-0.05)^2 - 4(0.00004)(15.125)}}{2(0.00004)}$$
$\cong 513.20,\ 736.80$

These x-values divide the number line into the intervals $(-\infty, 513.20)$, $(513.20, 736.80)$, and $(736.80, \infty)$.

Choose a test value in $(-\infty, 513.20)$:
$0.00004(0)^2 - 0.05(0) + 15.125 = 15.125 > 0$
The inequality is false.

Choose a test value in $(513.20, 736.80)$:
$0.00004(600)^2 - 0.05(600) + 15.125$
$= -0.475 < 0$
The inequality is true.

Choose a test value in $(-\infty, 513.20)$:
$0.00004(800)^2 - 0.05(800) + 15.125 = 0.725$
The inequality is false.

The solution is $(513.20, 736.80)$, so 223.60 feet of the roadway will be underwater.

51. $S(5) = -2.38$, indicating that a car traveling at an initial speed of 5 mph has a stopping distance of -2.38 feet. Because this result makes no physical sense, we can conclude that the function S does not describe the relationship between initial speed and stopping distance for speeds near 5 mph.

53. **a.** $Q(30) = 78$
$Q(40) = 128$
$Q(50) = 190$
$Q(60) = 264$

These values do not quite agree with those given in Table 19, but all are at least close.

b. The graphs are close over the entire interval, so the stopping distances predicted by S and Q are also close.

c. The graphs are not very close when x is near 0. Since the graph of Q has no negative y-values, it predicts more accurate stopping distances for low initial speeds.

d. Responses will vary. Most students will choose Q because it is a better model for low initial speeds, but some may argue that S is a better model for initial speeds between 30 and 60 mph, since it fits Table 19 exactly.

NUTSHELLS

Nutshell for Chapter 6 - QUADRATIC RELATIONS

What's familiar?

- As usual, we will discuss the numerical, analytical and graphical views. In this chapter, the analytical and graphical views dominate.
- Section 6-3 develops a comprehensive set of graphing transformation principles that unifies your work in Section 3-4 and 5-2 with the contents of this chapter.
- Section 6-6 uses methods to solve systems of quadratic equations and inequalities that are similar to those you used for linear systems in Chapter 4.

What's new?

- In this chapter, the relationships may or may not be functions. However, they all belong to the more inclusive category of *relations*.
- Section 6-1 shows you how to use your knowledge of functions to study more general relations.
- Section 6-4 uses geometry properties of the graphs of quadratic relations to construct mathematical models of physical problems.

List of hints for exercises in each chapter.

Chapter 6

- Use the strategy of *Making a Table* when doing graphical transformations.
- The terms *implicitly* and *explicitly* distinguish between the two ways to describe a relation analytically. The everyday definitions will help you understand the mathematical distinctions between the two forms.
- Note the similarity among the standard graphing forms of circles, parabolas, ellipses, and hyperbolas. The graphing transformation principles reflect what you learned in Sections 3-4 and 5-1. This observation will help you complete the exercises with ease.
- Although Graphical Transformations in the Context of Functions is a Supplementary Topic, you will probably want to complete these exercises. They reinforce the transformation techniques, in general, by pulling together Sections 3-4, 5-1, and 6-3.

Name:____________________

Chapter 6 Additional Exercises

Date:____________________

In Problems 1–3, find two functions defined implicitly by the relation given.

1. $x+3=-y^2$
2. $9x^2=4y^2+36$
3. $y^2+4x^2=16$

1. ____________________
2. ____________________
3. ____________________

In Problems 4–6, graph the relation.

4. $x+3=-y^2$ (This the relation from Problem 1.)

4.

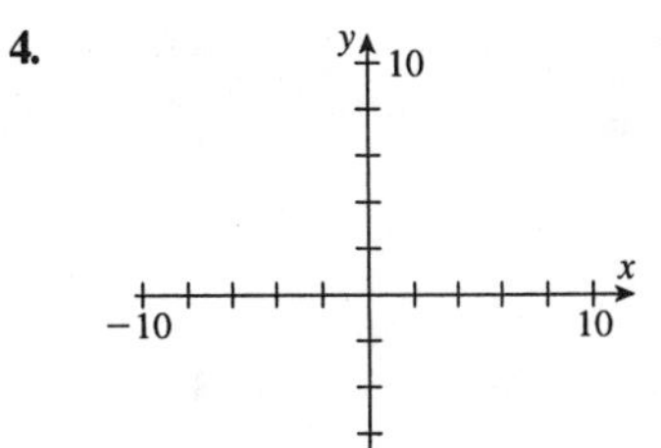

5. $9x^2=4y^2+36$ (This is the relation from Problem 2.)

5.

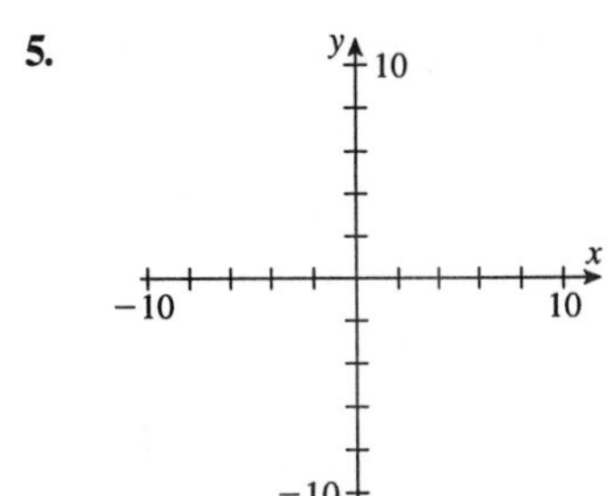

6. $y^2+4x^2=16$ (This is the relation from Problem 3.)

6.

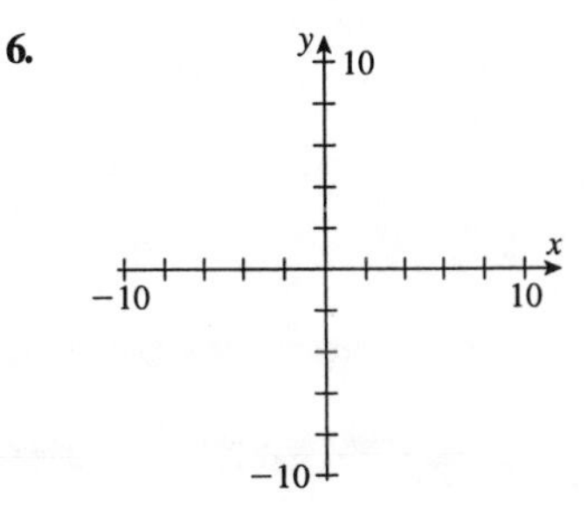

Chapter 6 Additional Exercises *(cont.)*

Name:____________________

In Problems 7–10, sketch the graph of each equation and identify any of the following features that apply: vertex, axis of symmetry, center, radius, vertices and covertices, major and minor axes, transverse and conjugate axes, asymptotes.

7. $x - 6y^2 + 12y = 6$

7. ____________________

____

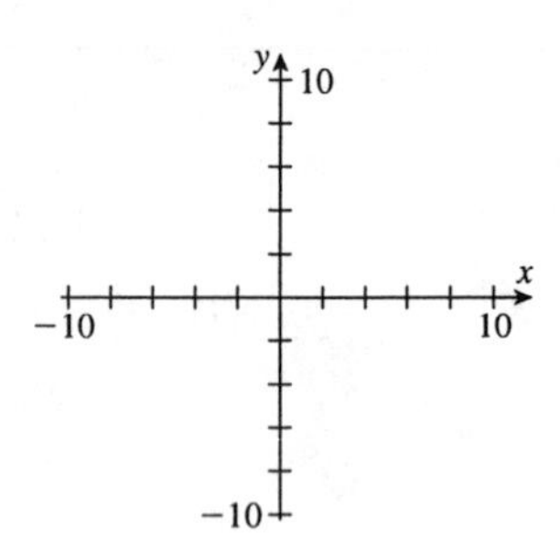

8. $9x^2 + 25y^2 - 54x - 100y = 44$

8. 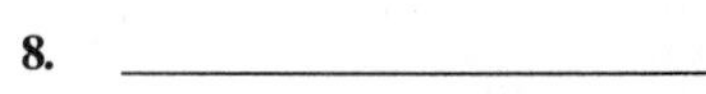

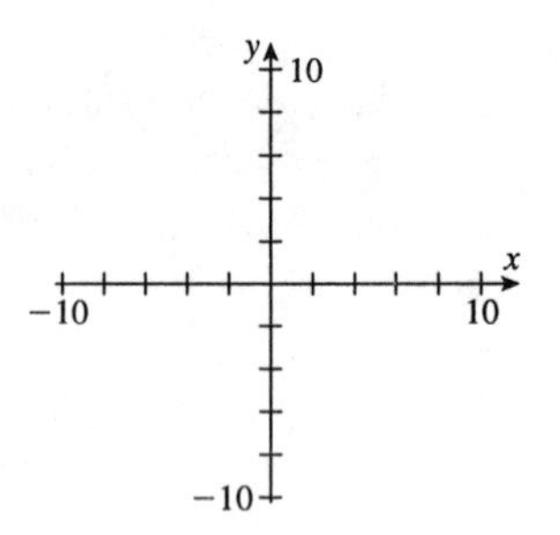

9. $-4x^2 + 9y^2 + 18y = 27$

9. 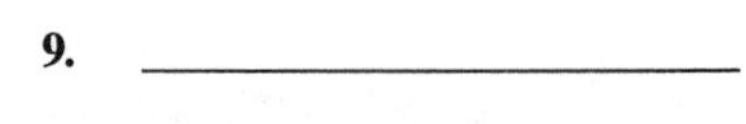

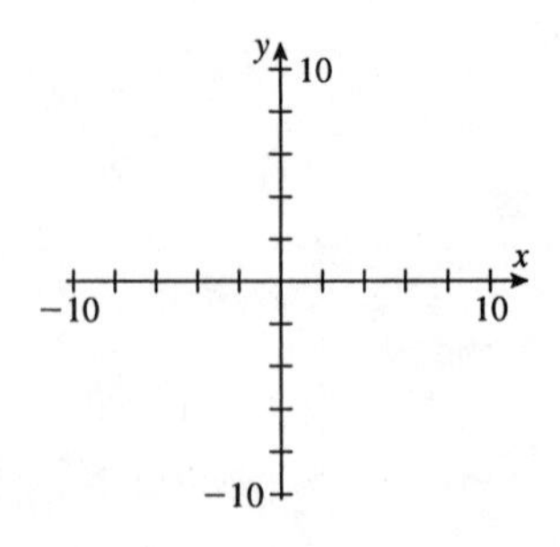

Chapter 6 Additional Exercises *(cont.)*

Name:____________________

10. $x^2 + y^2 + 4x + 2y = 20$

10. ____________________

____

y 10

x

−10 10

−10

11. Find the vertex and the axis of symmetry of $x - 4y^2 - 28y = 55.8$.

11. ____________________

12. Find the center and the asymptotes of $25x^2 - 16y^2 + 250x + 64y = -161$.

12. ____________________

13. Find the center, vertices, and covertices of $16x^2 + 9y^2 + 224x - 72y = -784$.

13. ____________________

14. How is the graph of $x^2 + y^2 = 1$ transformed to produce the graph of $16x^2 + 25y^2 = 1$? Identify the new vertices and covertices.

14. ____________________

15. How is the graph of $x^2 - y^2 = 1$ transformed to produce the graph of $9x^2 - (y+8)^2 = 81$? Identify the equations of the new asymptotes.

15. ____________________

16. How is the graph of $y = x^2$ transformed to produce the graph of $y = 6(x+5)^2 - 14$? Identify the new vertex.

16. ____________________

17. Is the graph of $y = \dfrac{x^2}{x^4 + 1}$ symmetric about the *x*-axis, *y*-axis, both, or neither?

17. ____________________

18. What equation results when the graph of $x^2 + y^2 = 1$ is compressed vertically by a factor of 7, stretched horizontally by a factor of 2, then shifted 1 unit right and 6 units down?

18. ____________________

19. Sketch the graph of the relation $\left(\frac{x+2}{9}\right)^2 - y^2 = 1$.

19.

y 10

x

−20 20

−10

20. Find the focus and directrix of the parabola $y = 4(x+5)^2 - 2$.

20. ____________________

Chapter 6 Additional Exercises *(cont.)*

Name:____________________

21. Find the foci and eccentricity of the ellipse.
$25x^2 + 9y^2 - 10x + 12y - 220 = 0$

21. ____________________

22. Find the foci of the hyperbola.
$y^2 - 4x^2 + 2y + 16x - 24 = 0$

22. ____________________

In Problems 23–25, sketch the conic section.

23. $y^2 + \frac{y}{3} + x = 8.$

23.

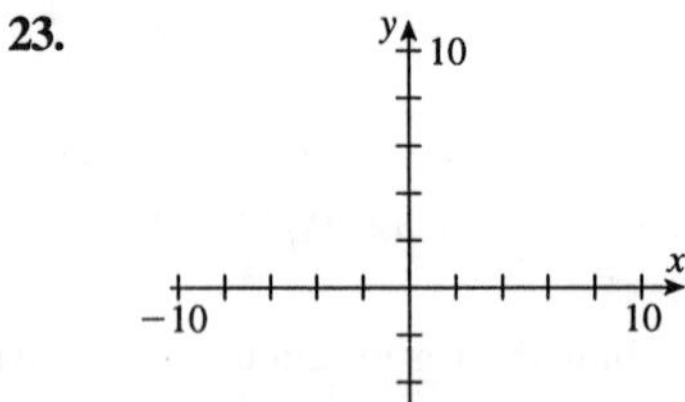
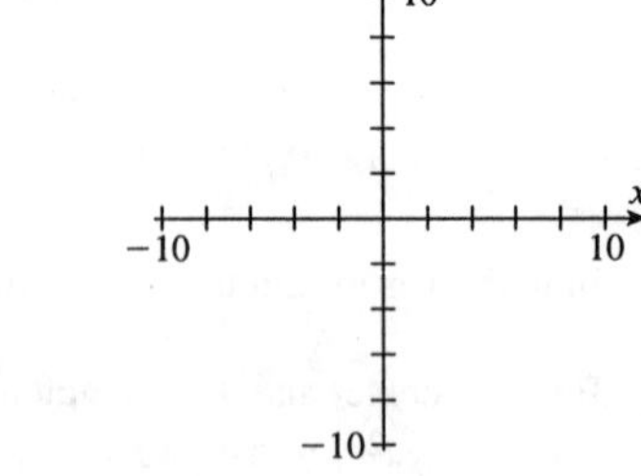

24. $4x^2 - 24x - 3y^2 - 36y - 108 = 0$

24.

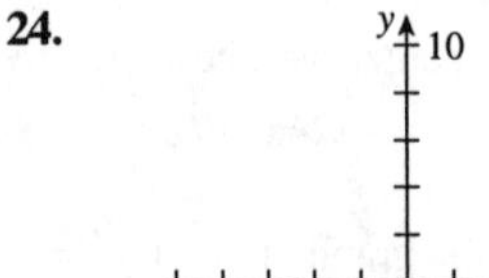
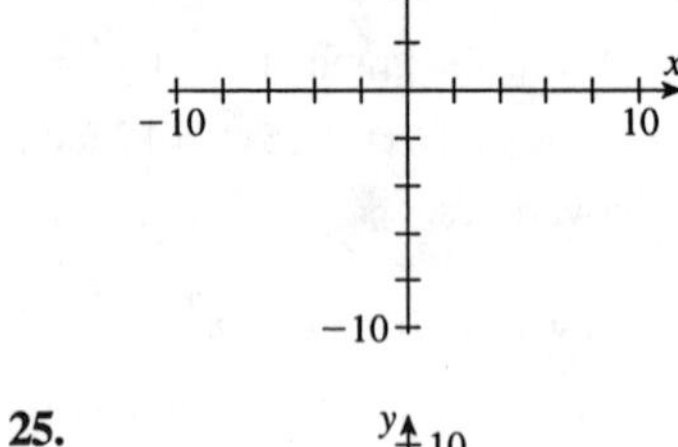

25. $5x^2 + 36y^2 + 20x + 216y + 164 = 0$

25.

y 10, x, −10, 10, −10

26. Find the domain and range of the function.
$y = -8\sqrt{3x}$

26. ____________________

27. Find the domain and range of the function.
$y = 8 + \sqrt{x - 7}$

27. ____________________

Chapter 6 Additional Exercises *(cont.)*

Name:____________________

28. Sketch the graph of the function.

$y = 2\sqrt{6 - 2.5x}$

28.

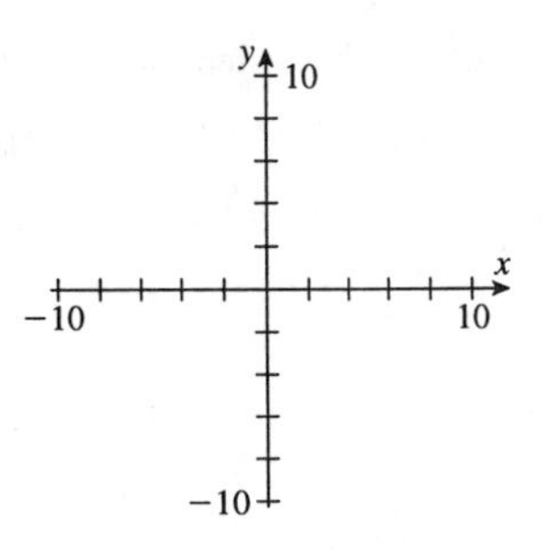

In Problems 29 and 30, find the domain and range. Then eliminate the radical to obtain the equation of a conic section that defines the function implicitly.

29. $y = 6\sqrt{x^2 + 9}$

29. ____________________

30. $y = 3 - \sqrt{16x^2 - 25}$

30. ____________________

31. Sketch the graph of the function.

$y = 6 - 2\sqrt{9 - (x - 3)^2}$

31.

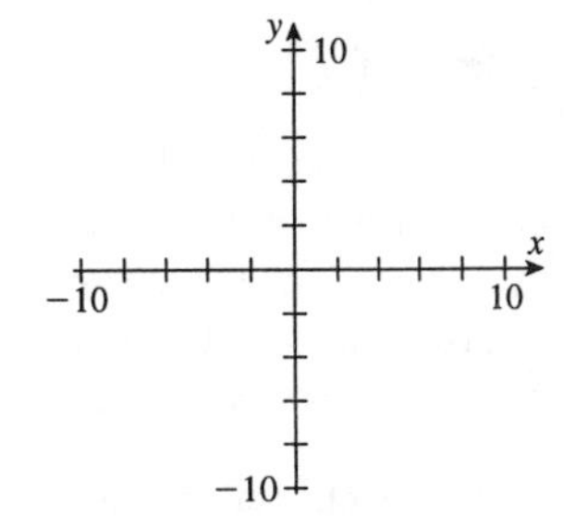

In Problems 32–35, solve the system of equations analytically.

32. $25x^2 + y^2 = 125$

$xy = -10$

32. ____________________

33. $x^2 - y^2 = 9$

$25x^2 + 4y^2 = 254$

33. ____________________

34. $x^2 + y^2 = 36$

$y = x^2 - 6$

34. ____________________

35. $x^2 + 8y^2 - x + 16y - 42 = 0$

$xy = 0$

35. ____________________

36. Solve the system of inequalities graphically.

$x^2 - y^2 < 4$

$y^2 - 2x^2 < 4$

36.

37. Solve the system of inequalities graphically.

$9x^2 + y^2 \le 81$

$y^2 - 9x^2 < 25$

$xy \le -3$

37.

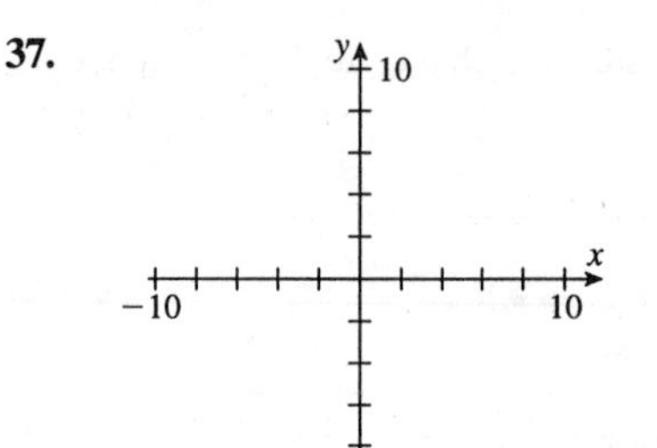

38. A computer-controlled lathe is to produce a hollow brass sphere 10 cm in diameter on the outside, one half at a time. Write an equation in terms of x and $y\,(y > 0)$ that describes the outside cutting profile.

38. ____________________

39. A ball is thrown directly up from ground level. Its height in meters as a function of time is given by $h(t) = -9.8t^2 + 39.2t$. Write two functions $t_1(h)$ and $t_2(h)$ that together describe t in terms of h.

39. ____________________

40. An electron is fired at a negatively charged sphere. It approaches within 8 m of the sphere, then flies away again. Model the electron's distance from the sphere as a function of time in seconds ($t = 0$ at instant of closest approach), using the upper half of a hyperbola with asymptotes having slopes of ± 1500 m/s.

40. ____________________

41. A parabolic reflector is 6 cm wide at a distance 8 cm from the vertex. Find the equation of the cross section, if the vertex is at the origin and the parabola opens upward.

41. ____________________

42. The outline of a 200-ft-tall pyramid can be described by the equation $y = -|x|$, with the origin at the tip of the pyramid. Describe the same outline when the origin is at the center of the pyramid's base.

42. ____________________

43. In relation to a street intersection, a sports stadium can be described by the equation $\frac{(x-5)^2}{16} + \frac{(y-3)^2}{4} = 1$ (units in city blocks). The original plans called for the equation to be 3 blocks farther west and 2 blocks farther north. What would its equation have been then?

43. ____________________

44. An ellipse is drawn using two tacks 7 in. apart and a 10-in. string between them. What is the ellipse's eccentricity?

44. ____________________

45. Using a hyperbolic mirror, light rays converging toward a single point are redirected toward another point 10 cm directly below the first. The second point is 9 cm from the mirror. Write an equation for the hyperbola (both halves). Take the two foci to be equal distances from the origin.

45. ____________________

46. How far from the vertex is the focal point of a parabolic mirror measuring 8 in. across at a distance of 18 in. from the vertex?

46. ____________________

Chapter 6 Additional Exercises *(cont.)*

Name:____________________

Problems 47 and 48 refer to the following scenario. A woman stands 20 ft from some railroad tracks as a train comes by, traveling at 90 mi/h.

47. If $t = 0$ when the locomotive is closest to the woman, write the locomotive's distance from the woman as a function of time in seconds. (1 mi = 5280 ft)

47. ____________________

48. When will the train whistle have the same pitch for the woman as it would if the train were standing still? (Sound travels at 1088 ft/s.)

48. ____________________

Problems 49 and 50 refer to the following situation: In an appropriate coordinate system, the orbit of the earth around the sun has the equation $\frac{x^2}{1.03} + \frac{y^2}{0.97} = 1$ (distances in astronomical units).

49. At what points would a comet with an orbit whose equation is $(x - 0.83)^2 + \frac{y^2}{0.66} = 1$ intersect the earth's orbit? Round your answer to the nearest hundredth.

49. ____________________

50. At what points would an asteroid with an orbit whose equation is $\frac{(x - 0.24)^2}{9} + \frac{y}{9} = 1$ intersect the earth's orbit?

50. ____________________

A Mathematical Looking Glass

Sonic Boom

(to follow Section 6-2)

An airplane flying at supersonic speeds emits sound waves that form the surface of a cone. The cone follows the airplane and always intersects the ground to form a conic section, as in Figure S-7. An observer hears a sonic boom at the instant the conic section moves through her location.

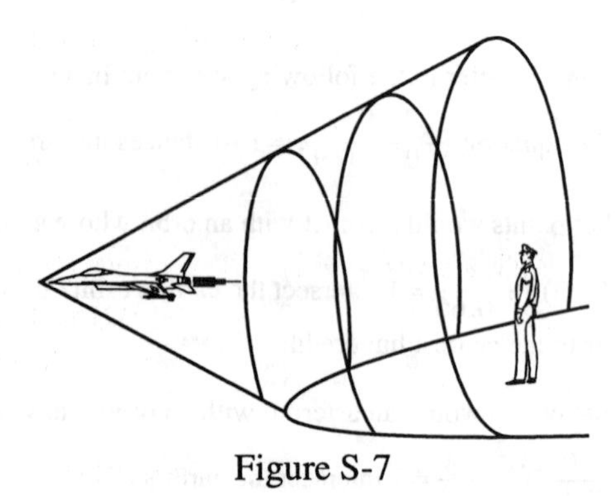

Figure S-7

What type of conic section is formed where the cone of sound waves intersects the ground? Let's find the equation of the conic section for a plane flying at an altitude of 30,000 feet with a speed of Mach 2 (that is, twice the speed of sound). Figure S-8 shows the airplane flying directly above the y-axis of a coordinate plane, on which distances are measured in feet. Let's construct the equation of the conic section at the instant the airplane passes over the origin.

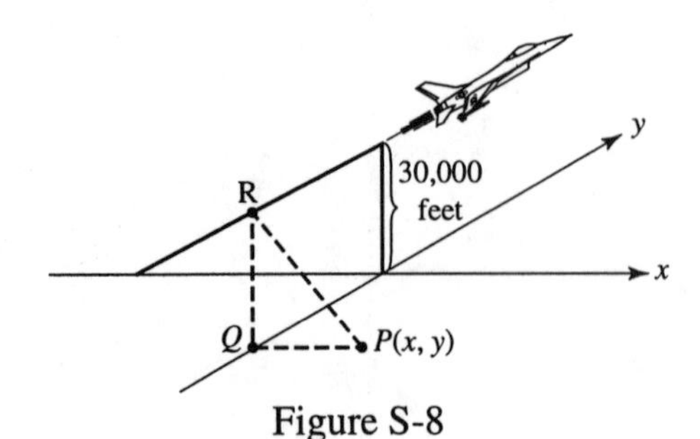

Figure S-8

Suppose an observer at $P = (x, y)$ hears the sonic boom at that instant. The sound she is hearing was emitted by the airplane when it passed the point R, directly above $Q = (0, y)$. Since that time, the airplane has flown a distance of y feet, and the sound has traveled from R to P. Because the airplane is traveling at twice the speed of sound, the length of RP must be $\frac{1}{2}y$.

Exercises

1. (Creating Models) Use the triangle PQR and the Pythagorean theorem to find the equation of the conic section. What type of conic section is it?

2. (Making Observations) What is the equation of the conic section if the airplane is traveling at Mach 3? Mach 4? Mach 5? What is the relationship between the Mach number and the equation?

A Mathematical Looking Glass
Whispering Galleries
(to follow Section 6-4)

Statuary Hall, in the U.S. Capitol Building in Washington, DC, has the shape of a quarter of a sphere, as shown in Figure S-9a.

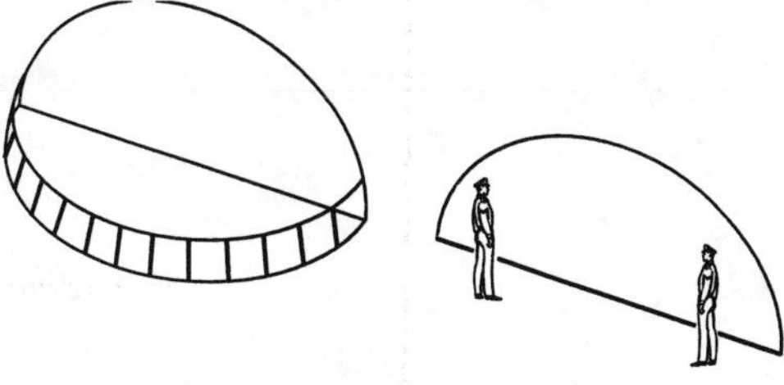

Figure S-9

If you stand at one end of the flat wall and whisper, you can be heard by someone standing at the other end, while a person standing in between cannot hear you at all. See Figure S-9b. St. Paul's Cathedral in London, and Union Station in St. Louis, also have the property that strategically placed people separated by some distance can hear each other's whispers. In his book, *Down by the Station*, Norman Temme says of Union Station's famed Whispering Arch, "During the War years many a departing GI whispered his fond farewells to his lady love only to discover they had a smiling and usually understanding audience."

Exercises

1. (Writing to Learn) Figure S-10a shows a room with whisperers at *A* and *B*. Cross sections through these points are semiellipses. Explain why this room is a more effective whispering gallery than the ones mentioned above, which all have semicircular cross sections.

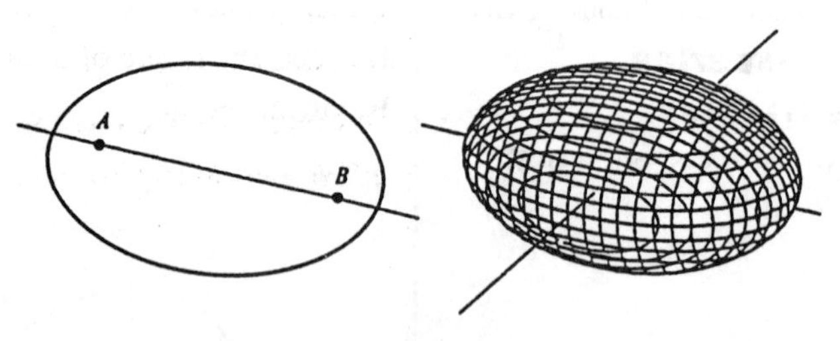

Figure S-10

2. a. (Problem Solving) In Figure S-10a, if the room is 120 feet long and 40 feet high, how far apart should *A* and *B* be?

b. (Problem Solving) How long would a sound wave from one person take to bounce off a wall and reflect to the other person? (Assume that the room is approximately at sea level, he the speed of sound is 1129 ft/sec.)

c. (Writing to Learn) Does the effectiveness of the whispering gallery depend on which way the people are facing? Explain your answer.

d. (Writing to Learn) Would the gallery be more effective if it could be designed as in Figure S-10b to have the shape of a full ellipsoid (that is, if all cross sections through *A* and *B* were full ellipses), with the whisperers suspended in space at the foci? Explain.

A Mathematical Looking Glass

Refraction

(to follow Section 6-5)

It is a physical curiosity that light rays travel between two points along a path which requires the least possible time. Since light travels through different media, such as air and water, at different speeds, the path of a light ray is not always a straight line. The bending of a light ray at the interface between two media, called **refraction**, is illustrated in Figure S-11.

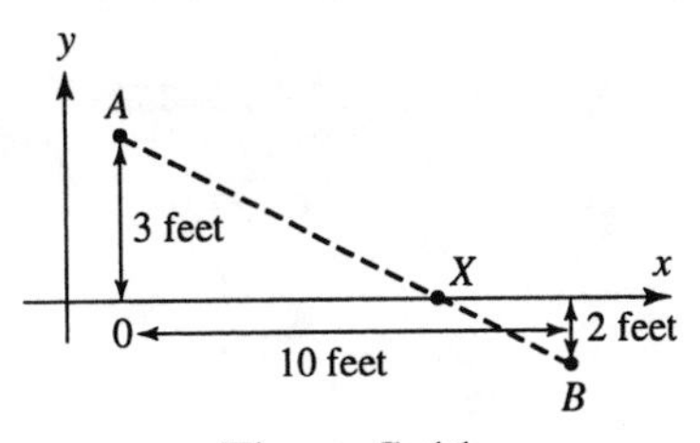

Figure S-11

Exercises

1. (Problem-Solving) Use the strategy of *Breaking the Problem into Parts* to find OX in Figure S-11.

   a. Use the Pythagorean Theorem to express AX and XB in terms of x.

   b. Light travels at 186,000 feet per second in air and 139,500 feet per second in water. Express the time needed to travel along each segment AX and XB as a function of x.

   c. Express the light ray's total time from A to B as a function of x.

   d. Find the minimum value of the function in part (c) graphically. How many feet from O does the light ray strike the water?

2. (Making Observations) How many feet from O would the light ray strike the water if it traveled in a straight line? How far is this from X?

Chapter 6

Section 6-1

1. Any vertical line between $x = -1$ and $x = 1$ intersects the graph twice.

3. The legs of the right triangle in the diagram have lengths x and y, and the hypotenuse has lnegth 1. By the Pythagorean Theorem, $x^2 + y^2 = 1$.

5. a. Any horizontal line between $y = -1$ and $y = 1$ intersects the graph twice.

b. The first and last rows have the same second entries and different first entries.

c. Solve for x yields $x = \pm\sqrt{1 - y^2}$. Some values of the independent variable y generate more than one value of the dependent variable x.

7. No two rows have the same first entry, so the table defines y as a function of x.

9. Two rows have the same first entry $\left(\sqrt{5}\right)$ and different second entries (2π and 3π), so the table does not define y as a function of x.

11. Many vertical lines intersect the graph more than once, so the graph does not define y as a function of x.

13. No vertical line intersects the graph more than once, so the graph defines y as a function of x.

15.

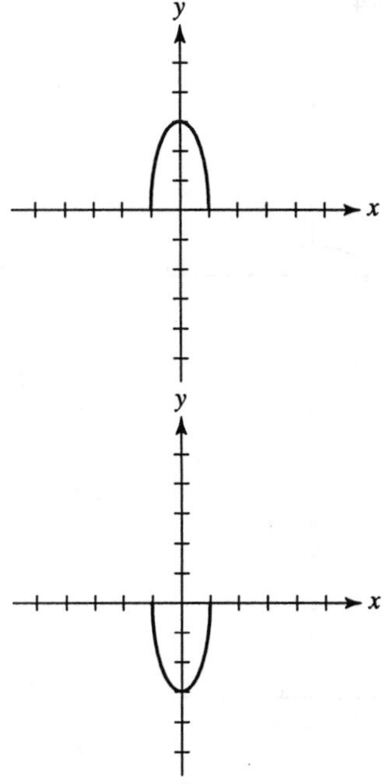

17.

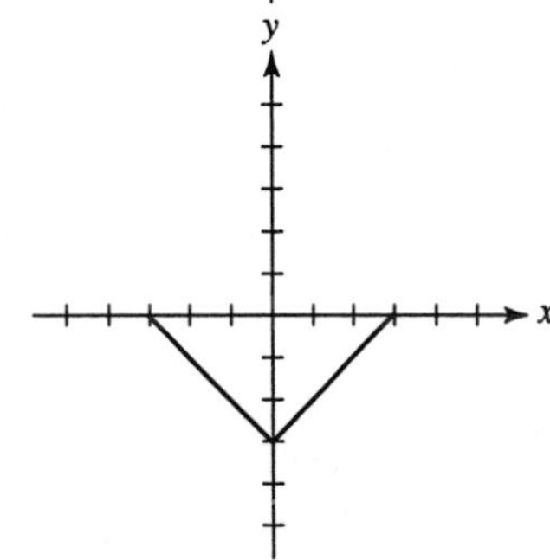

19. $x^2 + y = 4$
$y = 4 - x^2$
y is a function of x.

21. $(y-1)^2 - x^2 = 1$
$(y-1)^2 = x^2 + 1$
$y - 1 = \pm\sqrt{x^2 + 1}$
$y = 1 \pm \sqrt{x^2 + 1}$
y is not a function of x.

23. $x^2 + 4y^2 = 4$
$4y^2 = 4 - x^2$

$2y = \pm\sqrt{4 - x^2}$

$y = \pm\frac{1}{2}\sqrt{4 - x^2}$

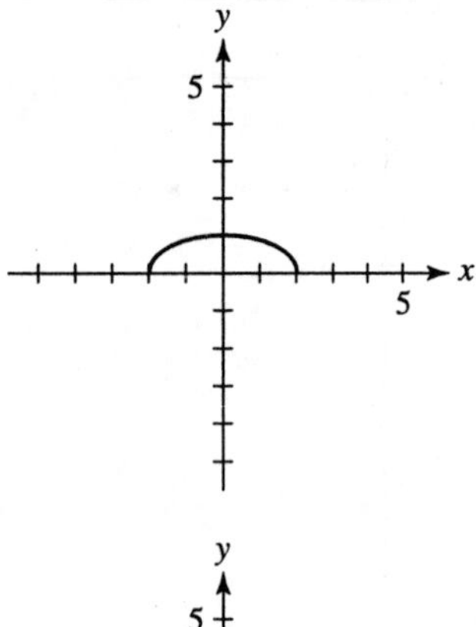

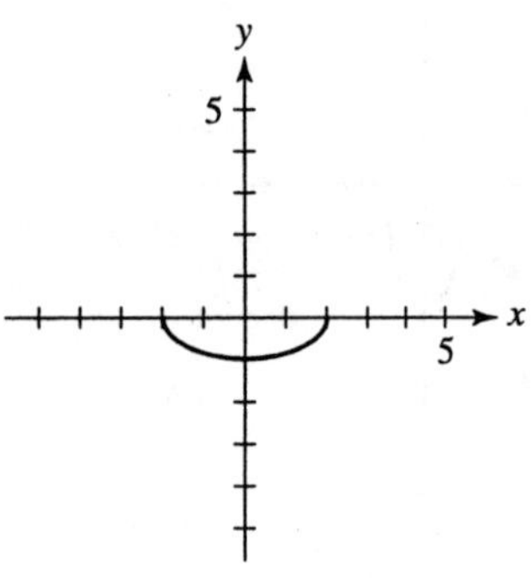

25. $4y^2 - x^2 = 4$

$4y^2 = 4 + x^2$

$2y = \pm\sqrt{4 + x^2}$

$y = \pm\frac{1}{2}\sqrt{4 + x^2}$

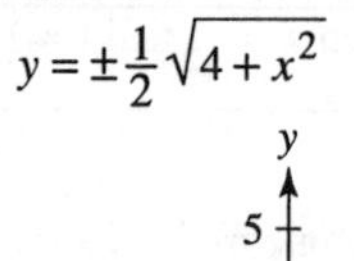

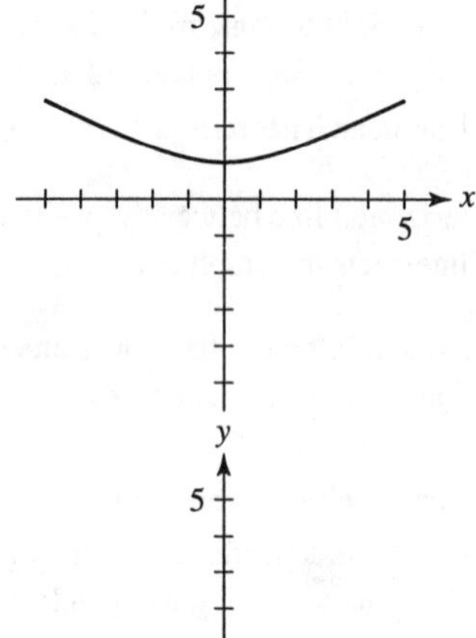

27. (Sample response) The graph of every quadratic relation consists of at most two pieces, each of which is the graph of a function. No vertical line can intersect the graph of either function more than once, so no vertical line can intersect the graph of the relation more than twice.

29. $x = [f(x)]^2$

$x = \left(\sqrt{x}\right)^2$

$x = x$

31. $9x^2 + [f(x) - 5]^2 = 36$

$9x^2 + \left[(5 - 3\sqrt{4 - x^2}) - 5\right]^2 = 36$

$9x^2 + \left(-3\sqrt{4 - x^2}\right)^2 = 36$

$9x^2 + 9(4 - x^2) = 36$

$9x^2 + 36 - 9x^2 = 36$

$36 = 36$

33. a. $4x^2 - 9y^2 = 36$

$9y^2 = 4x^2 - 36$

$3y = \pm\sqrt{4x^2 - 36}$

$y = \pm\frac{1}{3}\sqrt{4x^2 - 36}$

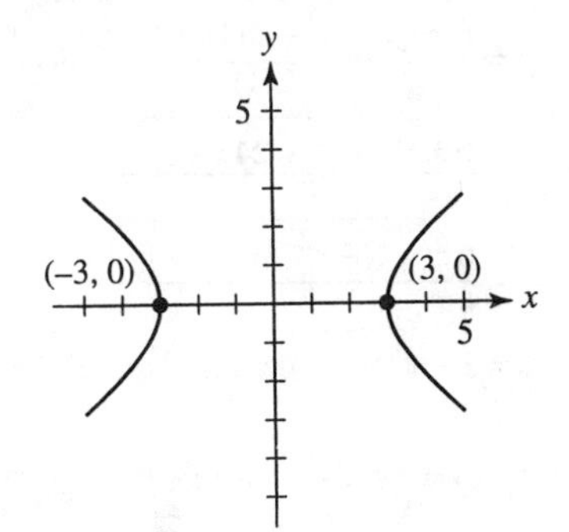

35. a. $x = 2y^2 - 6$

$2y^2 = x + 6$

$y^2 = \frac{x+6}{2}$

$y = \pm\sqrt{\frac{x+6}{2}}$

b.

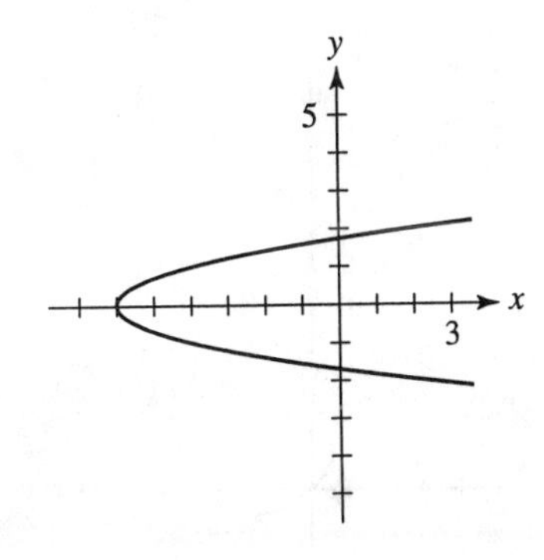

37.

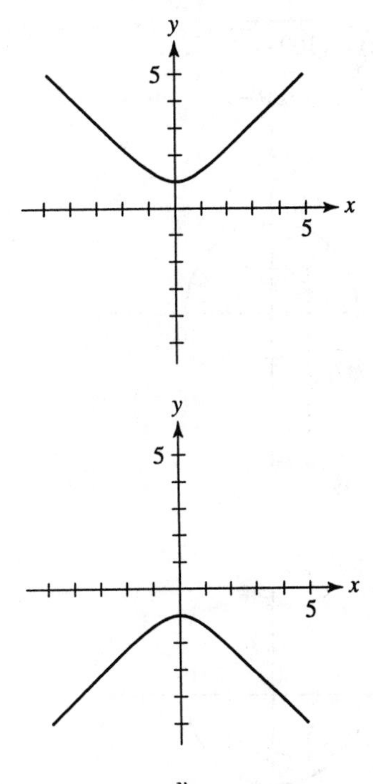

39.

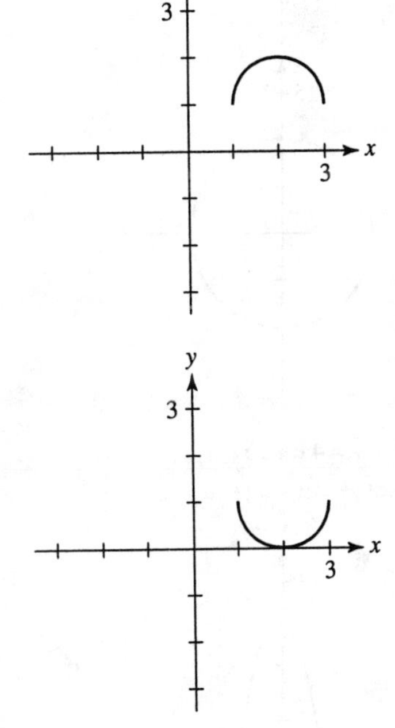

41. a. $x^2 + y^2 = 100$

$y^2 = 100 - x^2$

$y = \pm\sqrt{100 - x^2}$

b. $f_1(x) = \sqrt{100 - x^2}$

$f_2(x) = -\sqrt{100 - x^2}$

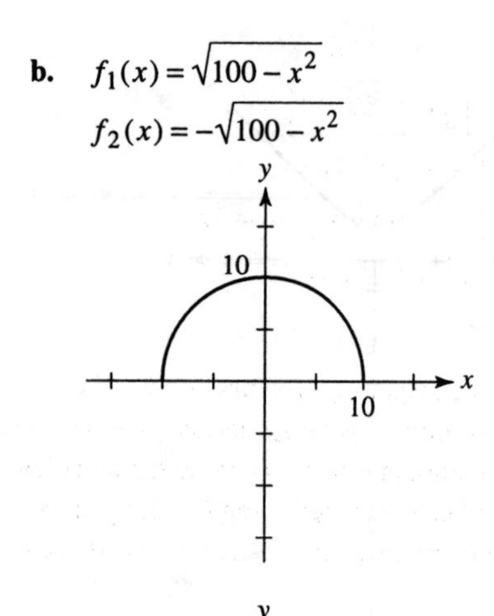

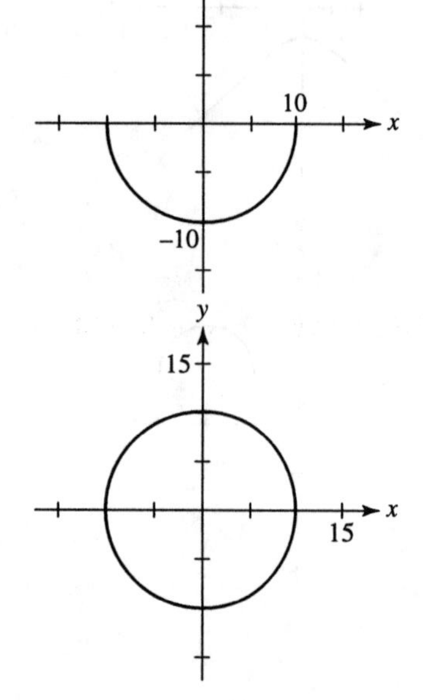

43. a. $4x^2 + y = 4x + 15$

$y = -4x^2 + 4x + 15$

y

18

5

x

45. a. $x^2 + y^2 = 6x + 6y$

$y^2 - 6y + (x^2 - 6x) = 0$

Regard the equation as $ay^2 + by + c = 0$ with $a = 1$, $b = -6$, and $c = x^2 - 6x$, and solve for y.

$$y = \frac{6 \pm \sqrt{(-6)^2 - 4(1)(x^2 - 6x)}}{2(1)}$$

$$y = \frac{6 \pm \sqrt{-4x^2 + 24x + 36}}{2}$$

$$y = \frac{6 \pm \sqrt{4(-x^2 + 6x + 9)}}{2}$$

$$y = 3 \pm \sqrt{-x^2 + 6x + 9}$$

b. $f_1(x) = 3 + \sqrt{-x^2 + 6x + 9}$

$f_2(x) = 3 - \sqrt{-x^2 + 6x + 9}$

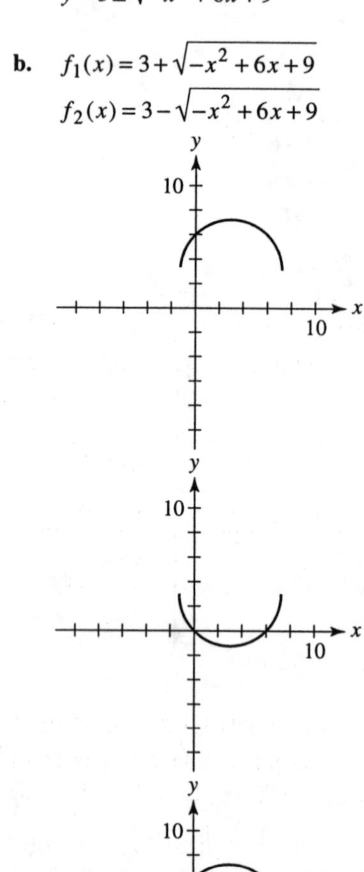

47. a. $x^2 = y^3 + 1$

$y^3 = x^2 - 1$

$y = \sqrt[3]{x^2 - 1}$

b.

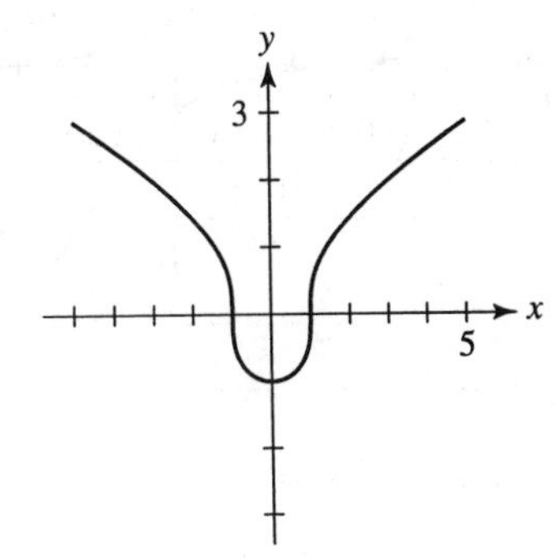

49. $3x - [f(x)]^2 = 6$

$3x - \left(\sqrt{3x-6}\right)^2 = 6$

$3x - (3x - 6) = 6$

$6 = 6$

51. $x^2 - 4[f(x) - 1] = 4x$

$x^2 - 4\left[\frac{1}{4}(x-2)^2 - 1\right] = 4x$

$x^2 - [(x-2)^2 - 4] = 4x$

$x^2 - \left[(x^2 - 4x + 4) - 4\right] = 4x$

$x^2 - (x^2 - 4x) = 4x$

$4x = 4x$

53. (Sample response) If a function $y = f(x)$ is defined implicitly by a relation, then the equation of the relation is true whenever the equation $y = f(x)$ is true. Substituting $f(x)$ for y in the relation produces an equation in x that is always true.

55. a. Rewrite the equation as $y^2 + (x+5)y + (x^2 + 4x + 10) = 0$.

Think of it as a quadratic equation $ay^2 + by + c = 0$ with $a = 1$, $b = x + 5$, and $c = x^2 + 4x + 10$.

Then use the quadratic formula to solve for y.

$$y = \frac{-(x+5) \pm \sqrt{(x+5)^2 - 4(1)(x^2 + 4x + 10)}}{2(1)}$$

$$y = \frac{-x - 5 \pm \sqrt{(x^2 + 10x + 25) - 4(x^2 + 4x + 10)}}{2}$$

$$y = \frac{1}{2}\left(-x - 5 \pm \sqrt{-3x^2 - 6x - 15}\right)$$

b. The graph is empty.

57. a. Rewrite the equation as $11y^2 + 10\sqrt{3}xy + (x^2 - 64) = 0$.

Think of it as a quadratic equation $ay^2 + by + c = 0$ with $a = 11$, $b = 10\sqrt{3}x$, and $c = x^2 - 64$.

Then use the quadratic formula to solve for y.

$$y = \frac{-10\sqrt{3}x \pm \sqrt{\left(10\sqrt{3}x\right)^2 - 4(11)(x^2 - 64)}}{2(11)}$$

$$y = \frac{-10\sqrt{3}x \pm \sqrt{300x^2 - 44(x^2 - 64)}}{2(11)}$$

$$y = \frac{1}{22}\left(-10\sqrt{3}x \pm \sqrt{256x^2 + 2816}\right)$$

b.

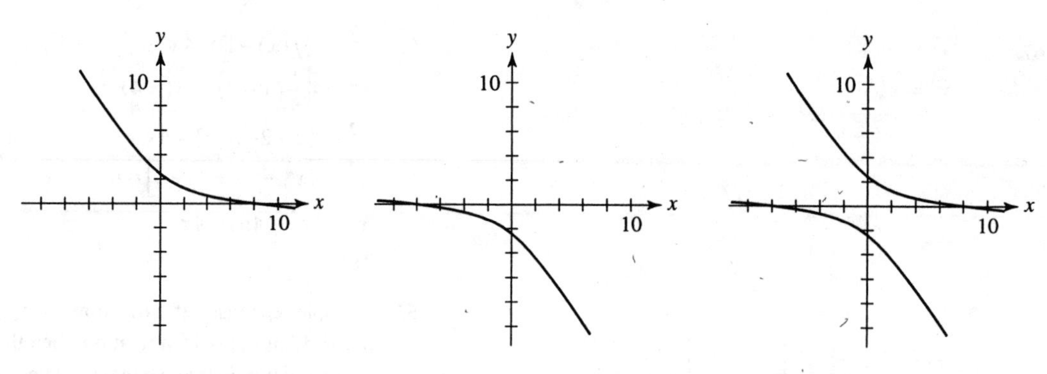

Section 6-2

1. Write the equation in standard graphing form.

$y = x^2 + 8x + 12$

$y = (x^2 + 8x) + 12$

$y = (x^2 + 8x + 16 - 16) + 12$

$y = (x^2 + 8x + 16) - 16 + 12$

$y = (x + 4)^2 - 4$

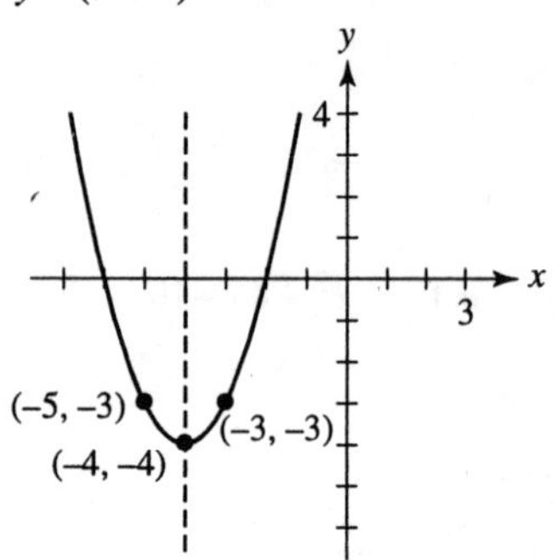

3. Write the equation in standard graphing form.

$x = -y^2 + 10y - 18$

$x = -(y^2 - 10y) - 18$

$x = -(y^2 - 10y + 25 - 25) - 18$

$x = -(y^2 - 10y + 25) + 25 - 18$

$x = -(y - 5)^2 + 7$

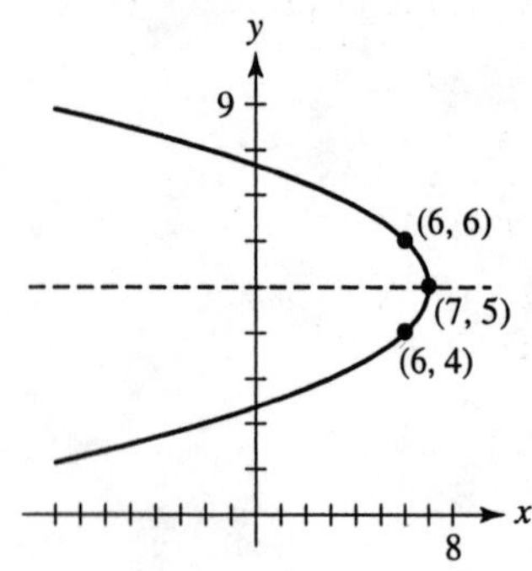

5. a. If $|x| > 3$, then $\left(\frac{x}{3}\right)^2 > 1$, so $\left(\frac{y}{2}\right)^2 < 0$ and there are no real solutions for y. This means that there are no points on the graph for which $|x| > 3$, so the entire graph lies on or between the lines $x = -3$ and $x = 3$.

b. If $|y| > 2$, then $\left(\frac{y}{2}\right)^2 > 1$, so $\left(\frac{x}{3}\right)^2 < 0$ and there are no real solutions for x. This means that there are no points on the graph for which $|y| > 2$, so the entire graph lies on or between the lines $y = -2$ and $y = 2$.

c. If $y = 0$, then

$$\left(\frac{x}{3}\right)^2 = 1$$
$$\frac{x}{3} = \pm 1$$
$$x = \pm 3$$

If $x = 0$, then

$$\left(\frac{y}{2}\right)^2 = 1$$
$$\frac{y}{2} = \pm 1$$
$$y = \pm 2$$

7. a. If $\left|\frac{x-10}{2}\right| > 1$, then $\left(\frac{x-10}{2}\right)^2 > 1$, so $\left(\frac{y-7}{3}\right)^2 < 0$, and there is no real solution for y. This means that there are no points on the graph for which $\left|\frac{x-10}{2}\right| > 1$. Thus all points on the graph have

$$\left|\frac{x-10}{2}\right| \le 1$$
$$-1 \le \frac{x-10}{2} \le 1$$
$$-2 \le x - 10 \le 2$$
$$8 \le x \le 12$$

b. If $\left|\frac{y-7}{3}\right| > 1$, then $\left(\frac{y-7}{3}\right)^2 > 1$, so $\left(\frac{x-10}{2}\right)^2 < 0$, and there is no real solution for x. This means that there are no points on the graph for which $\left|\frac{y-7}{3}\right| > 1$.

Thus all points on the graph have

$$\left|\frac{y-7}{3}\right| \le 1$$
$$-1 \le \frac{y-7}{3} \le 1$$
$$-3 \le y - 7 \le 3$$
$$4 \le y < 10$$

c. If $y = 7$, then

$$\left(\frac{x-10}{2}\right)^2 = 1$$
$$\frac{x-10}{2} = \pm 1$$
$$x - 10 = \pm 2$$
$$x = 8, 12$$

If $x = 10$, then

$$\left(\frac{y-7}{3}\right)^2 = 1$$
$$\frac{y-7}{3} = \pm 1$$
$$y - 7 = \pm 3$$
$$y = 4, 10$$

9. $4x^2 + y^2 = 64$

$$\frac{x^2}{16} + \frac{y^2}{64} = 1 \text{ or}$$
$$\left(\frac{x}{4}\right)^2 + \left(\frac{y}{8}\right)^2 = 1$$

The center is $(0, 0)$.
The vertices are $(0, \pm 8)$.
The covertices are $(\pm 4, 0)$.
The major axis is the y-axis.
The minor axis is the x-axis.

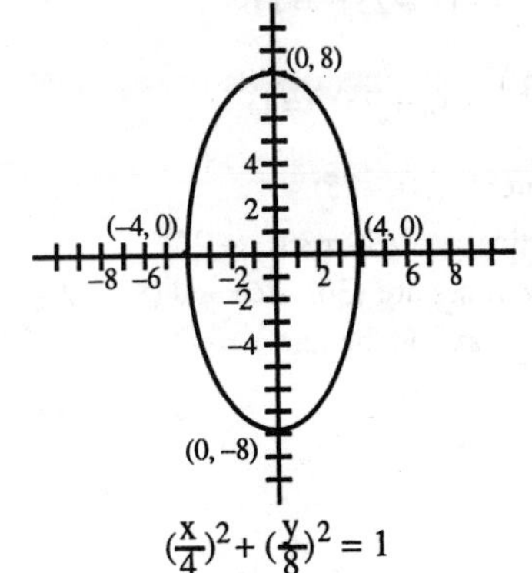

$\left(\frac{x}{4}\right)^2 + \left(\frac{y}{8}\right)^2 = 1$

11. $(x+4)^2+4(y-2)^2=16$

$\dfrac{(x+4)^2}{16}+\dfrac{(y-2)^2}{4}=1$ or

$\left(\dfrac{x+4}{4}\right)^2+\left(\dfrac{y-2}{2}\right)^2=1$

The center is (–4, 2).
the vertices are (–8, 2) and (0, 2).
The covertices are (–4, 0) and (–4, 4).
The major axis is the line $y=2$.
The minor axis is the line $x=-4$.

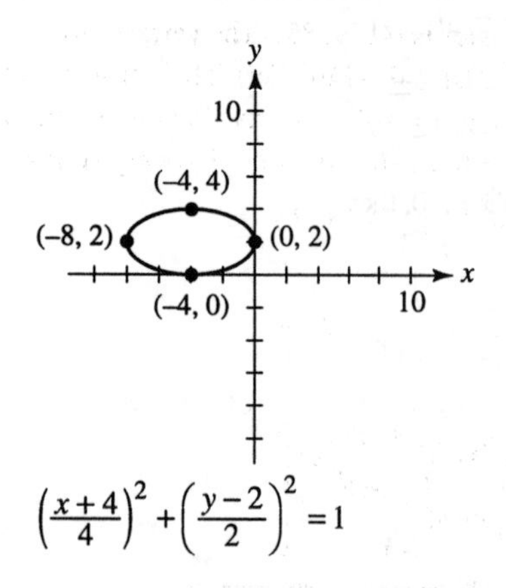

$\left(\dfrac{x+4}{4}\right)^2+\left(\dfrac{y-2}{2}\right)^2=1$

13. $(x^2-60x)+6(y^2+50y)=-4644$

$(x^2-60x+900-900)$
$\quad +6(y^2+50y+625-625)=-4644$

$(x^2-60x+900)-900$
$\quad +6(y^2+50y+625)-3750=-4644$

$(x-30)^2+6(y+25)^2=6$

$\dfrac{(x-30)^2}{6}+(y+25)^2=1$ or

$\left(\dfrac{x-30}{\sqrt{6}}\right)^2+(y+25)^2=1$

The center is (30, –25).
The vertices are $(30\pm\sqrt{6},\ -25)$.
The covertices are (30, –26) and (30, –24).
The major axis is the line $y=-25$.

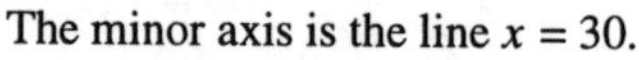
The minor axis is the line $x = 30$.

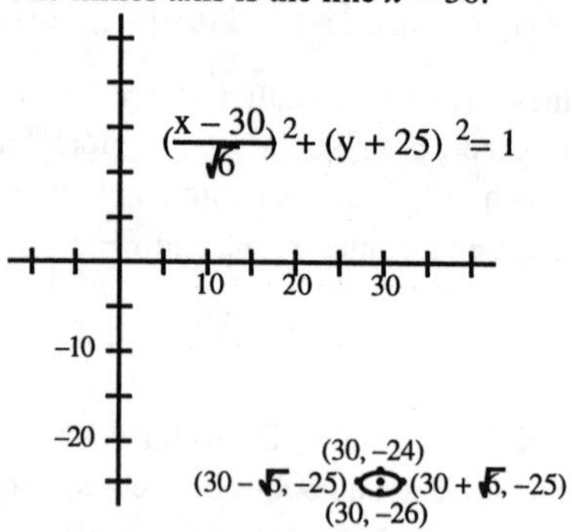

15. The center is (0, 0).
The radius is 5.

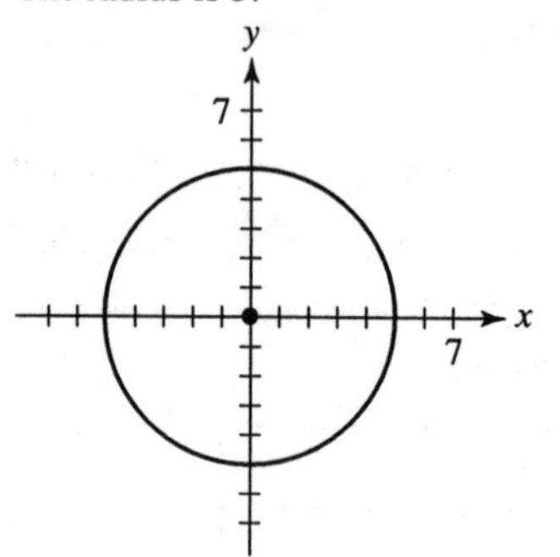

17. The center is (–6, 2).
The radius is 2.

19. Write the equation in standard graphing form.

$(x^2-8x)+(y^2-6y)+23=0$

$(x^2-8x+16-16)+(y^2-6y+9-9)+23=0$

$(x^2-8x+16)-16+(y^2-6y+9)-9+23=0$

$(x-4)^2+(y-3)^2=2$

The center is (4, 3).

The radius is $\sqrt{2}$.

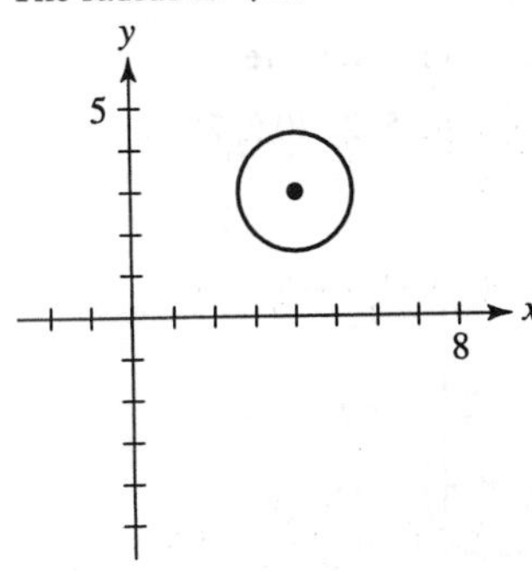

21. Solve the equation for y.

$x^2 - y^2 = 1$

$y^2 = x^2 - 1$

$y = \pm\sqrt{x^2 - 1}$.

In the first quadrant $y \geq 0$, so the last equation is equivalent to $y = \sqrt{x^2 - 1}$. In general, however, y can be negative, so the equations are not equivalent.

23.

x	y
1	0
2	$\sqrt{3} \cong 1.73$
10	$\sqrt{99} \cong 9.95$
100	$\sqrt{9999} \cong 99.99$

25. If (a, b) is on the graph of $x^2 - y^2 = 1$, then $a^2 - b^2 = 1$. In that case, it is also true that $(-a)^2 + b^2 = 1$ and $a^2 + (-b)^2 = 1$, so the points $(-a, b)$ and $(a, -b)$ are also on the graph.

27. The center is (–5, –6). The vertices are (–8, –6) and (–2, –6). The transverse axis is the line $y = -6$, and the conjugate axis is the line $x = -5$. The asymptotes have equations

$y + 6 = \pm\frac{4}{3}(x + 5)$.

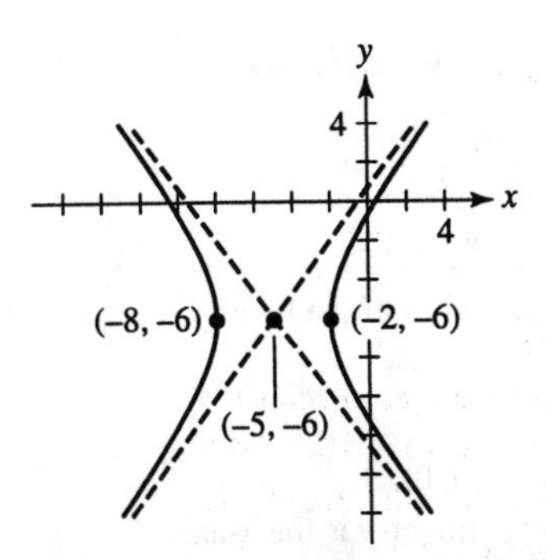

29. The center is (–129, 85). The vertices are (–129, 50) and (–129, 120). The transverse axis is the line $x = -129$, and the conjugate axis is the line $y = 85$. The asymptotes have equations

$y - 85 = \pm\frac{7}{10}(x + 129)$.

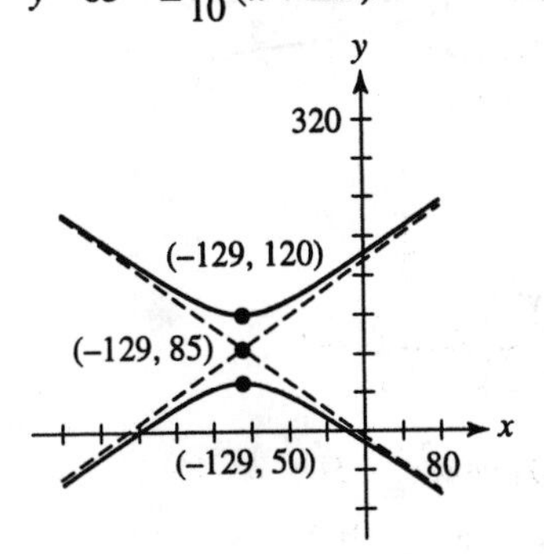

31. For any values of x and y, $x^2 \geq 0$ and $y^2 \geq 0$, so the equation $x^2 + y^2 = 0$ is true if and only if $(x, y) = (0, 0)$. The graph is a single point at the origin.

33. The equation $x^2 - y^2 = 0$ is equivalent to $y = \pm x$. The graph therefore consists of the two lines $y = x$ and $y = -x$.

35. Complete the square in x.

$(x^2 - 6x) + 4y^2 + 10 = 0$

$(x^2 - 6x + 9 - 9) + 4y^2 + 10 = 0$

$(x^2 - 6x + 9) - 9 + 4y^2 + 10 = 0$

$(x - 3)^2 + 4y^2 = -1$

The graph is empty.

37. Complete the square in x.

$(x^2 - 6x) - 4y^2 + 9 = 0$

$(x^2 - 6x + 9 - 9) - 4y^2 + 9 = 0$

$(x^2 - 6x + 9) - 9 - 4y^2 + 9 = 0$
$(x-3)^2 - 4y^2 = 0$
$4y^2 = (x-3)^2$
$2y = \pm(x-3)$
The graph consists of the lines $y = \pm\frac{1}{2}(x-3)$.

39. $x = 3y^2$
The vertex is (0, 0).
The axis of symmetry is the x-axis.

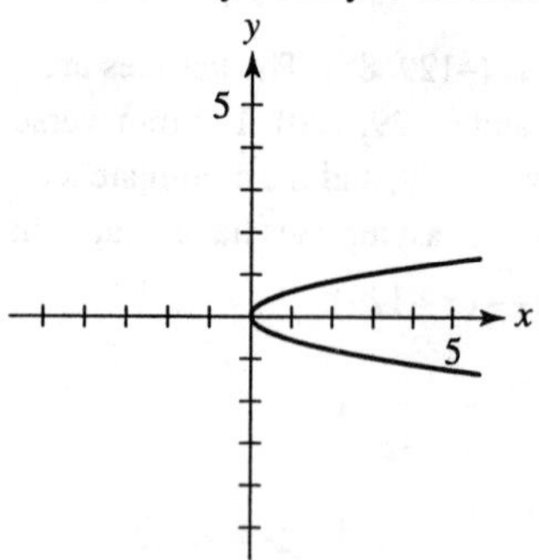

41. $2x^2 + 7y^2 = 28$
$\frac{x^2}{14} + \frac{y^2}{4} = 1$
$\left(\frac{x}{\sqrt{14}}\right)^2 + \left(\frac{y}{2}\right)^2 = 1$
The center is (0, 0).
The vertices are $(\pm\sqrt{14},\ 0)$.
The covertices are $(0, \pm 2)$.
The major axis is the x-axis.
The minor axis is the y-axis.

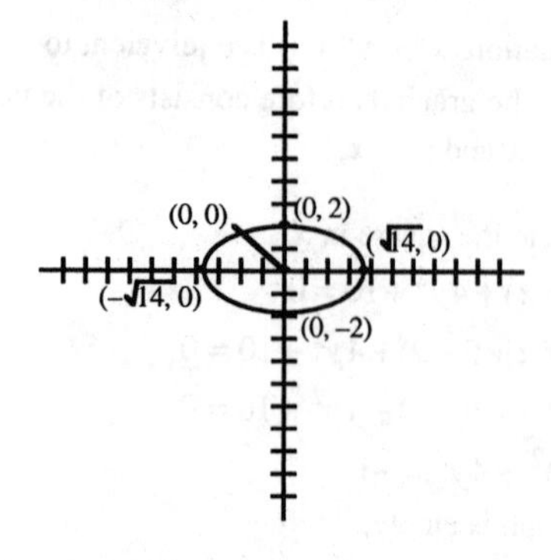

43. $4(x+1)^2 + 4(y-2)^2 = 9$
$(x+1)^2 + (y-2)^2 = \frac{9}{4}$
The graph is a circle.
The center is (–1, 2).

The radius is $\frac{3}{2}$.
Some points on the circle are:
$\left(-1 \pm \frac{3}{2},\ 2\right) = (-2.5, 2), (0.5, 2)$
$\left(-1,\ 2 \pm \frac{3}{2}\right) = (-1, 0.5), (-1, 3.5)$

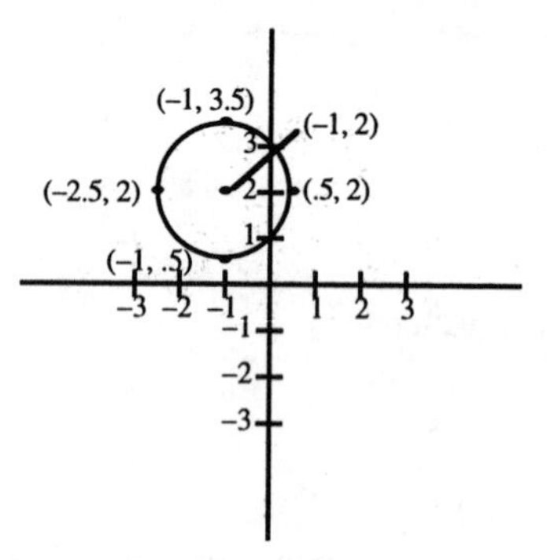

45. $4(x+1)^2 - 4(y-2) = 9$
$4(x+1)^2 = 4(y-2) + 9$
$y - 2 + \frac{9}{4} = (x+1)^2$
$y = (x+1)^2 - \frac{1}{4}$
The vertex is $\left(-1,\ -\frac{1}{4}\right)$.
The axis of symmetry is the line $x = -1$.

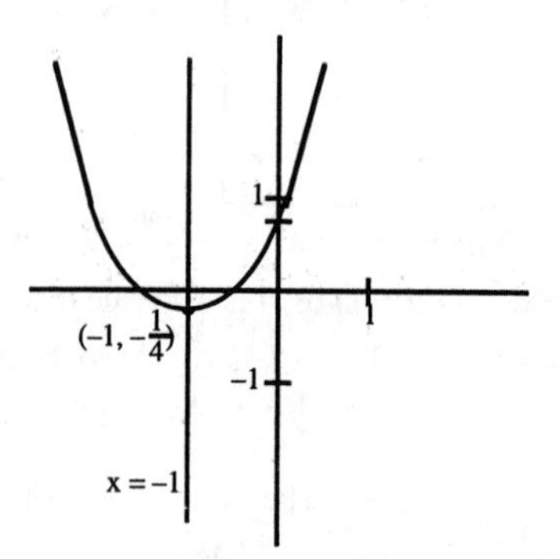

47. $5x^2 + y^2 - 70x + 225 = 0$
$5(x^2 - 14x) + y^2 = -225$
$5(x^2 - 14x + 49 - 49) + y^2 = -225$
$5(x^2 - 14x + 49) - 245 + y^2 = -225$
$5(x-7)^2 + y^2 = 20$
$\frac{(x-7)^2}{4} + \frac{y^2}{20} = 1$
The center is (7, 0).
The vertices are $(7,\ \pm\sqrt{20}) \cong (7,\ \pm 4.47)$.

The covertices are $(7 \pm 2, 0) = (5, 0), (9, 0)$.
The major axis is the line $x = 7$.
The minor axis is the x-axis.

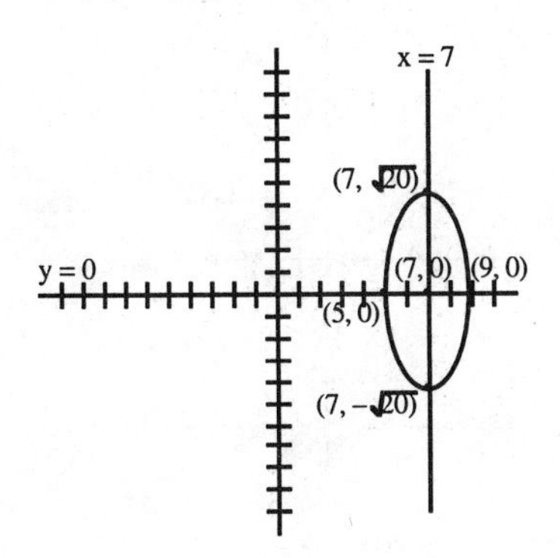

49. $2x^2 - 12x - y + 10 = 0$

$y = 2(x^2 - 6x) + 10$

$y = 2(x^2 - 6x + 9 - 9) + 10$

$y = 2(x^2 - 6x + 9) - 18 + 10$

$y = 2(x - 3)^2 - 8$

The vertex is $(3, -8)$.
The axis of symmetry is the line $x = 3$.

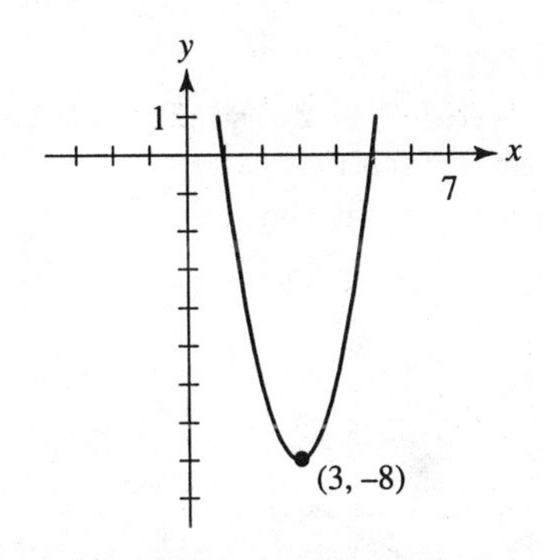

51. $4x^2 - y^2 + 24x - 8y + 20 = 0$

$4(x^2 + 6x) - (y^2 + 8y) = -20$

$4(x^2 + 6x + 9 - 9) - (y^2 + 8y + 16 - 16) = -20$

$4(x^2 + 6x + 9) - 36 + (y^2 + 8y + 16) + 16 = -20$

$4(x + 3)^2 - (y + 4)^2 = 0$

The equation cannot be written in standard form.

The graph is a pair of intersecting lines, with equations $y + 4 = \pm 2(x + 3)$.

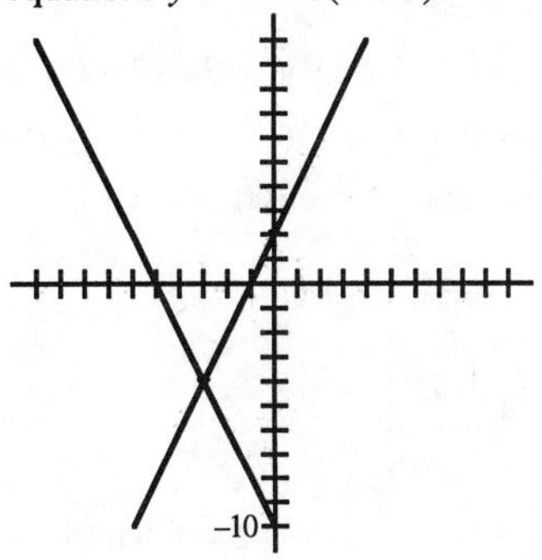

53. $100x^2 + 20x + 400y^2 - 480y - 9855 = 0$

$100(x^2 + 0.2x) + 400(y^2 - 1.2y) = 9855$

$100(x^2 + 0.2x + 0.01 - 0.01)$
$\quad +400(y^2 - 1.2y + 0.36 - 0.36) = 9855$

$100(x^2 + 0.2x + 0.01) - 1$
$\quad +400(y^2 - 1.2y + 0.36) - 144 = 9855$

$100(x + 0.1)^2 + 400(y - 0.6)^2 = 10,000$

$$\frac{(x + 0.1)^2}{100} + \frac{(y - 0.6)^2}{25} = 1$$

The center is $(-0.1, 0.6)$.
The vertices are
$(-0.1 \pm 10, 0.6) = (-10.1, 0.6), (9.9, 0.6)$.
The covertices are
$(-0.1, 0.6 \pm 5) = (-0.1, -4.4), (-0.1, 5.6)$.
The major axis is the line $y = 0.6$.
The minor axis is the line $x = -0.1$.

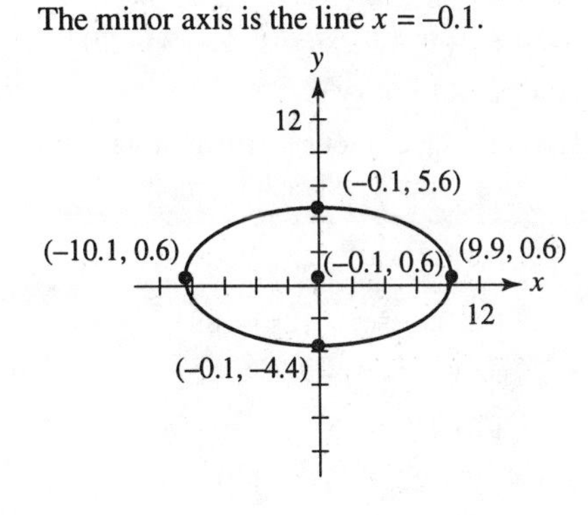

55. $4x^2 + 25y^2 + 24x - 50y - 39 = 0$

$4(x^2 + 6x) + 25(y^2 - 2y) = 39$

$4(x^2 + 6x + 9 - 9) + 25(y^2 - 2y + 1 - 1) = 39$

$4(x^2 + 6x + 9) - 36 + 25(y^2 - 2y + 1) - 25 = 39$

$4(x+3)^2 + 25(y-1)^2 = 100$

$\frac{(x+3)^2}{25} + \frac{(y-1)^2}{4} = 1$ or

$\left(\frac{x+3}{5}\right)^2 + \left(\frac{y-1}{2}\right)^2 = 1$

The center is (–3, 1).
The vertices are $(-3 \pm 5, 1) = (-8, 1), (2, 1)$.
The covertices are $(-3, 1 \pm 2) = (-3, -1), (-3, 3)$.
The major axis is the line $y = 1$.
The minor axis is the line $x = -3$.

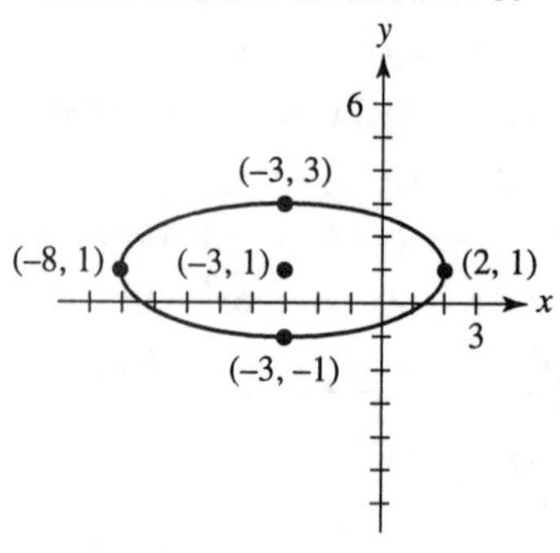

57. $x^2 + 5y^2 + 6x + 20y + 34 = 0$

$(x^2 + 6x) + 5(y^2 + 4y) = -34$

$(x^2 + 6x + 9 - 9) + 5(y^2 + 4y + 4 - 4) = -34$

$(x^2 + 6x + 9) - 9 + 5(y^2 + 4y + 4) - 20 = -34$

$(x+3)^2 + 5(y+2)^2 = -5$

The equation cannot be written in standard graphing form. The graph is empty.

59. $4x^2 - 25y^2 + 24x + 50y - 89 = 0$

$4(x^2 + 6x) - 25(y^2 - 2y) = 89$

$4(x^2 + 6x + 9 - 9) - 25(y^2 - 2y + 1 - 1) = 89$

$4(x^2 + 6x + 9) - 36 - 25(y^2 - 2y + 1) + 25 = 89$

$4(x+3)^2 - 25(y-1)^2 = 100$

$\frac{(x+3)^2}{25} - \frac{(y-1)^2}{4} = 1$ or

$\left(\frac{x+3}{5}\right)^2 - \left(\frac{y-1}{2}\right)^2 = 1$

The center is (–3, 1).
The vertices are $(-3 \pm 5, 1) = (-8, 1), (2, 1)$.
The transverse axis is the line $y = 1$.

The conjugate axis is the line $x = -3$.

The asymptotes are the lines $\frac{x+3}{5} = \pm\frac{y-1}{2}$.

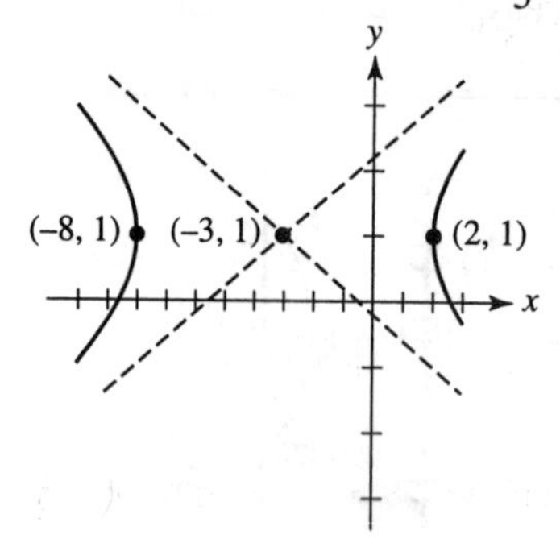

61. $x^2 - 5y^2 + 6x - 20y - 6 = 0$

$(x^2 + 6x) - 5(y^2 + 4y) = 6$

$(x^2 + 6x + 9 - 9) - 5(y^2 + 4y + 4 - 4) = 6$

$(x^2 + 6x + 9) - 9 - 5(y^2 + 4y + 4) + 20 = 6$

$(x+3)^2 - 5(y+2)^2 = -5$

$(y+2)^2 - \frac{(x+3)^2}{5} = 1$ or

$(y+2)^2 - \left(\frac{x+3}{\sqrt{5}}\right)^2 = 1$

The center is (–3, –2).
The vertices are $(-3, -2 \pm 1) = (-3, -3), (-3, -1)$.
The transverse axis is the line $x = -3$.
The conjugate axis is the line $y = -2$.
The asymptotes are the lines $y + 2 = \pm\frac{x+3}{\sqrt{5}}$.

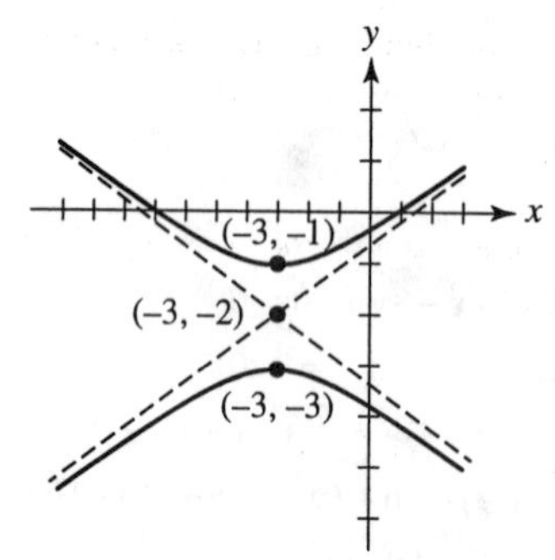

63. The cross section shown in Figure 6-20 is a parabola opening to the right, with its vertex at (0, 0). Its equation is $x = ay^2$ for some $a > 0$. Since the graph passes through $\left(3.88, \frac{6.52}{2}\right) = (3.88, 3.26)$, we have $3.88 = a(3.26)^2$. Therefore

$a = \frac{3.88}{(3.26)^2} \cong 0.365$. The equation is
$x = \frac{3.88}{(3.26)^2} y^2$.

Section 6-3

1. a. $\left(-\frac{2}{2}\right)^2 + (0)^2 = 1$; $\left(\frac{0}{2}\right)^2 + (1)^2 = 1$;
$\left(\frac{0}{2}\right)^2 + (-1)^2 = 1$; $\left(\frac{2}{2}\right)^2 + (0)^2 = 1$

b. (Sample response) The expression $\frac{x}{2}$ in the new equation plays the same role as x in the unit circle. Therefore if an ordered pair (a, b) satisfies the equation of the unit circle and (c, b) satisfies the new equation, then $\frac{c}{2} = a$, or equivalently, $c = 2a$.

c.

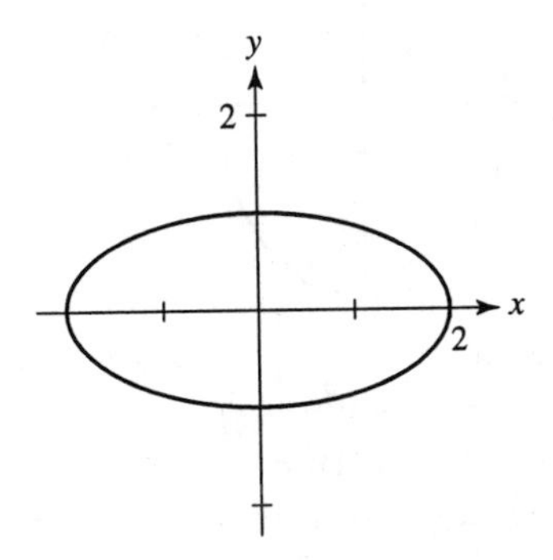

Each x-coordinate on the new graph is twice the corresponding x-coordinate on the unit circle.

d. The circle was stretched horizontally by a factor of 2.

e.

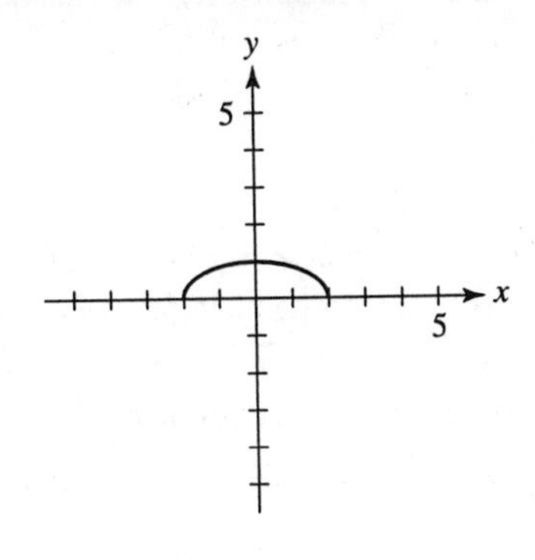

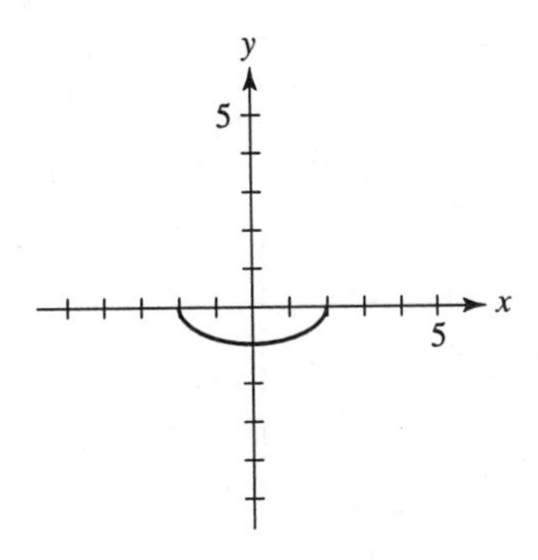

$$\left(\frac{x}{2}\right)^2 + y^2 = 1$$
$$y^2 = 1 - \left(\frac{x}{2}\right)^2$$
$$y = \pm\sqrt{1 - \left(\frac{x}{2}\right)^2}$$

3. a.

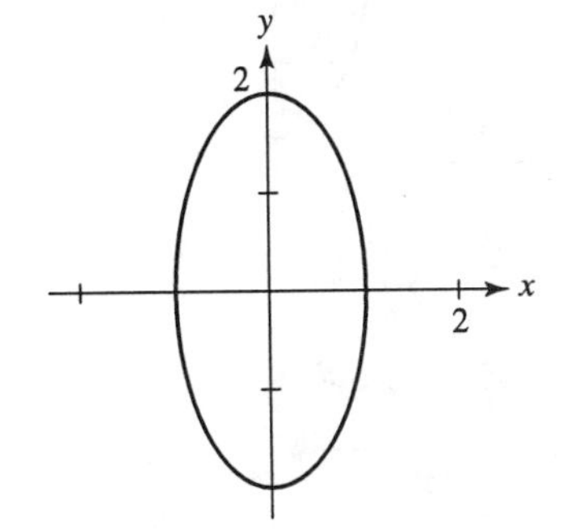

b. The circle was stretched vertically by a factor of 2.

c.

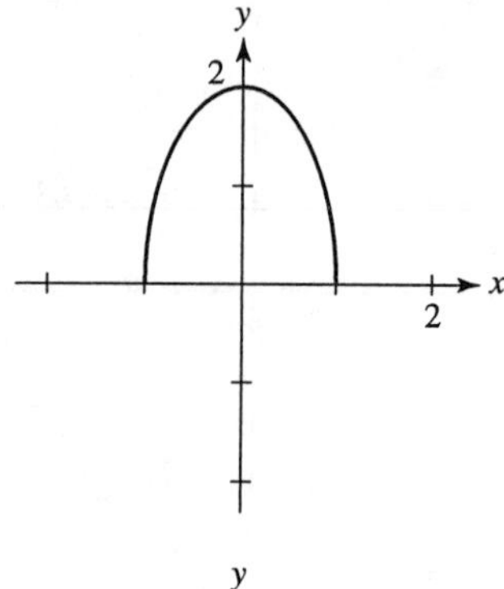

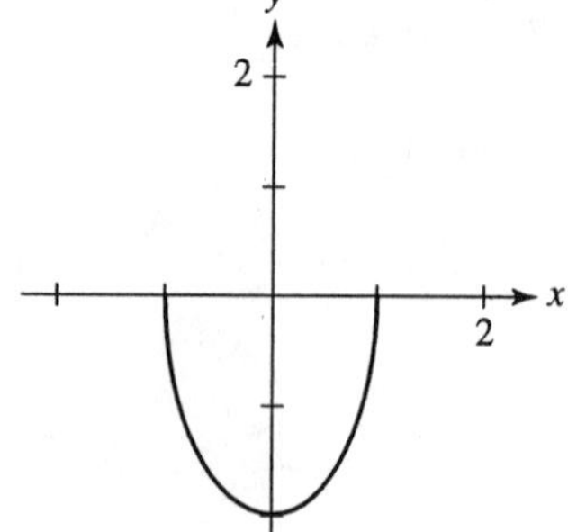

$$x^2 + \left(\frac{y}{2}\right)^2 = 1$$

$$\left(\frac{y}{2}\right)^2 = 1 - x^2$$

$$\frac{y}{2} = \pm\sqrt{1-x^2}$$

$$y = \pm 2\sqrt{1-x^2}$$

5. The graph is compressed horizontally by a factor of 3.

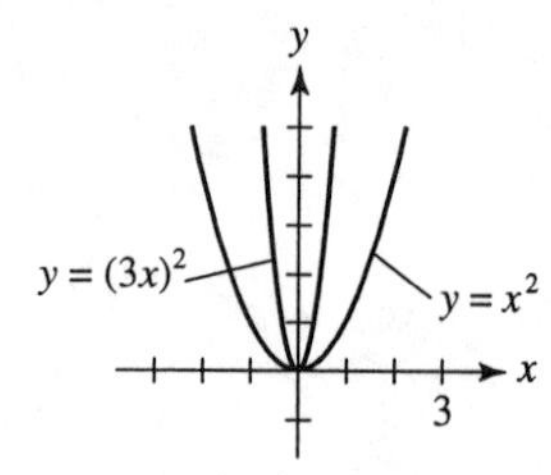

7. The graph is stretched horizontally by a factor of 2.

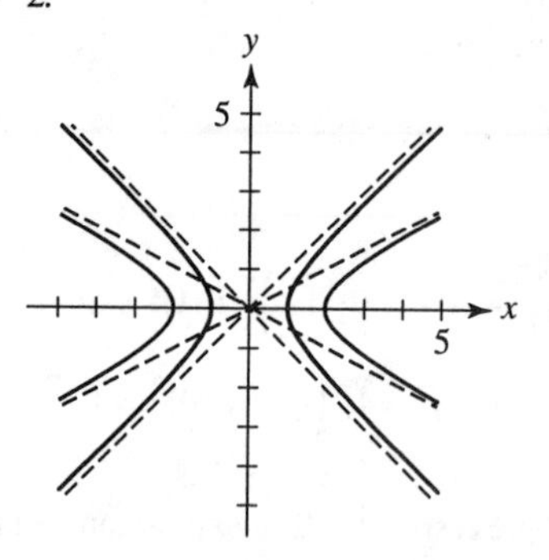

9. The graph is stretched horizontally by a factor of 3 and vertically by a factor of 4.

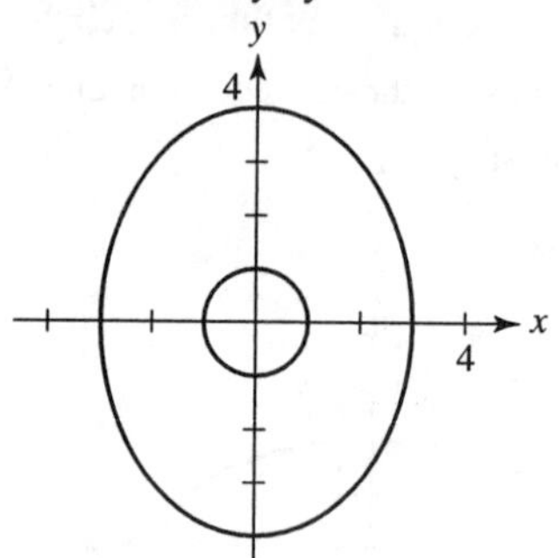

11. a. $(2-3)^2 + 0^2 = 1;$
$(3-3)^2 + (1)^2 = 1;$
$(3-3)^2 + (-1)^2 = 1;$
$(4-3)^2 + (0)^2 = 1$

b. (Sample response) The expression $x - 3$ in the new equation plays the same role as x in the unit circle. Therefore if an ordered pair (a, b) satisfies the equation of the unit circle and (c, b) satisfies the new equation, then $c - 3 = a$, or equivalently, $c = a + 3$.

c.

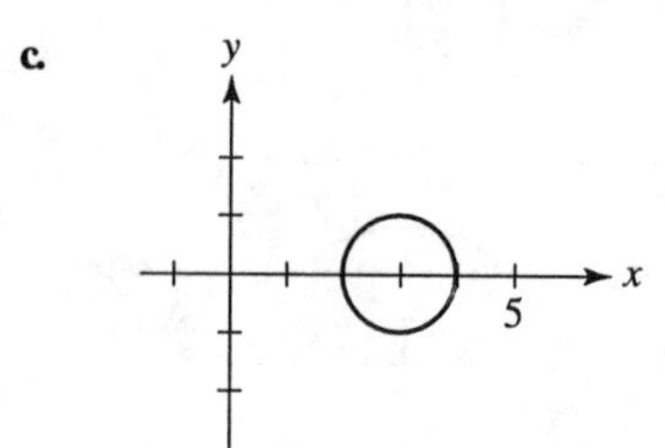

Each x-coordinate on the new graph is three more than the corresponding x-coordinate on the unit circle.

d. The circle was shifted 3 units to the right.

e.

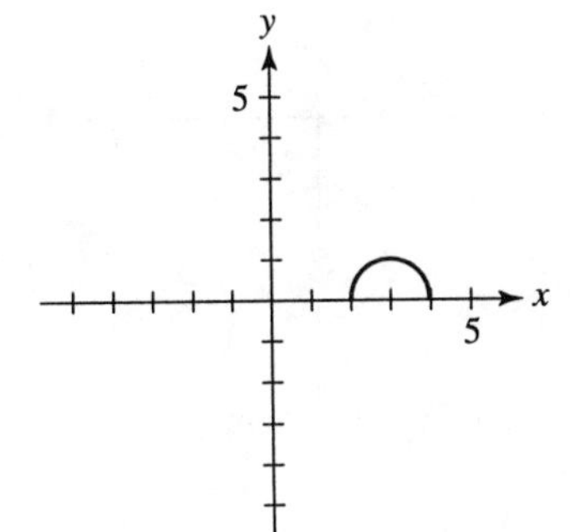

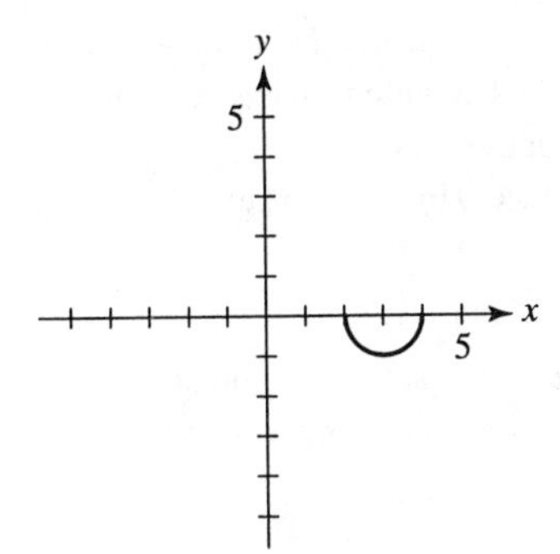

$$(x-3)^2+y^2=1$$
$$y^2=1-(x-3)^2$$
$$y=\pm\sqrt{1-(x-3)^2}$$

13. a.

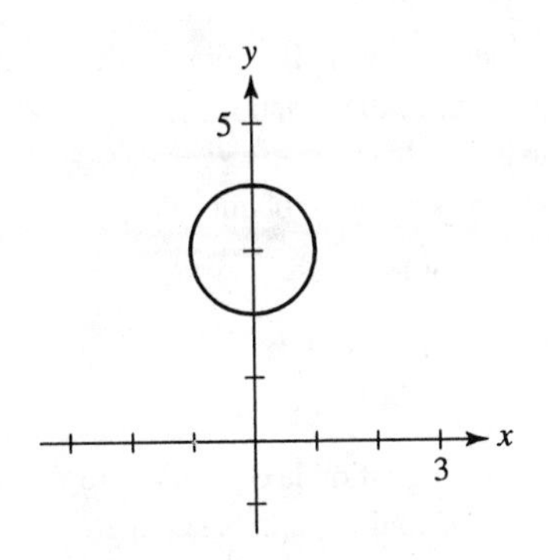

b. The circle was shifted 3 units up.

c.

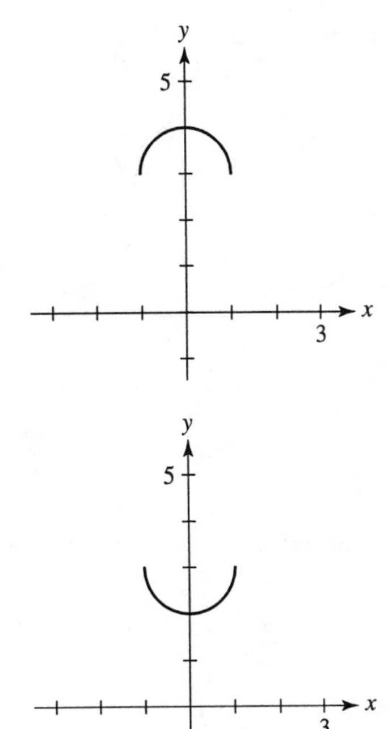

$$x^2+(y-3)^2=1$$
$$(y-3)^2=1-x^2$$
$$y-3=\pm\sqrt{1-x^2}$$
$$y=3\pm\sqrt{1-x^2}$$

15. The graph is shifted 3 units to the right.

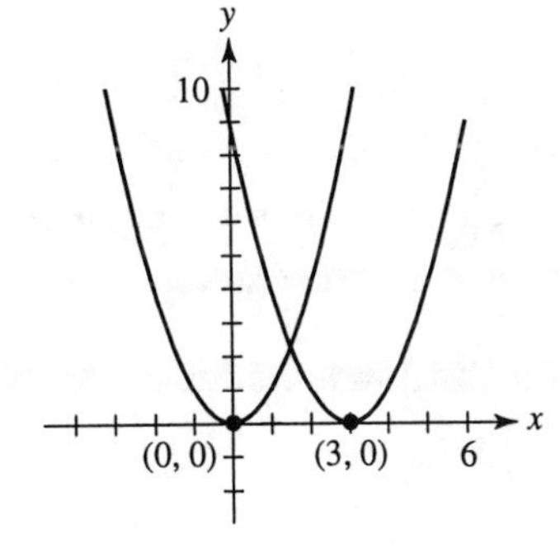

17. The graph is shifted 2 units to the left.

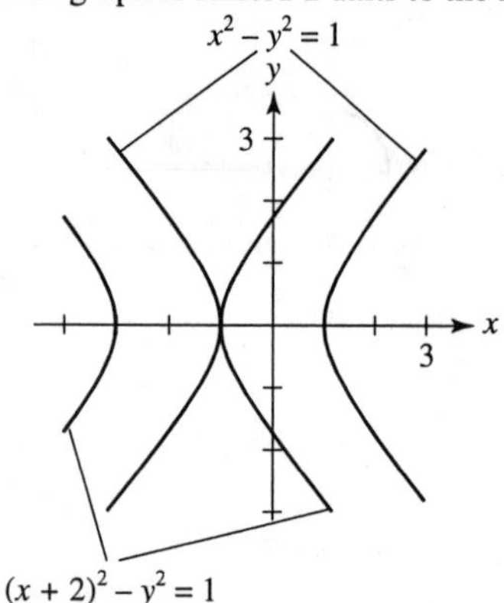

19. The graph is shifted 3 units to the left and 4 units down.

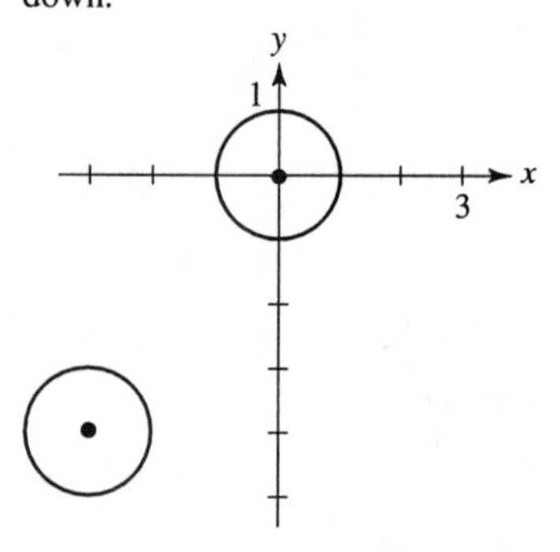

21. $(-x)^2 + y^2 = 1$

$x^2 + y^2 = 1$

$x^2 + (-y)^2 = 1$

$x^2 + y^2 = 1$

The graph of each equation is the unit circle.

23. **a.** $x^2 + (-y-2)^2 = 1$

$x^2 + [-(y+2)]^2 = 1$

$x^2 + (y+2)^2 = 1$

b. The second graph is the reflection of the first graph in the x-axis.

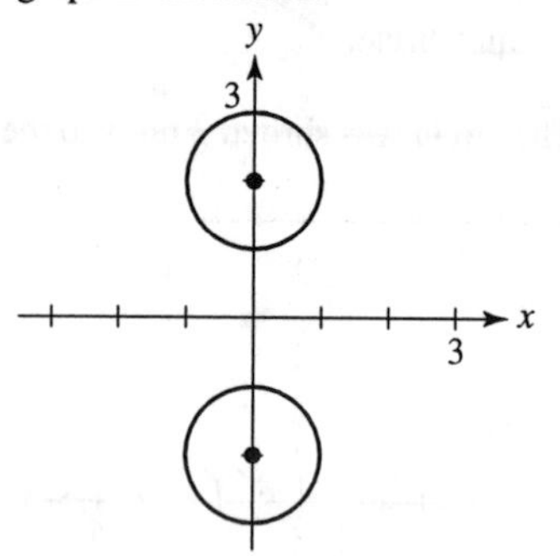

25. Replace x by $-x$ to obtain

$(-x) - y^2 = 1$

$-x - y^2 = 1$

The new equation is not equivalent to the original equation, so the graph is not symmetric about the y-axis.

Replace y by $-y$ to obtain

$x - (-y)^2 = 1$

$x - y^2 = 1$

The new equation is equivalent to the original equation, so the graph is symmetric about the x-axis.

27. Replace x by $-x$ to obtain

$(-x)^2 + \left(\frac{y}{4}\right)^2 = 1$

$x^2 + \left(\frac{y}{4}\right)^2 = 1$

The new equation is equivalent to the original equation, so the graph is symmetric about the y-axis.

Replace y by $-y$ to obtain

$x^2 + \left(-\frac{y}{4}\right)^2 = 1$

$x^2 + \left(\frac{y}{4}\right)^2 = 1$

The new equation is equivalent to the original equation, so the graph is symmetric about the x-axis.

29. a.

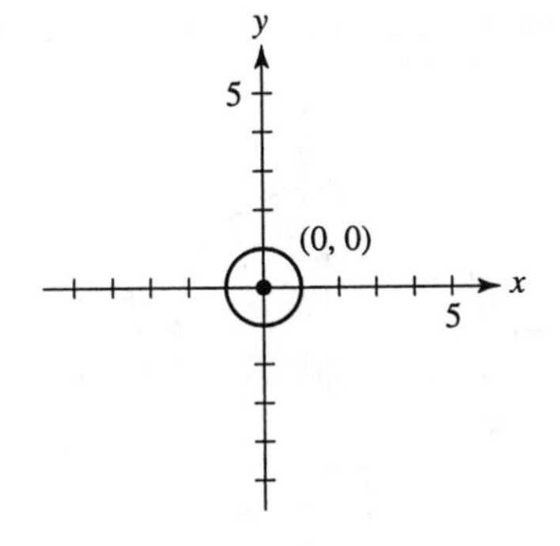

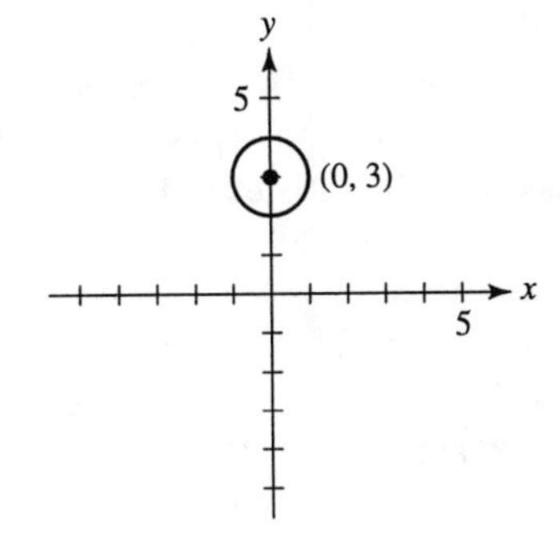

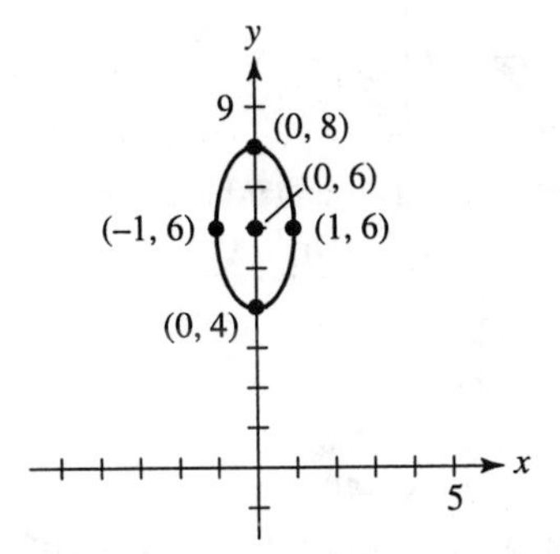

b. $x^2 + y^2 = 1$

$x^2 + (y-3)^2 = 1$

$x^2 + \left(\frac{y}{2} - 3\right)^2 = 1$

31. The center is (–129, 85).
The vertices are $(-129 \pm 50, 85)$
= (–179, 85), (–79, 85).
The covertices are $(-129, 85 \pm 35)$
= (–129, 50), (–129, 120).
The major axis is the line $y = 85$.
The minor axis is the line $x = -129$.

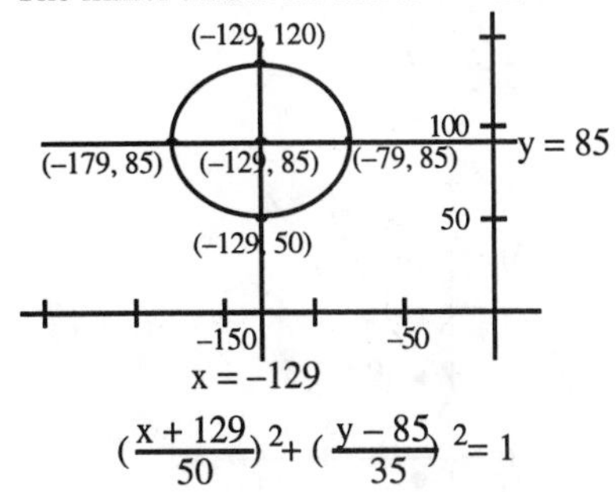

$$\left(\frac{x+129}{50}\right)^2 + \left(\frac{y-85}{35}\right)^2 = 1$$

33. The center is (0.5, 1.5).
The vertices are (0.5, –0.5) and (0.5, 3.5).
The covertices are (–0.5, 1.5) and (1.5, 1.5).
The major axis is the line $x = 0.5$.
The minor axis is the line $y = 1.5$.

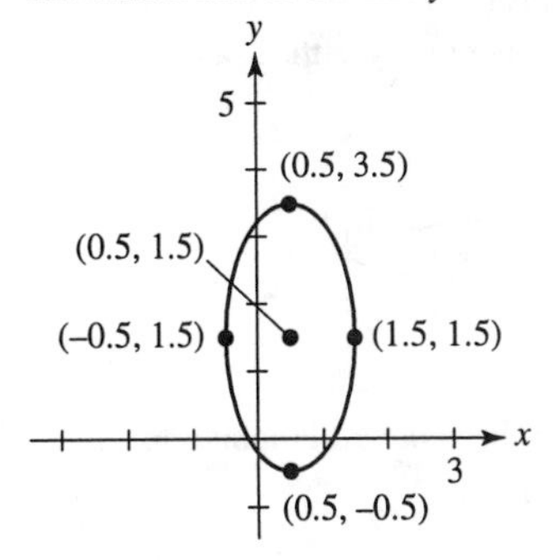

35. The center is (–129, 85).
The vertices are (–179, 85), (–79, 85).
The transverse axis is the line $y = 85$.
The conjugate axis is the line $x = -129$.
The asymptotes are the lines
$\frac{y-85}{35} = \pm\frac{x+129}{50}$.

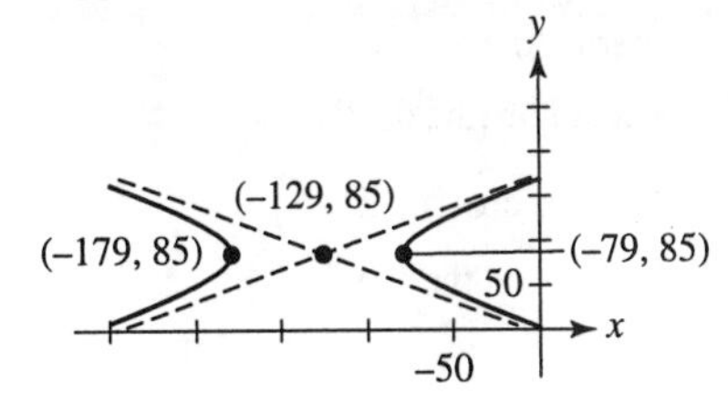

37. The center is (0.5, 1.5).
The vertices are (0.5, –0.5) and (0.5, 3.5).
The transverse axis is the line $x = 0.5$.
The conjugate axis is the line $y = 1.5$.

The asymptotes are the lines
$\frac{y-1.5}{2} = \pm(x-0.5)$.

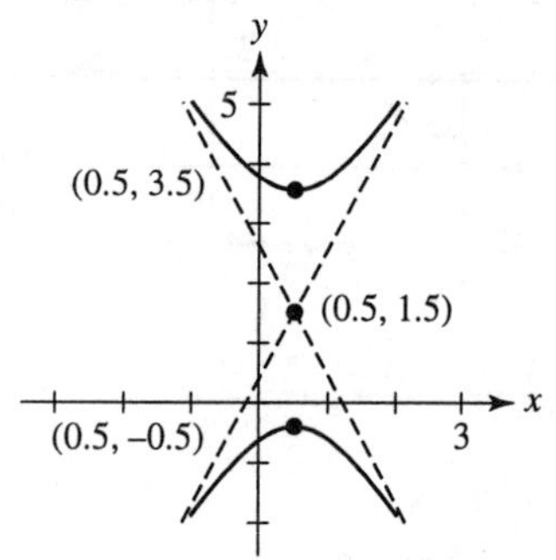

39. The graph of $x = y^2$ is reflected in the y-axis and stretched horizontally by a factor of 5.
The vertex is (0, 0).
The axis of symmetry is the x-axis.

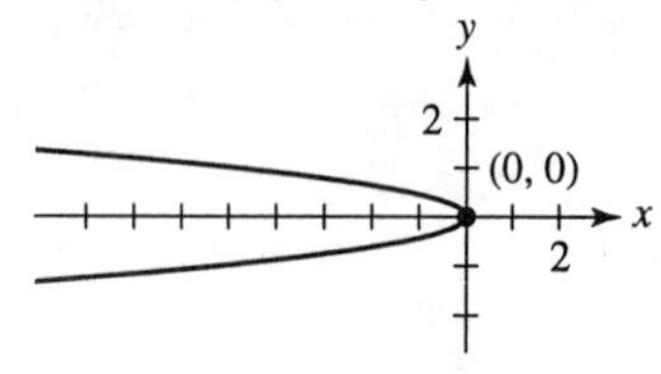

41. First obtain the standard graphing form.
$5x^2 + 6y^2 = 30$
$\frac{x^2}{6} + \frac{y^2}{5} = 1$
$\left(\frac{x}{\sqrt{6}}\right)^2 + \left(\frac{y}{\sqrt{5}}\right)^2 = 1$
The graph of $x^2 + y^2 = 1$ is stretched horizontally by a factor of $\sqrt{6}$ and vertically by a factor of $\sqrt{5}$.
The center is (0, 0).
The vertices are $\left(\pm\sqrt{6},\ 0\right)$.
The covertices are $\left(0,\ \pm\sqrt{5}\right)$.
The major axis is the x-axis.

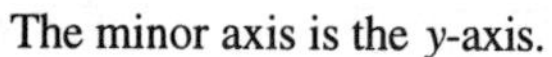
The minor axis is the y-axis.

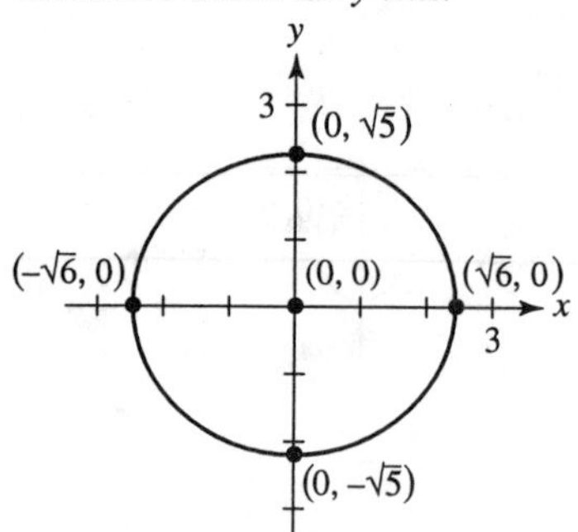

43. The graph of $x^2 + y^2 = 1$ is stretched both horizontally and vertically by a factor of 5, then shifted 10 units to the left and 20 units up.
The center is (–10, 20).
The radius is 5.

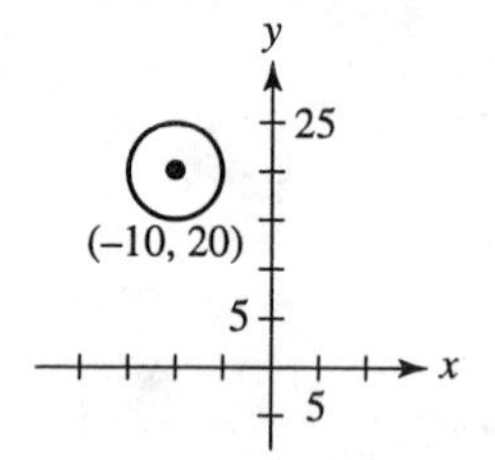

45. First obtain the standard graphing form.
$(x+10)^2 - (y-20) = 25$
$y = (x+10)^2 - 5$
The graph of $y = x^2$ is shifted 10 units to the left and 5 units down.
The vertex is (–10, –5).
The axis of symmetry is the line $x = -10$.

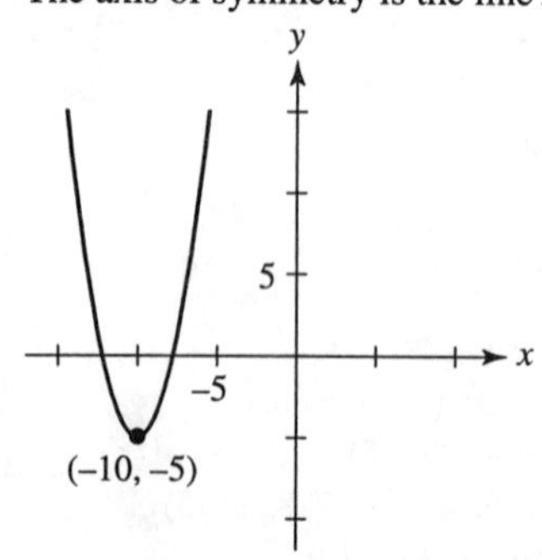

47. The graph of $x^2 + y^2 = 1$ is stretched horizontally by a factor of 4 and vertically by a factor of 6, then shifted 2 units to the left and 3 units down.

The center is (–2, –3).
The vertices are (–2, –9) and (–2, 3).
The covertices are (–6, –3) and (2, –3).
The major axis is the line $x = -2$.
The minor axis is the line $y = -3$.

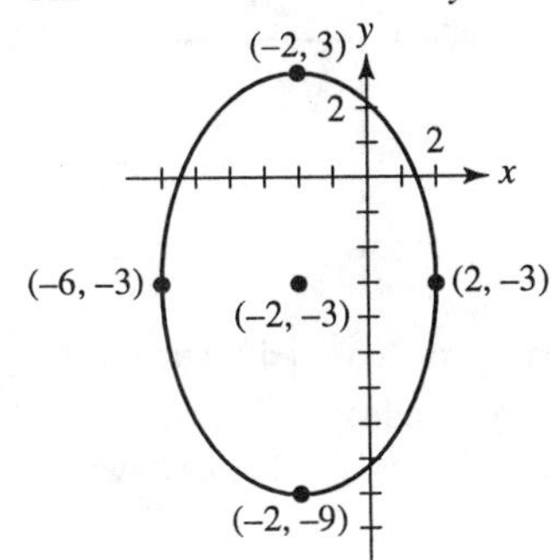

49. The graph of $x^2 - y^2 = 1$ is stretched horizontally by a factor of 4 and vertically by a factor of 6, then shifted 2 units to the left and 3 units down.
The center is (–2, –3).
The vertices are (–6, –3) and (2, –3).
The transverse axis is the line $y = -3$.
The conjugate axis is the line $x = -2$.
The asymptotes are the lines $\frac{y+3}{6} = \pm\frac{x+2}{4}$.

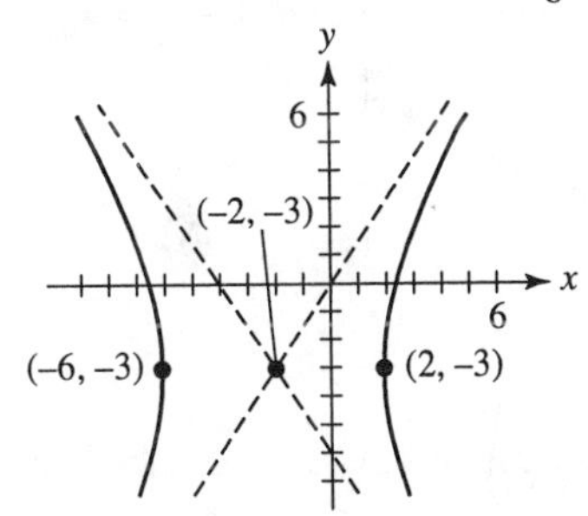

51. Replace x by $-x$ in the equation to obtain
$y = (-x)^2 - 4$
$y = x^2 - 4$
The resulting equation is equivalent to the original equation, so the graph is symmetric about the y-axis.
Replace y by $-y$ in the equation to obtain
$-y = x^2 - 4$
$y = -x^2 + 4$
The resulting equation is not equivalent to the original equation, so the graph is not symmetric about the x-axis.

53. Replace x by $-x$ in the equation to obtain
$y = (-x)^3 - 4(-x)^2$
$y = -x^3 - 4x^2$
The resulting equation is not equivalent to the original equation, so the graph is not symmetric about the y-axis.
Replace y by $-y$ in the equation to obtain
$-y = x^3 - 4x^2$
$y = -x^3 + 4x^2$
The resulting equation is not equivalent to the original equation, so the graph is not symmetric about the x-axis.

55. Replace x by $-x$ in the equation to obtain
$(-x)^3 + 4(-x)^2 - y^2 = 0$
$-x^3 + 4x^2 - y^2 = 0$
The resulting equation is not equivalent to the original equation, so the graph is not symmetric about the y-axis.
Replace y by $-y$ in the equation to obtain
$x^3 + 4x^2 - (-y)^2 = 0$
$x^3 + 4x^2 - y^2 = 0$
The resulting equation is equivalent to the original equation, so the graph is symmetric about the x-axis.

57. a. $(x-5)^2 + (y+2)^2 = 1$

b. $\left(\frac{x}{2}\right)^2 + (2y)^2 = 1$

c. $\left(\frac{x+1}{3}\right)^2 + y^2 = 1$

d. $\left(\frac{x}{3} + 1\right)^2 + y^2 = 1$

59.

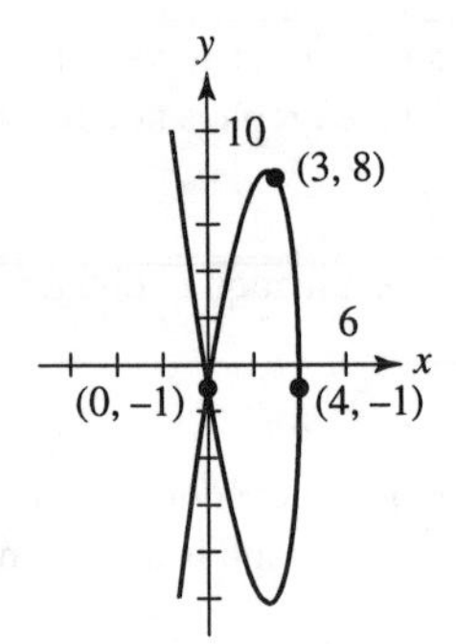

61.

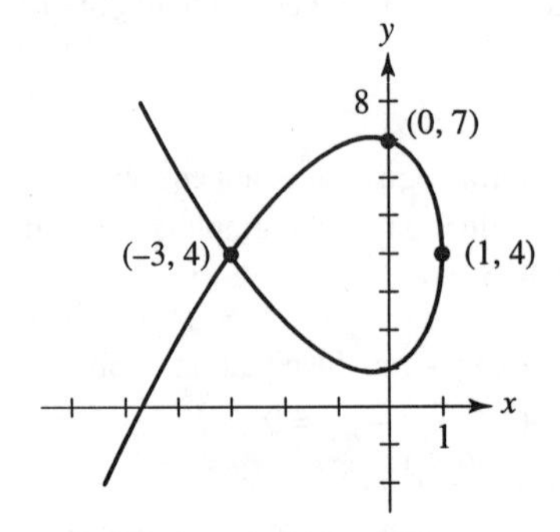

63. The crest of the hill is at (5000, 200). To move this point to the origin of your system, shift the graph 5000 units left and 200 units down. The equation is

$$y = (2.4\times10^{-5})(x+5000)^2 - (3.2\times10^{-9})(x+5000)^3 - 200,$$

which can be simplified to

$$y = (-2.4\times10^{-5})x^2 - (3.2\times10^{-9})x^3.$$

65. a.

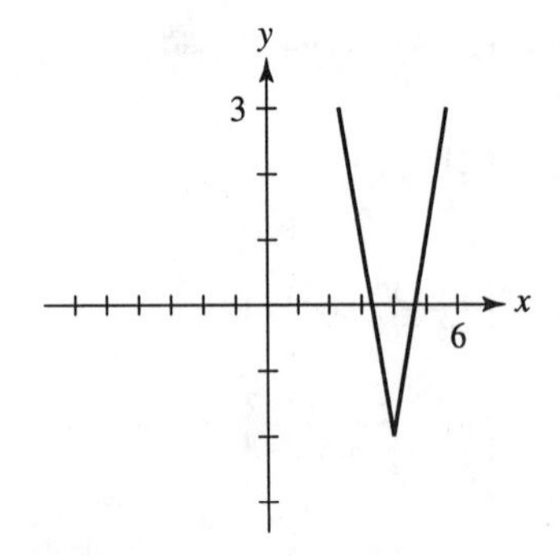

The vertex of the graph is (4, –2). The slopes of the two linear pieces are ±3.

b.

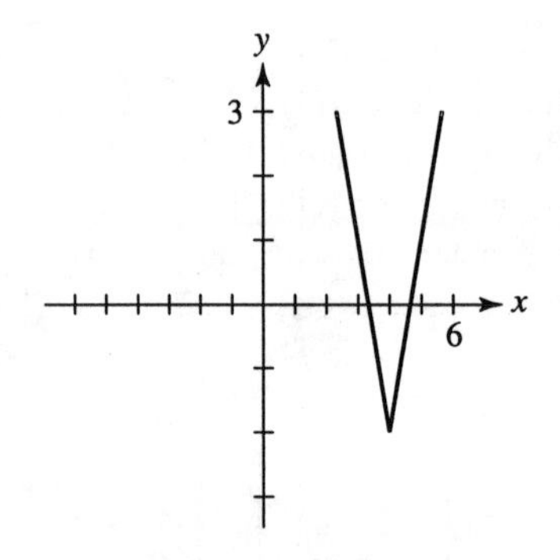

The basic graph $y = |x|$ is stretched vertically by a factor of 3, then shifted 4 units to the right and 2 units down.

c. Begin with

$$\frac{y+2}{3} = |x-4|$$
$$y+2 = 3|x-4|$$
$$y = 3|x-4| - 2$$

67.

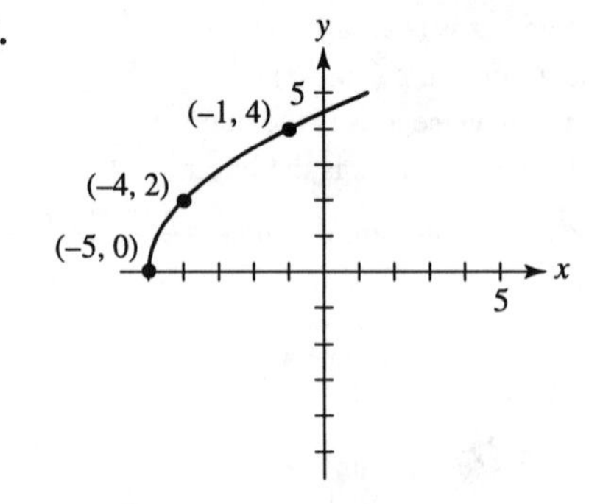

69.

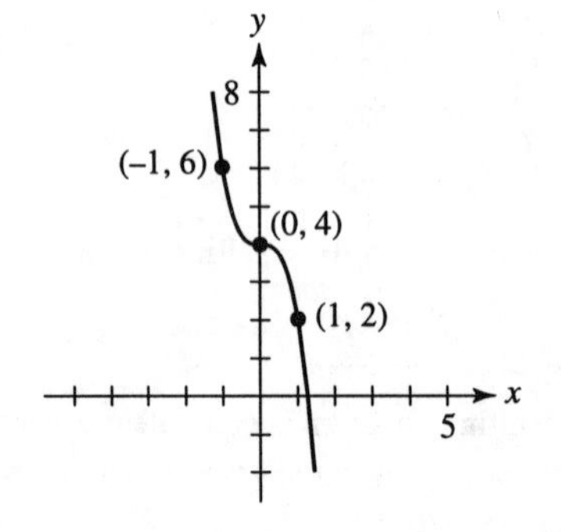

71. a.

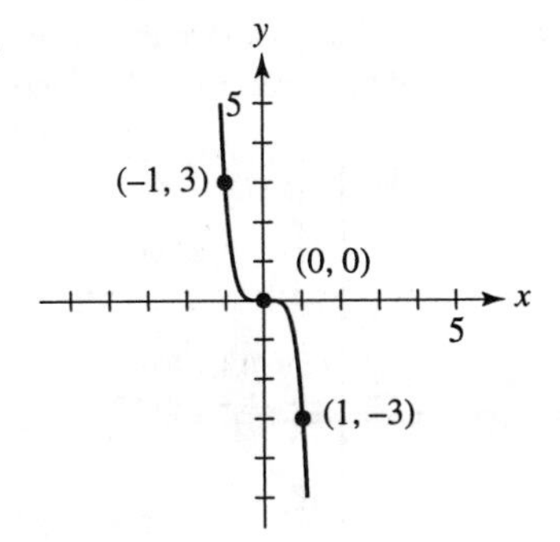

b.

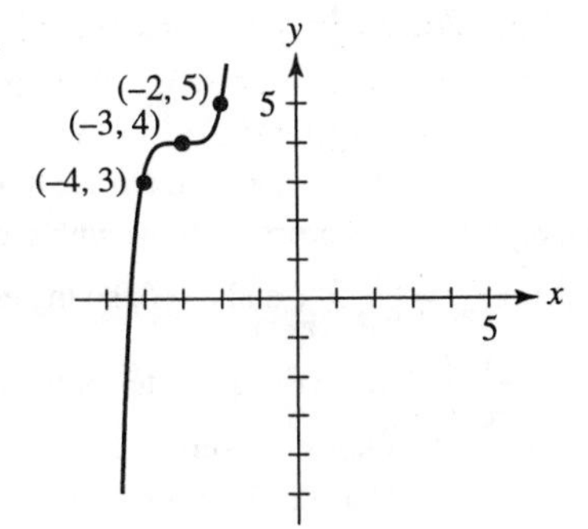

c.

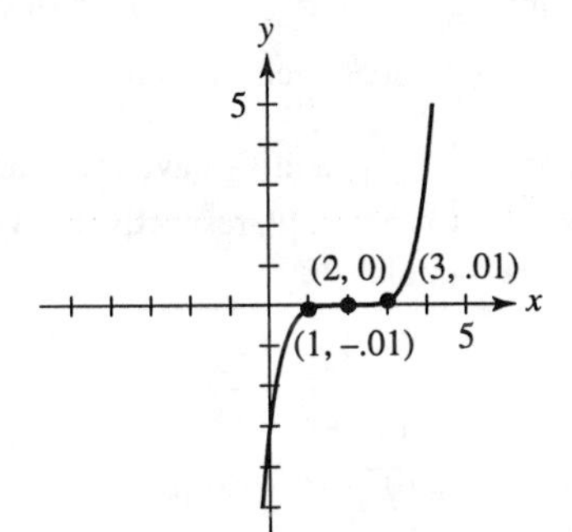

d.

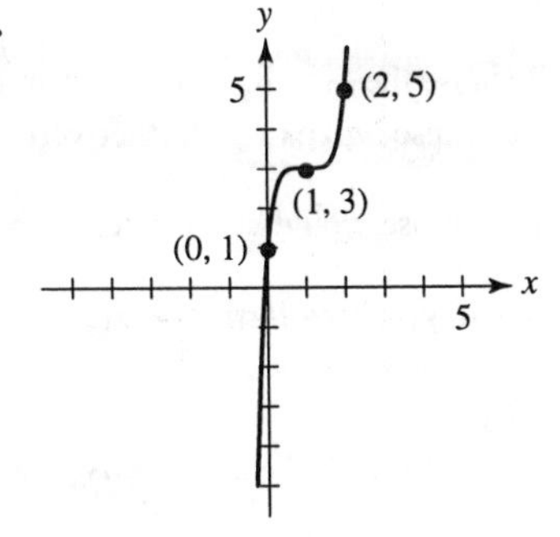

73. a.

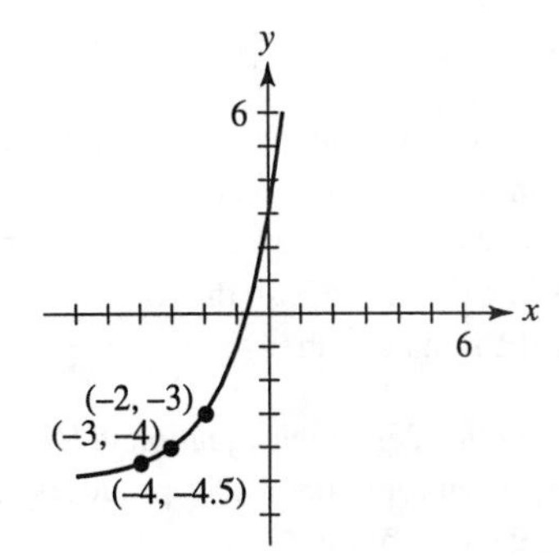

b.

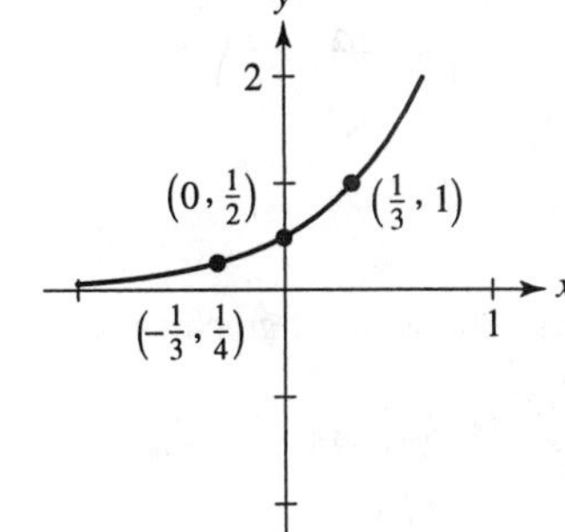

c.

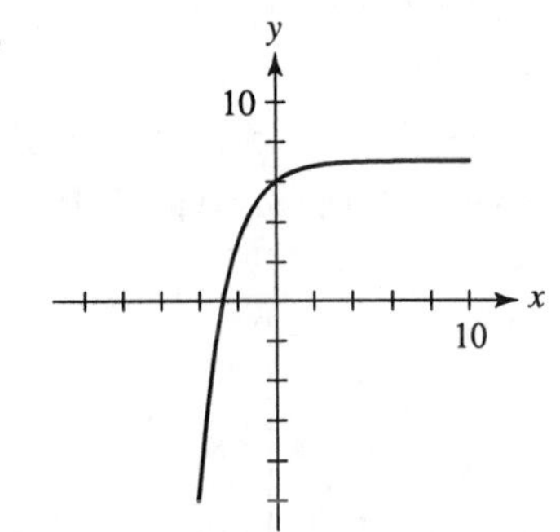

d.

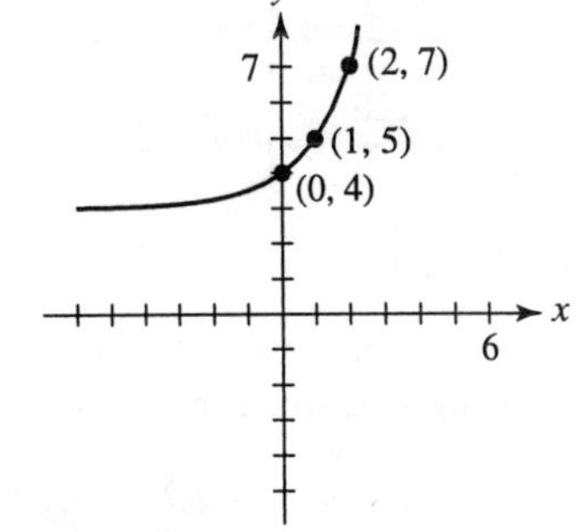

Section 6-4

1. Folding the paper as shown in Figure 128 superimposes point D on point F, while leaving point P stationary. Thus the line segment PD is superimposed on PF, so the two segments must have the same length.

3. Since both sides of the equation in Exercise 2 are positive, squaring both sides produces an equivalent equation:

$$x^2+(y-p)^2=(y+p)^2$$
$$x^2+y^2-2py+p^2=y^2+2py+p^2$$
$$x^2=4py$$
$$y=\frac{1}{4p}x^2$$

5. To obtain the standard graphing form of the equation:

$$-8x=(y^2-2y)-15$$
$$-8x=(y^2-2y+1-1)-15$$
$$-8x=(y^2-2y+1)-1-15$$
$$-8x=(y-1)^2-16$$
$$x=-\frac{1}{8}(y-1)^2+2$$

The vertex is (2, 1), and the parabola opens to the left.

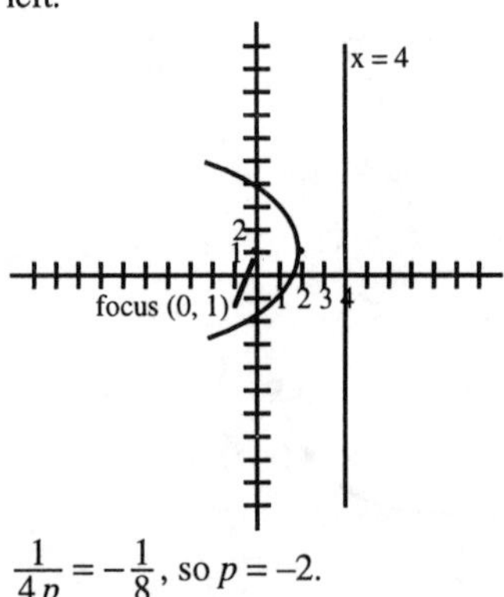

$\frac{1}{4p}=-\frac{1}{8}$, so $p=-2$.

The focus is 2 units to the left of the vertex, at (0, 1), and the directrix is the line $x=4$.

7. The equation $y=5(x-0.4)^2+0.03$ is in standard graphing form. The vertex is (0.4, 0.03), and the parabola opens upward.

$\frac{1}{4p}=5$, so $p=0.05$.

The focus is 0.05 units above the vertex, at (0.4, 0.08), and the directrix is the line $y=-0.02$.

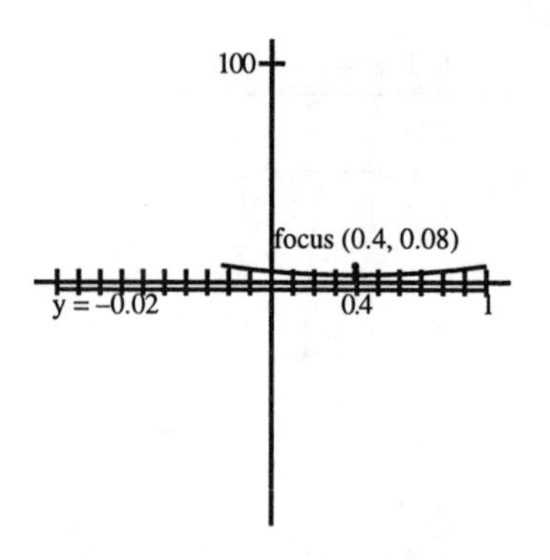

9. In the coordinate system of Exercise 8, Dr. Harvey's dish has cross sections with the equation $y=\frac{1}{4p}x^2$, and $p=6$, so the equation is $y=\frac{1}{24}x^2$. Since the diameter of the dish is 20 feet, the outer ends of the cross section have x-coordinates of ± 10. Their y-coordinates are therefore $\frac{1}{24}(10)^2\cong 4.17$, so the dish should be 4.17 feet (about 50 inches) deep.

11. The points P, F_1, and F_2 have coordinates (x, y), $(-c, 0)$ and $(c, 0)$, respectively. By the distance formula:

$\overline{PF}1=\sqrt{(x-(-c))^2+(y-0)^2}$, and

$\overline{PF}2=\sqrt{(x-c)^2+(y-0)^2}$

Since $\overline{PF_1}+\overline{PF_2}=2a$, we have

$$\sqrt{(x+c)^2+y^2}+\sqrt{(x-c)^2+y^2}=2a$$

13. **a.** Because the thumbtacks are closer together, the ellipse will extend farther vertically.

b. The ellipse would be a circle.

15. In standard graphing form, the equation is

$$\frac{x^2}{25}+\frac{y^2}{16}=1$$

The center is (0, 0), the major axis is horizontal, and $a=5$, $b=4$, $c=\sqrt{5^2-4^2}=3$.

The foci are located 3 units to the left and right of the center, at (±3, 0). The eccentricity is $\frac{3}{5}$.

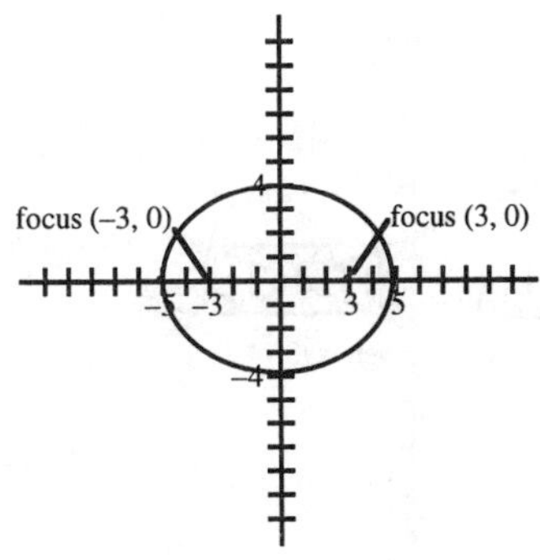

17. To obtain standard graphing form:

$9(x^2+2x)+4(y^2+6y)=99$

$9(x^2+2x+1-1)+4(y^2+6y+9-9)=99$

$9(x^2+2x+1)-9+4(y^2+6y+9)-36=99$

$9(x+1)^2+4(y+3)^2=144$

$\frac{(x+1)^2}{16}+\frac{(y+3)^2}{36}=1$

The center is (–1, –3), the major axis is vertical, and $a=6$, $b=4$, $c=\sqrt{36-16}=\sqrt{20}$. The foci are $\sqrt{20}$ units above and below the center, at $(-1,\ -3\pm\sqrt{20})\cong(-1,-7.47),(-1,1.47)$.

The eccentricity is $\frac{\sqrt{20}}{6}\cong 0.75$.

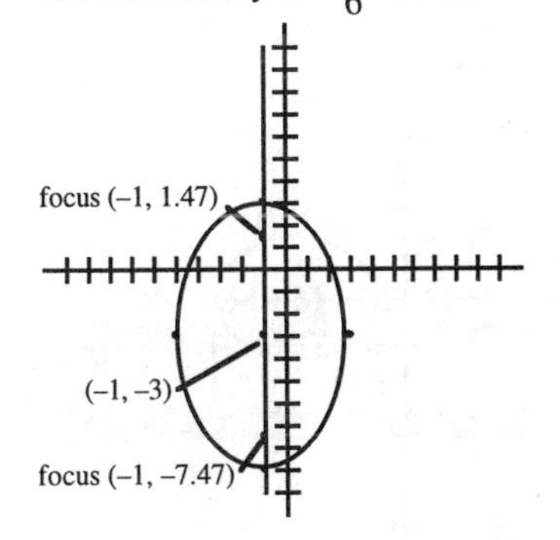

19. The vertices of the ellipse are 15 feet apart, and the covertices are 5 feet apart, so $a=7.5$ and $b=2.5$. The value of c is then

$\sqrt{7.5^2-2.5^2}\cong 7.07$. The patient should be at one focus and the ultrasound source at the other, each 7.07 feet from the center of ellipse.

21. The points P, F_1, F_2 have coordinates (x, y), $(-c, 0)$, and $(c, 0)$, respectively. By the distance formula:

$\overline{PF_1}=\sqrt{(x-(-c))^2+(y-0)^2}$, and

$\overline{PF_2}=\sqrt{(x-c)^2+(y-0)^2}$

Since $\overline{PF_1}-\overline{PF_2}=2a$, we have

$\sqrt{(x+c)^2+y^2}-\sqrt{(x-c)^2+y^2}=2a$

23. In standard graphing form, the equation is

$\frac{x^2}{16}-\frac{y^2}{25}=1$

The center is (0, 0), the transverse axis is horizontal and $a=4$, $b=5$,

$c=\sqrt{4^2+5^2}=\sqrt{41}$. The foci are located $\sqrt{41}$ units to the left and right of the center, at

$(\pm\sqrt{41},\ 0)\cong(\pm 6.40,\ 0)$.

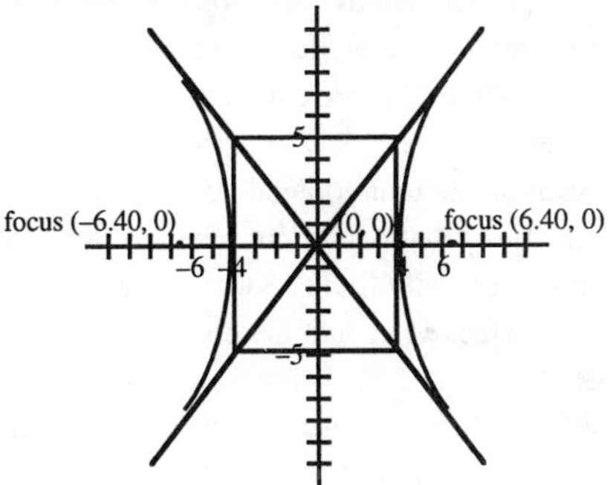

25. To obtain standard graphing form:

$9(x^2+2x)-4(y^2+6y)=171$

$9(x^2+2x+1-1)-4(y^2+6y+9-9)=171$

$9(x^2+2x+1)-9-4(y^2+6y+9)+36=171$

$9(x+1)^2-4(y+3)^2=144$

$\frac{(x+1)^2}{16}-\frac{(y+3)^2}{36}=1$

The center is (–1, –3), the transverse axis is horizontal, and $a = 4$, $b = 6$, $c = \sqrt{36+16} = \sqrt{52}$. The foci are $\sqrt{52}$ units to the left and right of the center, at $\left(-1 \pm \sqrt{52},\ -3\right) \cong (-8.21, -3)$ and (6.21, –3).

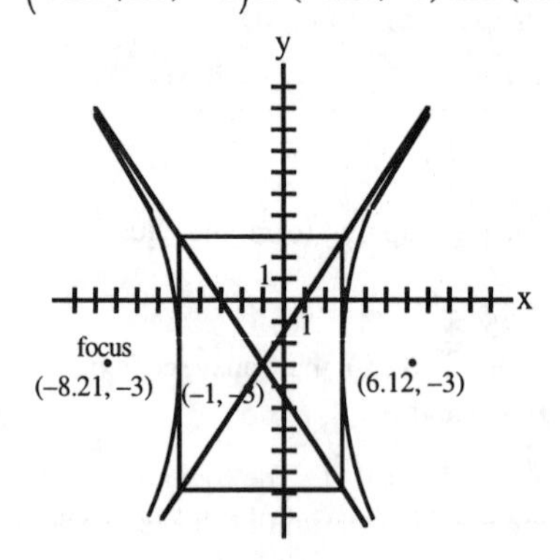

27. The primary shock wave took 0.5 seconds longer to reach Petaluma than to reach Sacramento. Since the shock wave travels at 25 km/sec, this means that the epicenter was 12.5 km closer to Sacramento than to Petaluma. The points satisfying that condition lie on one branch of a hyperbola with foci at Sacramento and Petaluma. The value of c is half the distance between the two cities, or 55. The value of a is half the difference in the distances from the epicenter, or 6.25. This value could also be calculated as 12.5 times the difference of 0.5 seconds in arrival times.

29. In standard form, the equation is

$$\frac{(x-5)^2}{14} - \frac{y^2}{2} = 1$$

The center is (5, 0), the transverse axis is horizontal, and

$a = \sqrt{14},\ b = \sqrt{2},\ c = \sqrt{14+2} = 4$. The vertices are $\sqrt{14}$ units to the left and right of the center, at $\left(5 \pm \sqrt{14},\ 0\right) \cong (1.26, 0), (8.74, 0)$.

The foci are 4 units to the left and right of the center, at $(5 \pm 4, 0) = (1, 0), (9, 0)$.

The asymptotes have equations

$$\frac{x-5}{\sqrt{14}} = \pm\frac{y}{\sqrt{2}}$$

$$y = \pm\frac{1}{\sqrt{7}}(x-5)$$

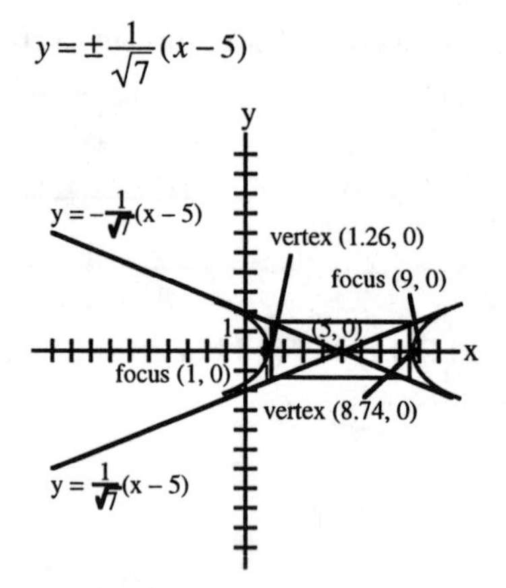

31. In standard form, the equation is

$$\frac{(x-5)^2}{14} + \frac{y^2}{2} = 1$$

The center is (5, 0), the major axis is horizontal, and $a = \sqrt{14},\ b = \sqrt{2},\ c = \sqrt{14-2} = \sqrt{12}$.

The vertices are $\sqrt{14}$ units to the left and right of the center, at $\left(5 \pm \sqrt{14},\ 0\right) \cong (1.26, 0), (8.74, 0)$.

The covertices are $\sqrt{2}$ units above and below the center, at $\left(5,\ \pm\sqrt{2}\right) \cong (5, -1.41), (5, 1.41)$.

The foci are $\sqrt{12}$ units to the left and right of the center, at $\left(5 \pm \sqrt{12},\ 0\right) \cong (1.54, 0), (8.46, 0)$.

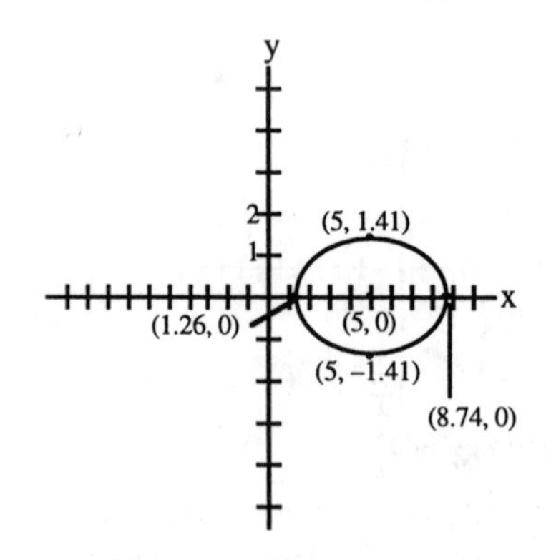

33. If we attempt to write the equation in standard form, we obtain

$9(x-500)^2 - 4(y-300)^2 = 0$

Thus, the equation has no standard form. It can be rewritten as

$3(x-500) = \pm 2(y-300)$

$y - 300 = \pm 1.5(x-500)$

Its graph is a pair of lines through (500, 300), with slopes of ±1.5.

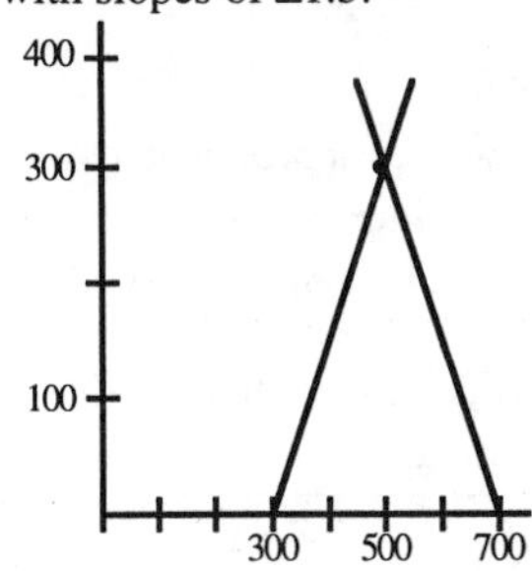

35. To obtain the standard form of the equation:

$(x^2 - x) + 9(y^2 + y) = 38$

$\left(x^2 - x + \frac{1}{4} - \frac{1}{4}\right) + 9\left(y^2 + y + \frac{1}{4} - \frac{1}{4}\right) = 38$

$\left(x^2 - x + \frac{1}{4}\right) - \frac{1}{4} + 9\left(y^2 + y + \frac{1}{4}\right) - \frac{9}{4} = 38$

$\left(x - \frac{1}{2}\right)^2 + 9\left(y + \frac{1}{2}\right)^2 = \frac{81}{2}$

$\dfrac{\left(x - \frac{1}{2}\right)^2}{\frac{81}{2}} + \dfrac{\left(y + \frac{1}{2}\right)^2}{\frac{9}{2}} = 1$

The center is $\left(\frac{1}{2}, -\frac{1}{2}\right)$, the major axis is horizontal, $a = \sqrt{\frac{81}{2}}$, $b = \sqrt{\frac{9}{2}}$,

$c = \sqrt{\frac{81}{2} - \frac{9}{2}} = 6.$

The vertices are $\sqrt{\frac{81}{2}}$ units to the left and right of the center, at

$\left(\frac{1}{2} \pm \sqrt{\frac{81}{2}}, -\frac{1}{2}\right) \cong (-5.86, -0.5), (6.86, -0.5).$

The covertices are $\sqrt{\frac{9}{2}}$ units above and below the center, at

$\left(\frac{1}{2}, -\frac{1}{2} \pm \sqrt{\frac{9}{2}}\right) \cong (0.5, -2.62), (0.5, 1.62).$

The foci are 6 units to the left and right of the center, at

$\left(\frac{1}{2} \pm 6, -\frac{1}{2}\right) = (-5.5, -0.5), (6.5, -0.5).$

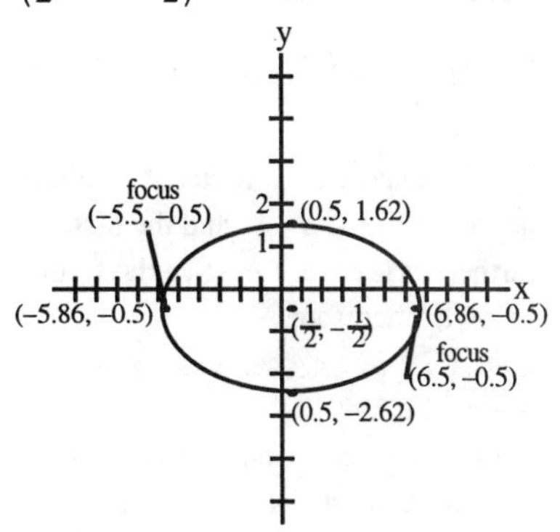

37. To obtain the standard form of the equation:

$x^2 + y^2 - 139x - 93y = 0$

$(x^2 - 139x) + (y^2 - 93y) = 0$

$(x^2 - 139x + 4830.25 - 4830.25)$
$\quad + (y^2 - 93y + 2162.25 - 2162.25) = 0$

$(x^2 - 139x + 4830.25) - 4830.25$
$\quad + (y^2 - 93y + 2162.25) - 2162.25 = 0$

$(x - 69.5)^2 + (y - 46.5)^2 = 6992.5$

Since the coefficients of x^2 and y^2 are equal, the graph is a circle. The center is (69.5, 46.5), and the radius is $\sqrt{6992.5} \cong 83.62$.

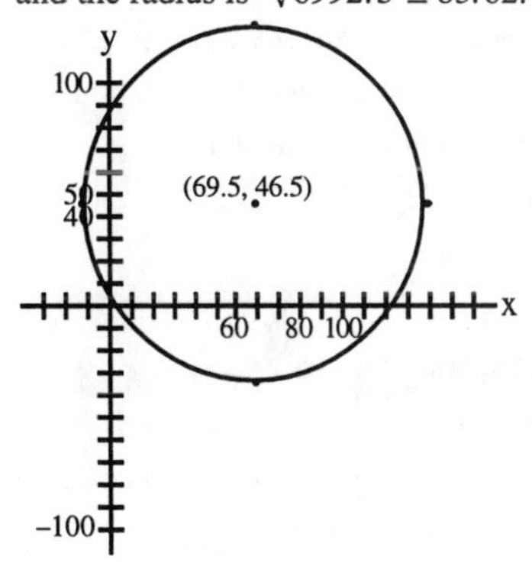

39. a. If the bulb is at the focus of the parabola and the axis of symmetry points straight ahead of the car, then all light rays coming from the bulb will strike the reflective surface and will be reflected straight ahead of the car, resulting in maximum illumination.

b. If $\frac{1}{4p} = 0.37$, then $p = \frac{1}{1.48} \cong 0.68$. The bulb should be placed about 0.68 inches in front of the vertex.

41. a. $e = \frac{30.33 - 29.79}{30.33 + 29.79} \cong 0.0090$

b. The distance from a focus of an ellipse to the nearer vertex is $a - c$, and the distance to the farther vertex is $a + c$. (See the figure below.) Therefore,

$a - c = 29.79$

$a + c = 30.33$

Add the two equations to obtain $a = 30.06$. Substitute 30.06 for a in either equation to obtain $c = 0.27$.

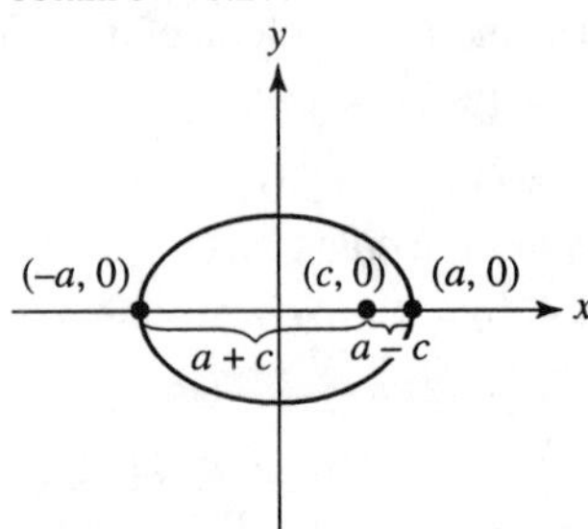

c. The sun must be c units to the left of the center, so the center is (0.27, 0). The value of b is

$\sqrt{a^2 - c^2} = \sqrt{30.06^2 - 0.27^2} \cong 30.06.$

Thus the orbit is very nearly circular. Its equation is approximately

$(x - 0.27)^2 + y^2 = 30.06^2$

43.

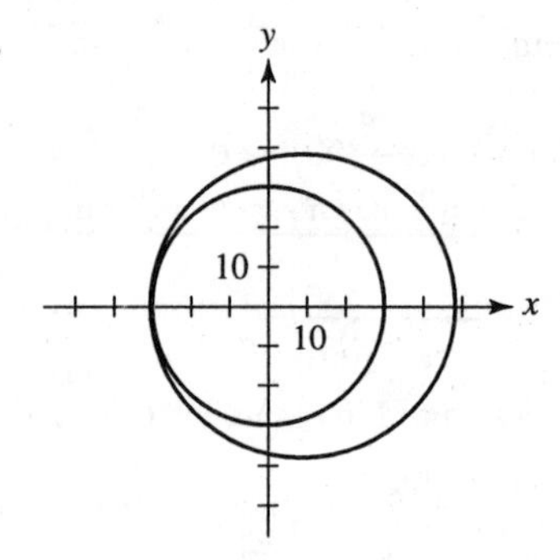

The graphs intersect at approximately (–28.80, ±7.64). Estimates of the fraction of Pluto's orbit inside Neptune's orbit will vary.

45. Since PD is vertical, its length is $y - (k - p)$. By the distance formula,

$\overline{PF} = \sqrt{(x-h)^2 + [y-(k+p)^2]}.$

Therefore:

$\sqrt{(x-h)^2 + [y-(k+p)^2]} = y - (k - p)$

Since both sides of the equation represent positive distances, squaring both sides produces an equivalent equation:

$(x-h)^2 + [y-(k+p)]^2 = y^2 - 2(k-p)y + (k-p)^2$

$(x-h)^2 + y^2 - 2(k+p)y + (k+p)^2$

$= y^2 - 2(k-p)y + (k-p)^2$

$(x-h)^2 + y^2 - 2ky - 2py + k^2 + 2pk + p^2$

$= y^2 - 2ky + 2py + k^2 - 2pk + p^2$

$(x-h)^2 = 4py - 4pk$

$4py = (x-h)^2 + 4pk$

$y = \frac{1}{4p}(x-h)^2 + k$

47. By the distance formula:

$\overline{PF_1} = \sqrt{[x-(h-c)]^2+(y-k)^2} = \sqrt{[(x-h)+c]^2+(y-k)^2}$, and

$\overline{PF_2} = \sqrt{[x-(h+c)]^2+(y-k)^2} = \sqrt{[(x-h)-c]^2+(y-k)^2}$. Thus:

$$\sqrt{[(x-h)+c]^2+(y-k)^2} - \sqrt{[(x-h)-c]^2+(y-k)^2} = 2a$$
$$\sqrt{[(x-h)+c]^2+(y-k)^2} = 2a + \sqrt{[(x-h)-c]^2+(y-k)^2}$$
$$\left(\sqrt{[(x-h)+c]^2+(y-k)^2}\right)^2 = \left(2a + \sqrt{[(x-h)-c]^2+(y-k)^2}\right)^2$$
$$[(x-h)+c]^2+(y-k)^2 = 4a^2 + 4a\sqrt{[(x-h)-c]^2+(y-k)^2} + [(x-h)-c]^2+(y-k)^2$$
$$4a\sqrt{[(x-h)-c]^2+(y-k)^2} = -4a^2 - [(x-h)-c]^2 + [(x-h)+c]^2$$
$$4a\sqrt{[(x-h)-c]^2+(y-k)^2} = -4a^2 - (x-h)^2 + 2c(x-h) - c^2 + (x-h)^2 + 2c(x-h) + c^2$$
$$4a\sqrt{[(x-h)-c]^2+(y-k)^2} = -4a^2 + 4c(x-h)$$
$$a\sqrt{[(x-h)-c]^2+(y-k)^2} = -a^2 + c(x-h)$$
$$\left(a\sqrt{[(x-h)-c]^2+(y-k)^2}\right)^2 = \left[-a^2 + c(x-h)\right]^2$$
$$a^2\left\{[(x-h)-c]^2+(y-k)^2\right\} = a^4 - 2a^2c(x-h) + c^2(x-h)^2$$
$$a^2\left[(x-h)^2 - 2c(x-h) + c^2 + (y-k)^2\right] = a^4 - 2a^2c(x-h) + c^2(x-h)^2$$
$$a^2(x-h)^2 - 2a^2c(x-h) + a^2c^2 + a^2(y-k)^2 = a^4 - 2a^2c(x-h) + c^2(x-h)^2$$
$$a^2c^2 - a^4 = c^2(x-h)^2 - a^2(x-h)^2 - a^2(y-k)^2$$
$$(c^2-a^2)(x-h)^2 - a^2(y-k)^2 = a^2(c^2-a^2)$$
$$\frac{(x-h)^2}{a^2} - \frac{(y-k)^2}{c^2-a^2} = 1$$

Section 6-5

1. The ratio $\frac{d}{s^2} = \frac{1}{15}$ for any data point.

3. The first differences of the s-values are not constant.

5. As the water becomes shallower, the value of d decreases, so the value of $s = \sqrt{15d}$ also decreases.

7.

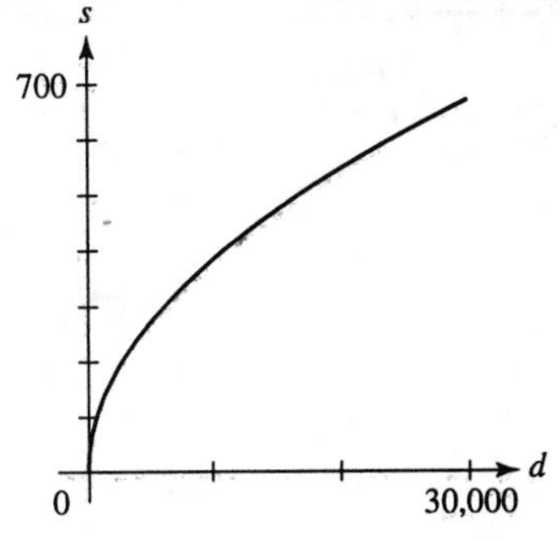

The graph is steeper when d is small.

9. a. The domain is $[0, \infty)$ because some point on the graph lies on the vertical line $x = x_0$ if and only if $x_0 \geq 0$. Similarly, the range is $[0, \infty)$ because some point on the graph lies on the horizontal line $y = y_0$ if and only if $y_0 \geq 0$.

b. The expression $y = \sqrt{x}$ is defined if and only if $x \geq 0$, so the domain is $[0, \infty)$. The value of $\sqrt{x}$ can be any nonnegative real number, so the range is also $[0, \infty)$.

11. a.

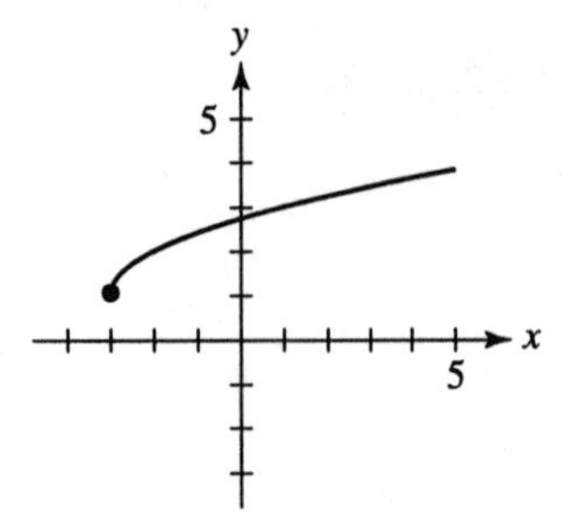

b. The domain is $[-3, \infty)$ and the range is $[1, \infty)$.

13. a.

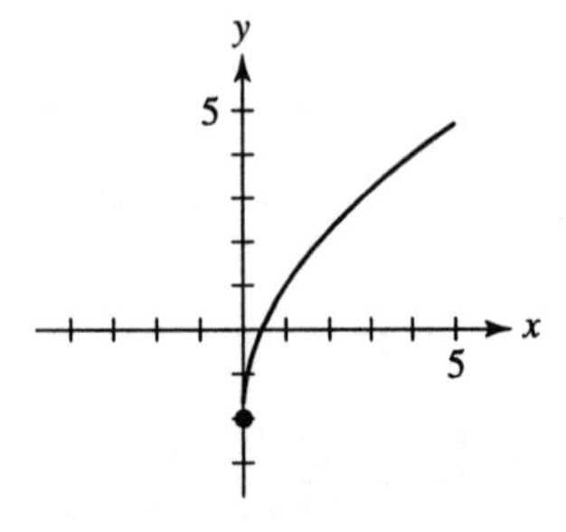

b. The domain is $[0, \infty)$ and the range is $[-2, \infty)$.

15. Each half of the rail has a length of 1026.5 feet before it expands. If it expands to $1026.5 + x$ feet, then $y = \sqrt{(x+1026.5)^2 - 1026.5^2}$. Equivalently, $y = \sqrt{x^2 + 2053x}$.

17. $y^2 = (x-1026)^2 - 1026.5^2$

$(x-1026.5)^2 - y^2 = 1026.5^2$

$\left(\frac{x-1026.5}{1026.5}\right)^2 - \left(\frac{y}{1026.5}\right)^2 = 1$

a.

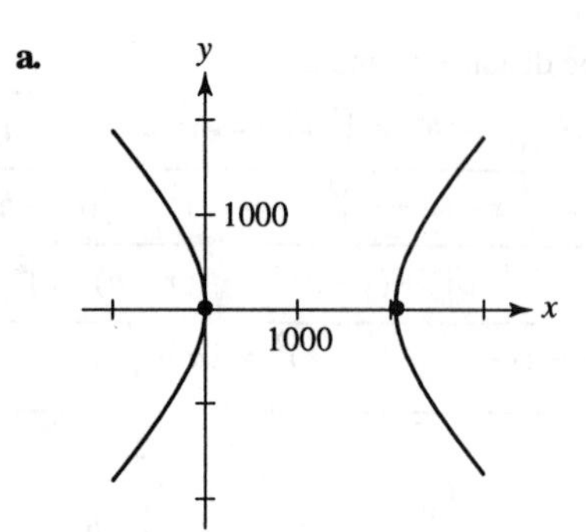

b.

c

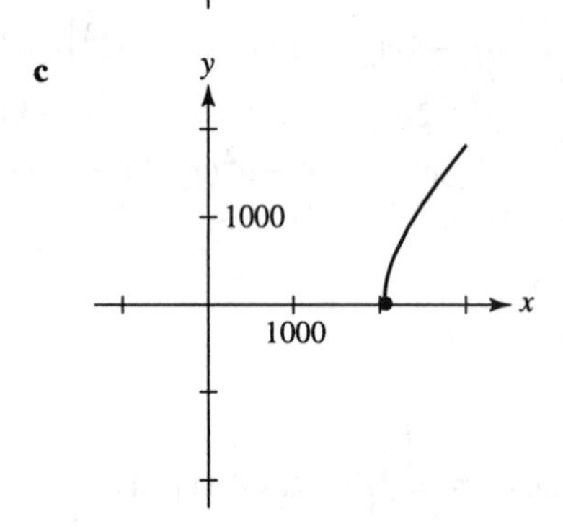

19. a. $y^2 = 16 - 4x^2$

$4x^2 + y^2 = 16$

$\frac{x^2}{4} + \frac{y^2}{16} = 1$

$\left(\frac{x}{2}\right)^2 + \left(\frac{y}{4}\right)^2 = 1$

b.

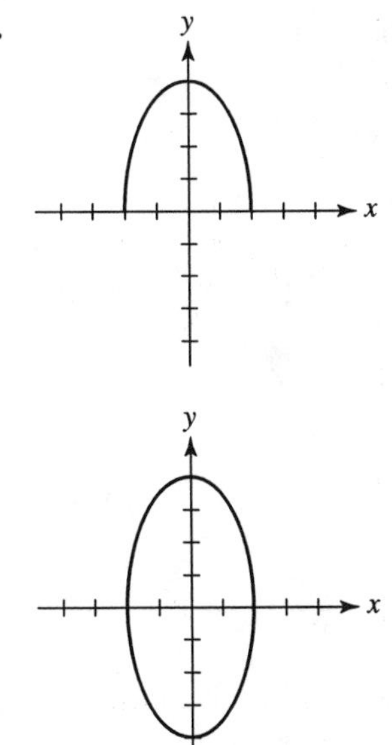

c. The domain is [–2, 2]. The range is [0, 4].

21. a.

$$y^2 = 4(25 + x^2)$$
$$y^2 = 100 + 4x^2$$
$$y^2 - 4x^2 = 100$$
$$\frac{y^2}{100} - \frac{x^2}{25} = 1$$
$$\left(\frac{y}{10}\right)^2 - \left(\frac{x}{5}\right)^2 = 1$$

b.

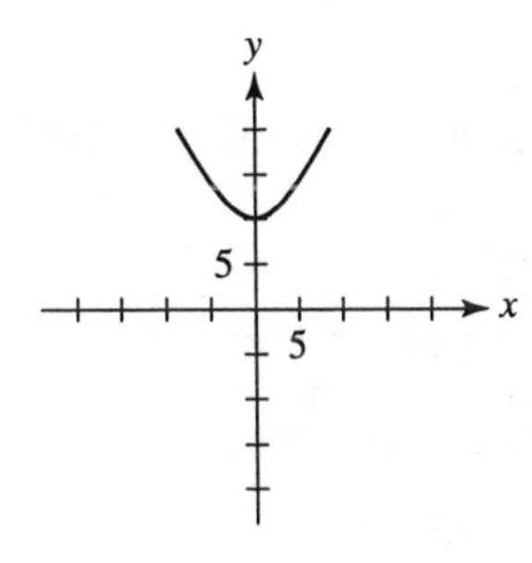

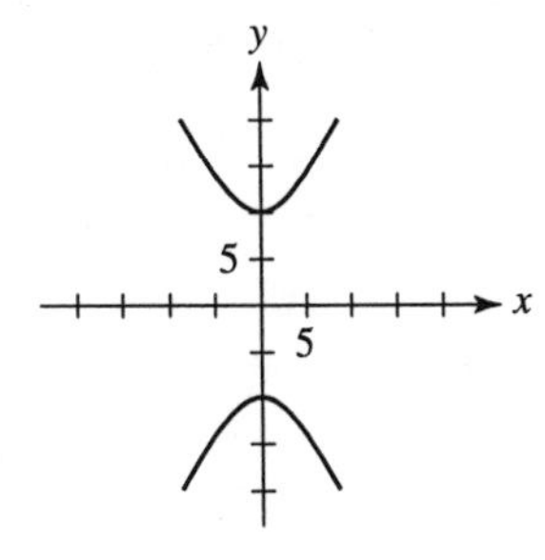

c. The domain is (–∞, ∞).
The range is [10, ∞).

23. a.

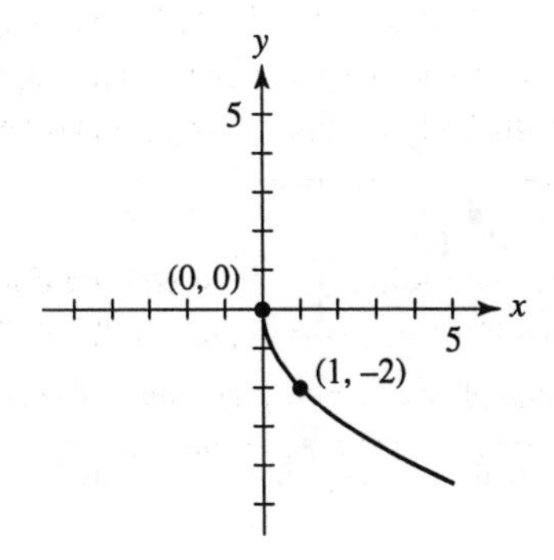

b. The domain is [0, ∞).
The range is (–∞, 0].

25. a.

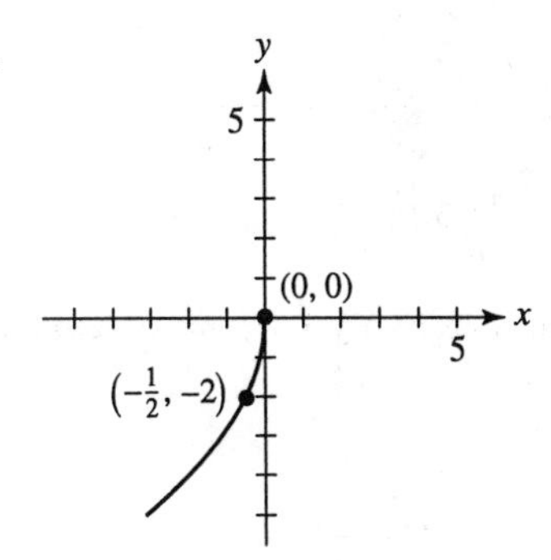

b. The domain is (–∞, 0]. The range is (–∞, 0].

27. a.

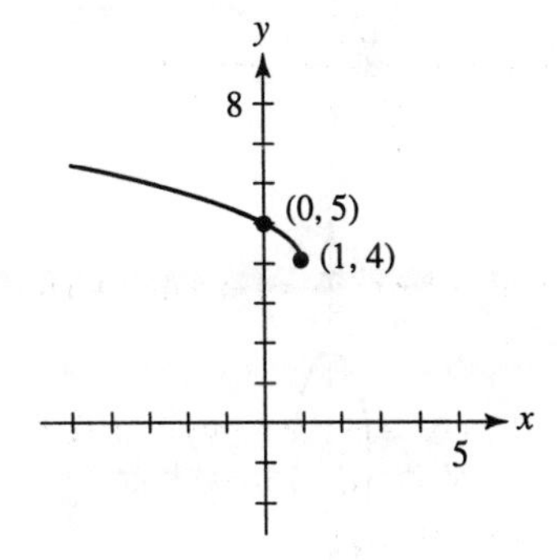

b. The domain is (–∞, 1]. The range is [4, ∞).

29. a.

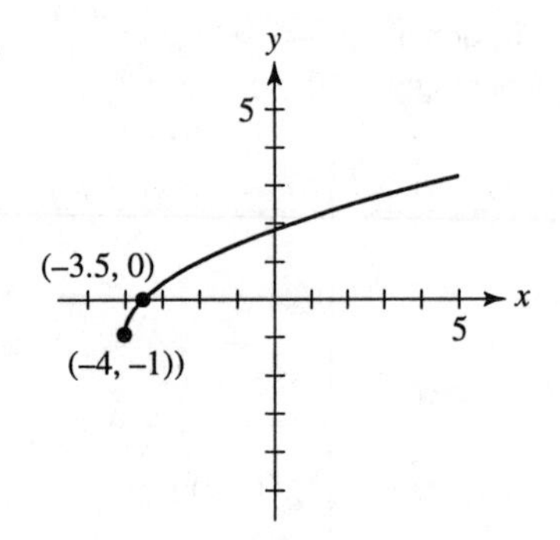

b. The domain is [−4, ∞). The range is [−1, ∞).

31. a.
$$y^2 = 9(1 - x^2)$$
$$y^2 = 9 - 9x^2$$
$$9x^2 + y^2 = 9$$
$$x^2 + \frac{y^2}{9} = 1$$
$$x^2 + \left(\frac{y}{3}\right)^2 = 1$$

b.

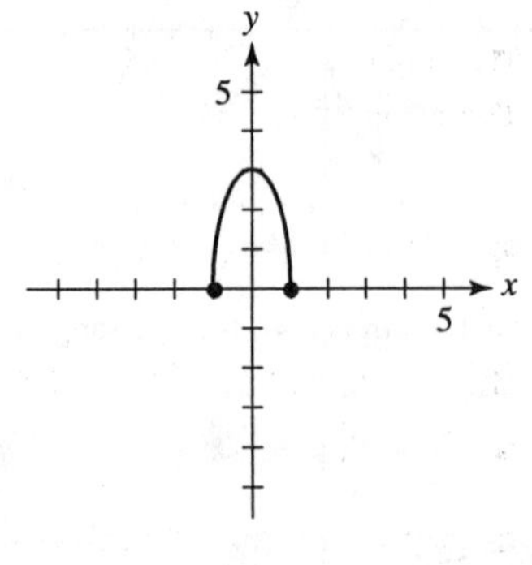

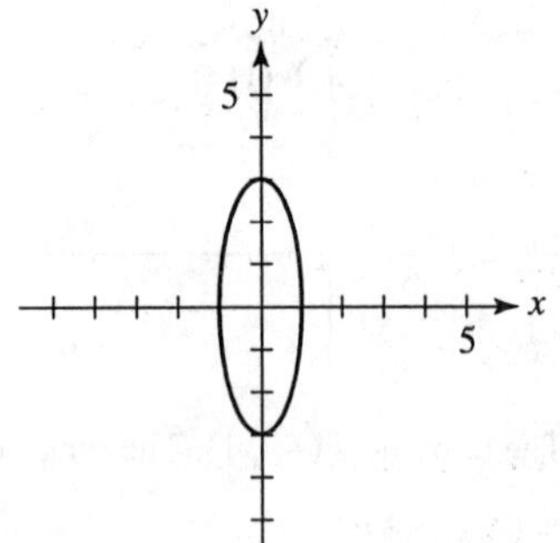

c. The domain is [−1, 1]. The range is [0, 3].

33. a.
$$y^2 = 9(x^2 + 1)$$
$$y^2 = 9x^2 + 9$$
$$9x^2 - y^2 = -9$$
$$\frac{y^2}{9} - x^2 = 1$$
$$\left(\frac{y}{3}\right)^2 - x^2 = 1$$

b.

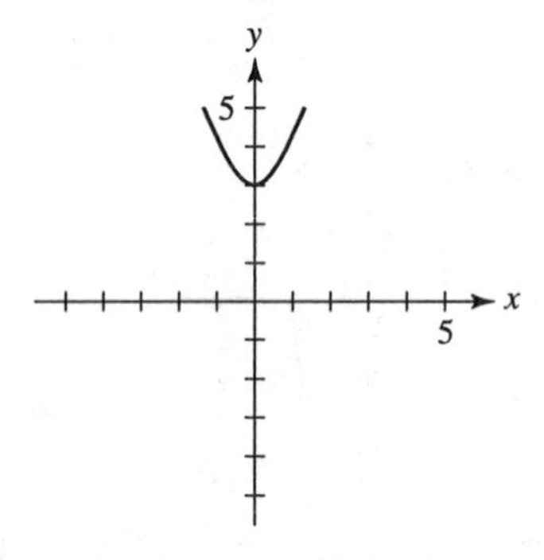

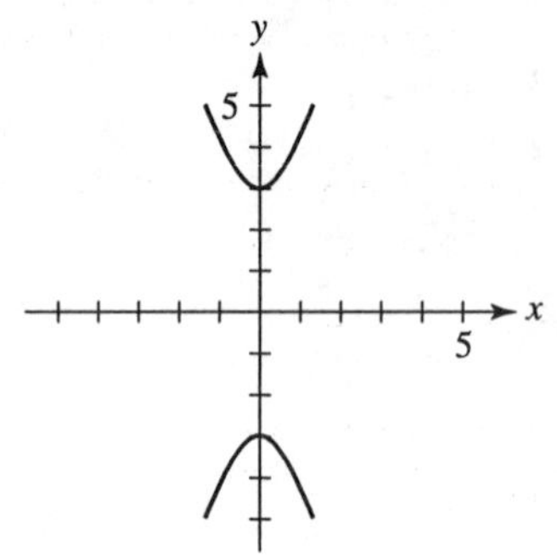

c. The domain is (−∞, ∞). The range is [3, ∞).

35. a.
$$y - 5 = \sqrt{9x^2 - 36}$$
$$(y-5)^2 = 9x^2 - 36$$
$$9x^2 - (y-5)^2 = 36$$
$$\frac{x^2}{4} - \frac{(y-5)^2}{36} = 1$$
$$\left(\frac{x}{2}\right)^2 - \left(\frac{y-5}{6}\right)^2 = 1$$

b.

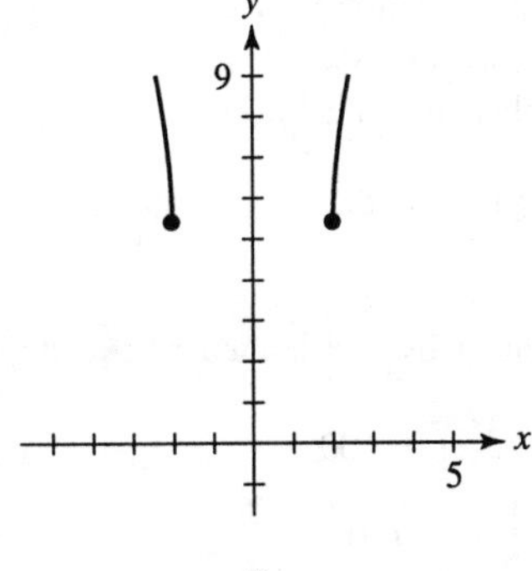

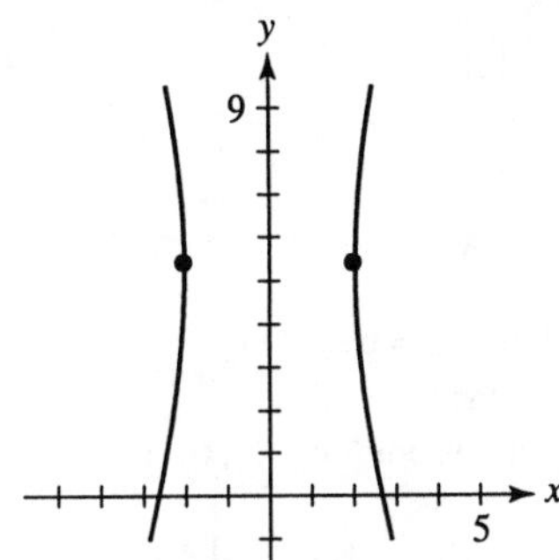

c. The domain is $(-\infty, -2] \cup [2, \infty)$.
The range is $[5, \infty)$.

37. a.

$$y + 6 = \sqrt{4 - (x-5)^2}$$
$$(y+6)^2 = 4 - (x-5)^2$$
$$(x-5)^2 + (y+6)^2 = 4$$

b.

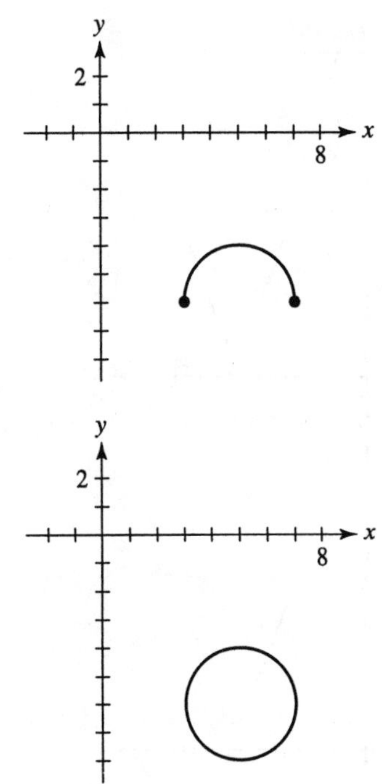

c. The domain is [3, 7].
The range is [–6, –4].

39. a. If x is an abstract variable, then $h(x)$ is defined whenever $x + 9 \geq 0$, so the domain is $[-9, \infty)$. Since $\sqrt{x+9} \geq 0$ for all x, the value of $h(x)$ is at least $\frac{0-3}{15} = -0.2$, so the range is $[-0.2, \infty)$.

b. In the physical context $x \geq 0$, so the domain is $[0, \infty)$. If $x \geq 0$, then $\sqrt{x+9} \geq 3$, so the value of $h(x)$ is at least $\frac{3-3}{15} = 0$. The range is $[0, \infty)$.

41. By the Pythagorean Theorem,
$x^2 + 30^2 = y^2$, so $y = \sqrt{x^2 + 900}$.

43.

$$y^2 = (450 - 90t)^2 + 900$$
$$y^2 - (450 - 90t)^2 = 900$$
$$y^2 - [90(5-t)]^2 = 900$$
$$y^2 - 8100(t-5)^2 = 900$$
$$\frac{y^2}{900} - 9(t-5)^2 = 1$$

$$\frac{y^2}{900}-\frac{(t-5)^2}{1/9}=1$$

$$\left(\frac{y}{30}\right)^2-\left(\frac{t-5}{1/3}\right)^2=1$$

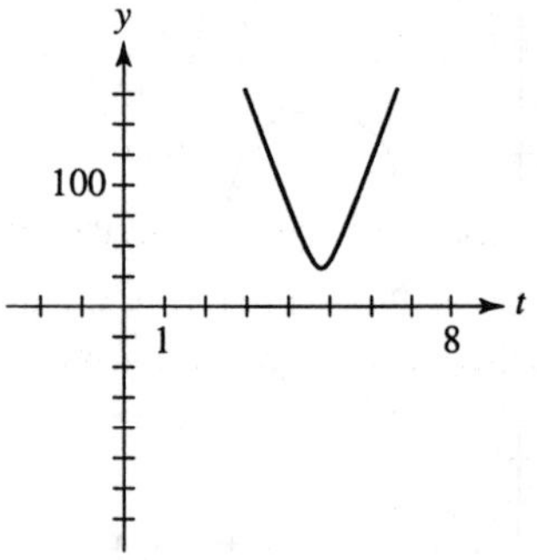

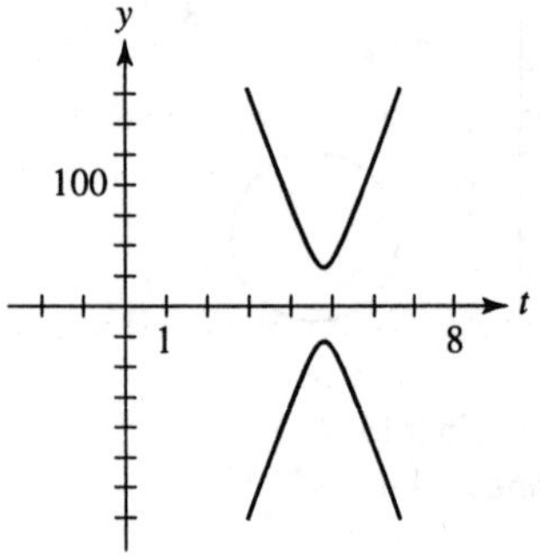

Section 6-6

1. The foci of the hyperbola are located at Sacramento and Modesto, so the center is halfway between them, at (0, –55), and the value of c is 55. As in Exercises 31 and 32 of Section 6-3, the value of a is 12.5 times the difference in arrival times, that is (12.5)(4) = 50. Therefore, $b=\sqrt{55^2-50^2}\cong 22.91$. The equation is $\frac{(y+55)^2}{50^2}-\frac{x^2}{22.91^2}=1$. Since the primary shock wave arrived earlier at Modesto (the lower focus), the epicenter lies on the lower branch of the hyperbola.

3. Solving for x in the second equation:
   $x=2-2y$
   Replacing x by $2-2y$ in the first equation:
   $(2-2y)+y^2=1$
   $y^2-2y+1=0$
   $(y-1)^2=0$
   $y=1$
   Using the equation $x=2-2y$ to find x:
   If $y=1$, then $x=0$.
   The only solution is (0, 1).

5. Solving for y in the second equation:
   $y=\frac{3}{2}x$
   Replacing y by $\frac{3}{2}x$ in the first equation:
   $4x^2+\left(\frac{3}{2}x\right)^2=100$
   $4x^2+\frac{9}{4}x^2=100$
   $\frac{25}{4}x^2=100$
   $x^2=16$
   $x=\pm 4$
   Since $y=\frac{3}{2}x$:
   If $x=4$, then $y=6$
   If $x=-4$, then $y=-6$.
   The solutions are (–4, –6), (4, 6).

7. The first equation is solved for x^2:
   $x^2=2y^2-14$
   Replace x^2 by $2y^2-14$ in the second equation.
   $y^2=2(2y^2-14)-5$
   $3y^2=33$
   $y^2=11$
   $y=\pm\sqrt{11}\cong\pm 3.32$
   $x^2=2y^2-14$
   $=2(11)-14=8$
   $x=\pm\sqrt{8}\cong\pm 2.83$
   The solutions are (–2.83, –3.32), (–2.83, 3.32), (2.83, –3.32), (2.83, 3.32).

9. Multiply the second equation by 2, and add:
   $$\begin{aligned} x^2+2y^2&=11\\ 4x^2-2y^2&=14\\ \hline 5x^2\quad\;\;&=25 \end{aligned}$$
   $x^2=5$
   $x=\pm\sqrt{5}\cong\pm 2.24$
   Replace x^2 by 5 in the first equation:
   $5+2y^2=11$
   $y^2=3$
   $y=\pm\sqrt{3}\cong 1.73$
   The solutions are (–2.24, –1.73), (–2.24, 1.73), (2.24, –1.73), (2.24, 1.73).

11. Solving the second equation for *y*:

$y = \frac{156}{x}$

Replacing *y* by $\frac{156}{x}$ in the first equation:

$x^2 + \left(\frac{156}{x}\right)^2 = 25$

$x^4 + 156 = 25x^2$

$x^4 - 25x^2 + 156 = 0$

$(x^2 - 12)(x^2 - 13) = 0$

$x = \pm\sqrt{12},\ \pm\sqrt{13} \cong \pm 3.46,\ \pm 3.61$

Since $y = \frac{156}{x}$:

If $x = \pm\sqrt{12}$, then $y = \pm\frac{156}{\sqrt{12}} \cong \pm 45.03$.

If $x = \pm\sqrt{13}$, then $y = \pm\frac{156}{\sqrt{13}} \cong \pm 43.27$.

The solutions are (–3.61, –43.27), (–3.46, –45.03), (3.46, 45.03), (3.61, 43.27).

13. Solving the second equation for *y*:

$y = -\frac{6}{x}$

Replacing *y* by $-\frac{6}{x}$ in the first equation:

$x^2 + 4\left(-\frac{6}{x}\right)^2 = 40$

$x^2 + \frac{144}{x^2} = 40$

$x^4 + 144 = 40x^2$

$x^4 - 40x^2 + 144 = 0$

$(x^2 - 36)(x^2 - 4) = 0$

$(x + 6)(x - 6)(x + 2)(x - 2) = 0$

$x = \pm 6, \pm 2.$

Since $y = -\frac{6}{x}$:

If $x = -6$, then $y = -\frac{6}{-6} = 1$

If $x = -2$, then $y = -\frac{6}{-2} = 3$

If $x = 2$, then $y = -\frac{6}{2} = -3$

If $x = 6$, then $y = -\frac{6}{6} = -1$

The solutions are (–6, 1), (–2, 3), (2, –3), (6, –1).

15. **a.**

b.

17. Solve the first equation for *y* to obtain $y = \frac{13 - x^2}{3x}$. Graph this equation together with $y = x + 5$. The solutions of the system are the points where the graphs intersect, at about (–4.48, 0.52), (0.73, 5.73).

19. Solve each equation for *y* to obtain $y = \pm\sqrt{\frac{2 - 5x^2}{4}}$ and $y = \pm\sqrt{\frac{9 - 6x^2}{7}}$. Graph both curves (that is, all four equations). Since the curves do not intersect, the system has no real solutions.

21. **a.** The ship lies on or inside a circle of radius 20, centered at (0, –40).

b. The equation of the circle is $x^2 + (y + 40)^2 = 400$.

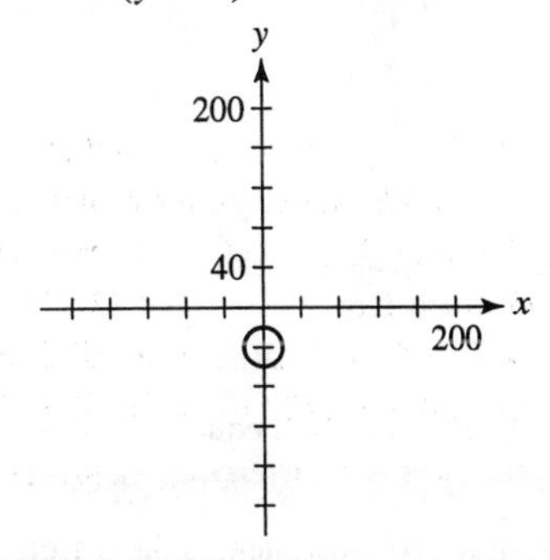

23. No, since the region in Exercise 22 does not include the entire region in Exercise 21.

25. Begin by graphing the relation $3x - y^2 = 6$. (The standard graphing form is $x = \frac{1}{3}y^2 + 2$.)

Choose a point not on the graph, say (0, 0). At (0, 0), the inequality asserts that $3(0)-(0)^2 \le 6$, which is true. The solution is therefore the region indicated below.

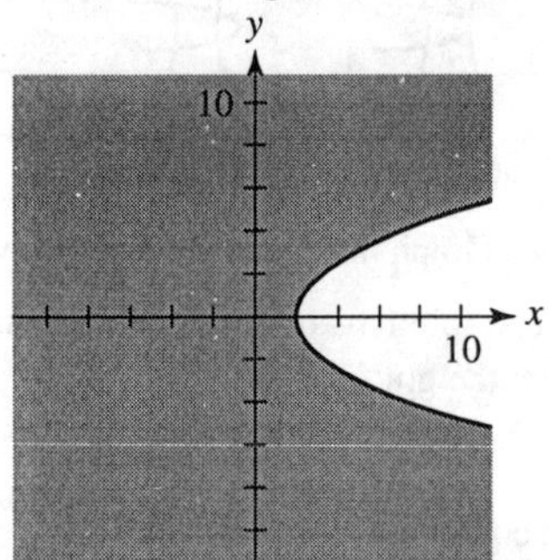

27. Begin by graphing the relation $y^2 - 16x^2 = 96$, as in Exercise 26. Again, choose any point not on the graph, say (0, 0). At (0, 0), the inequality asserts that $(0)^2 - 16(0)^2 \ge 96$, which is false. The solution is therefore the region indicated below.

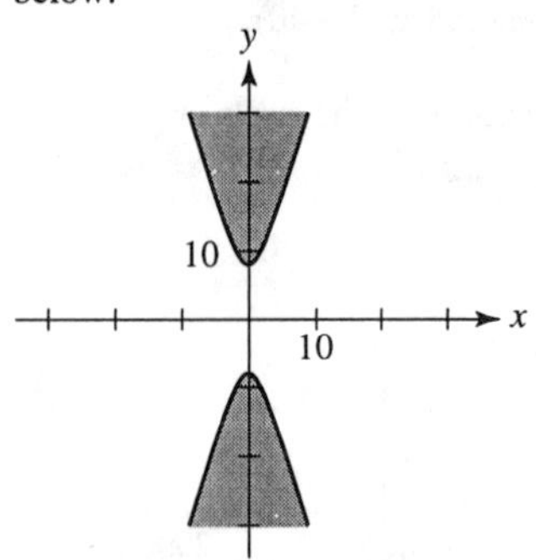

29. Begin by solving the inequalities $3x - y^2 \le 6$ and $y^2 - 16x^2 < 96$, as in Exercises 25 and 26. The solution to the system is the region common to the solutions to both inequalities, as shown below.

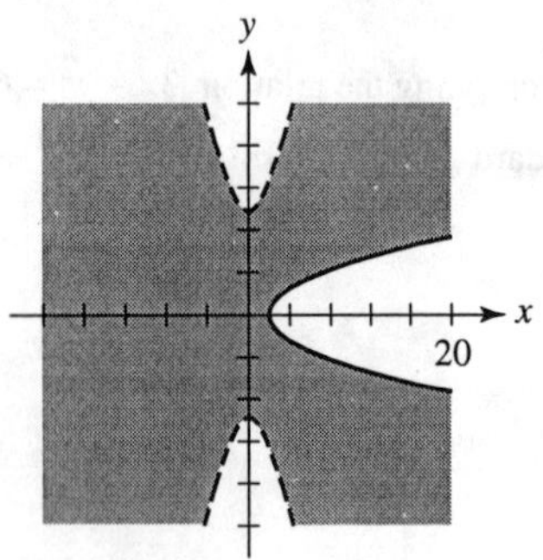

31. First, graph $x^2 + y^2 = 64$, $x^2 + y^2 = 100$, and $y = x^2 + 8$. Their graphs are respectively the smaller circle, the larger circle, and the parabola shown below. Choose a point not on any of the graphs, say (0, 0). At (0, 0), the first inequality asserts that $(0)^2 + (0)^2 > 64$, which is false. The second asserts that $(0)^2 + (0)^2 < 100$, which is true. The third asserts that $(0) - (0)^2 \ge 8$, which is false. The solution to each inequality is indicated below. The solution to the system is the region common to the solutions to all three inequalities.

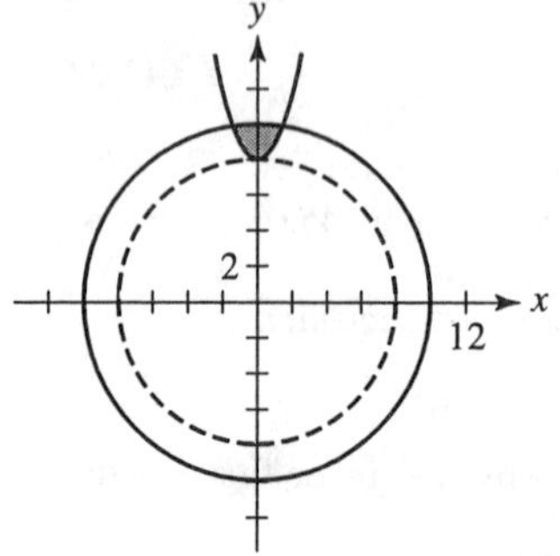

33. A helicopter based in Nassau can search within a circle of radius 100, centered at (100, 0). Its equation is $(x-100)^2 + y^2 = 10{,}000$, and the circle together with its interior is described the inequality $(x-100)^2 + y^2 \le 10{,}000$.

The system is

$x^2 + (y+40)^2 \le 400$

$(x-100)^2 + y^2 \le 10{,}000$

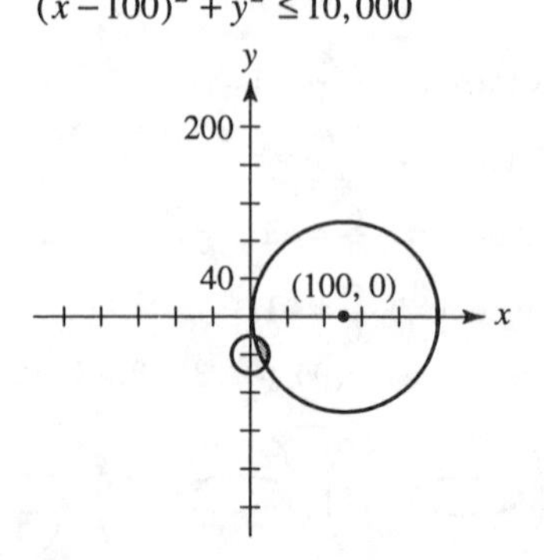

35. Responses will vary. Most students will choose to have the helicopter fly on to Miami because this plan allows the total area covered by the two helicopters to be greater.

37. Solving the second equation for *y*:

$y = -\frac{4}{x}$

Replacing *y* by $-\frac{4}{x}$ in the first equation:

$x^2 + 5\left(-\frac{4}{x}\right)^2 = 24$

$x^2 + \frac{80}{x^2} = 24$

$x^4 + 80 = 24x^2$

$x^4 - 24x^2 + 80 = 0$

$(x^2 - 4)(x^2 - 20) = 0$

$x^4 = 4$ or $x^2 = 20$

$x = \pm 2$ or $x = \pm\sqrt{20} \cong \pm 4.47$

Since $y = -\frac{4}{x}$:

If $x = -\sqrt{20}$, then $y = \frac{4}{\sqrt{20}} \cong 0.89$.

If $x = -2$, then $y = \frac{4}{2} = 2$.

If $x = 2$, then $y = -\frac{4}{2} = -2$.

If $x = \sqrt{20}$, then $y = -\frac{4}{\sqrt{20}} \cong -0.89$.

The solutions are (–4.47, 0.89), (–2, 2), (2, –2) (4.47, –0.89).

39. Multiply the first equation by 9, and add:

$$\begin{aligned} 9x^2 + 9y^2 &= 36 \\ 9x^2 + 16y^2 &= 144 \\ \hline 25y^2 &= 180 \end{aligned}$$

$y^2 = 7.2$

$y = \pm\sqrt{7.2} \cong \pm 2.68$

Replace y^2 by 7.2 in the first equation:

$x^2 + 7.2 = 4$

$x^2 = -3.2$

$x = \pm\sqrt{3.2}i \cong \pm 1.79i$

The solutions are $(-1.79i, -2.68)$, $(-1.79i, 2.68)$, $(1.79i, -2.68)$, $(1.79i, 2.68)$.

41. Solving for x^2 in the second equation:

$x^2 = y + 6$

Replacing x^2 by $y + 6$ in the first equation:

$(y + 6) + y^2 = 36$

$y^2 + y - 30 = 0$

$(y + 6)(y - 5) = 0$

$y = -6, 5$

Since $x^2 = y + 6$:

If $y = -6$, then $x^2 = 0$, so $x = 0$.

If $y = 5$, then $x^2 = 11$, so $x = \pm\sqrt{11} \cong \pm 3.32$.

The solutions are (–3.32, 5), (0, –6), (3.32, 5).

43. From the second equation, either $x = 0$ or $y = 0$.

If $x = 0$, then substitution into the first equation produces

$4y^2 + 16y + 16 = 0$

$4(y + 2)^2 = 0$

$y = -2$

If $y = 0$, then substitution into the first equation produces

$x^2 - 8x + 16 = 0$

$(x - 4)^2 = 0$

$x = 4$

The solutions are (0, –2), (4, 0).

45. Begin by graphing the relations $3x^2 + 5y^2 = 15$ and $3x^2 + 5y^2 = 90$. The standard graphing forms are

$\frac{x^2}{5} + \frac{y^2}{3} = 1$ and $\frac{x^2}{30} + \frac{y^2}{18} = 1$.

Choose any point not on either graph, say (0, 0). At (0, 0), the first inequality asserts that $3(0)^2 + 5(0)^2 \geq 15$, which is false. The second inequality asserts that $3(0)^2 + 5(0)^2 < 90$, which is true. The solution to the system is the shaded region below.

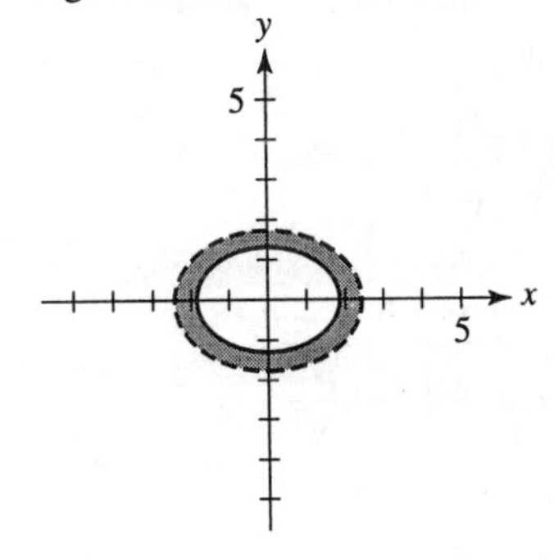

47. Begin by graphing the relations

$25x^2 + 4y^2 = 100$, $y^2 - 9x^2 = 9$, and $xy = 4$.

The standard graphing forms for the first two are

$\frac{x^2}{4} + \frac{y^2}{25} = 1$ and $\frac{y^2}{9} - x^2 = 1$.

Solving the third equation for *y* yields $y = \frac{4}{x}$.

Choose any point not on any of the graphs, say (0, 0). At (0, 0), the first inequality asserts that $25(0)^2 + 4(0)^2 \le 100$, which is true. The second asserts that $(0)^2 - 9(0)^2 \ge 9$, which is false. The third asserts that $(0)(0) < 4$, which is true. The solution to the system is the heavily shaded region below.

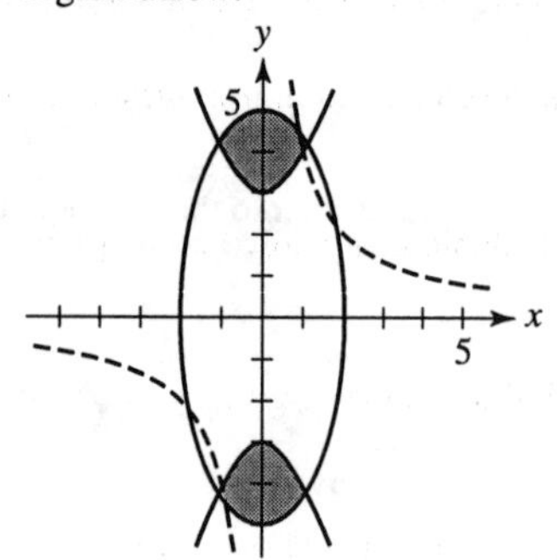

49. a. Solve the comet's equation for y to obtain $y = -\frac{x}{x+1}$. From the graph, it appears that the comet's orbit is a hyperbola. (The comet actually traces out only one branch of the hyperbola). Therefore, it will never be seen again. (Also, since the equation of the comet's orbit represents y as a function of x, the graph cannot be an ellipse.)

b. The comet will cross the earth's orbit at the points representing the solutions to the system

$$\frac{x^2}{1.03} + \frac{y^2}{0.97} = 1$$
$$xy + x + y = 0$$

To solve the system graphically, solve each equation for y to obtain:

$$y = \pm\sqrt{0.97\left(1 - \frac{x^2}{1.03}\right)}$$
$$y = -\frac{x}{x+1}$$

The comet will cross the earth's orbit twice, at about (–0.47, 0.87) and (0.89, –0.47).

51. a. The time for the batted ball to reach the fielder plus the time for the thrown ball to reach first base must be no more than 3 seconds. Since the ball always travels at 120 feet per second, the total distance traveled by the ball must be 360 feet. Thus $\overline{AB} + \overline{BC} = 360$. This means that the fielder is on an ellipse with foci at A and C and with $a = 180$.

b. For the ellipse in part (a), the center is (45, 0). From part (a), $a = 180$ and $c = 45$ (the distance from the center to either focus). Then $b = \sqrt{180^2 - 45^2} \cong 174.28$. The equation of the ellipse is

$$\left(\frac{x-45}{180}\right)^2 + \left(\frac{y}{174.28}\right)^2 = 1$$

The fielder must play on or inside the ellipse, and in the first quadrant of the coordinate system, so

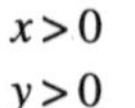

$$\left(\frac{x-45}{180}\right)^2 + \left(\frac{y}{174.28}\right)^2 \le 1$$
$$x > 0$$
$$y > 0$$

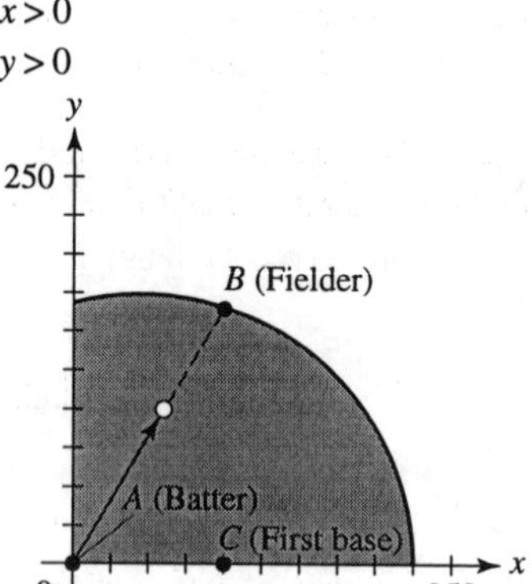

NUTSHELLS

Nutshell for Chapter 7 - POLYNOMIAL FUNCTIONS

What's familiar?

- Linear and quadratic functions are polynomial functions. Therefore, the study of polynomials numerically, analytically, and graphically will give us a broader framework in which to place the ideas of Chapter 3 and 5.
- Section 7-1 discusses the idea of *polynomial variation*, which generalizes the concept of linear variation from Chapter 3.
- Section 7-1 continues the work on graphical transformations. You learn to obtain many polynomial graphs by transforming a few basic one, including $y = x$ and $y = x^2$.
- Sections 7-2 and 7-3 relate the factors of a polynomial $f(x)$ both to the solutions of the equation $f(x) = 0$ and to the x–intercepts on the graph of $f(x)$. We also did this for quadratic equations in Chapter 5.
- Section 7-4 explores a general method for fitting a polynomial function to a table. In Section 5-2, you used a special case of this method.
- Section 7-5 discusses solving polynomial equations and inequalities. You use methods that build on those you used for quadratic equations and inequalities in Section 5-2.

What's new?

- Polynomial functions exhibit fewer common features than either linear or quadratic functions.
- Section 7-1 begins by looking at the simplest polynomial functions. There are the *power functions* $y = x$, $y = x^2$, $y = x^3$, $y = x^4$, and so on. Then we take a look at their numerical, graphical, and analytical properties before considering more general polynomial functions.
- Section 7-2 looks at the analytical properties of polynomial functions historically. Much of their history comes through the fascinating mathematicians who discovered them.
- Section 7-3 lets you discover why the behavior of the "tails" on a polynomial graph depend only on the equation's term of highest degree.
- Section 7-4 contains a numerical method for solving polynomial equations. This complements the analytical and graphical methods you already show.

List of hints for exercises in each chapter.

Chapter 7

- Before completing the exercises in Section 7-1, review the transformation principles in Sections 3-4, 5-1 and 6-3.
- Review the connection between x-intercepts on quadratic graphs and solutions to quadratic equations. Section 7-2 expands the connection to polynomials of higher degree.
- Review the first and second differences of Sections 3-2 and 5-1. Section 7-4 generalizes this concept.

Name:____________________

Chapter 7 Additional Exercises

Date:____________________

1. Sketch the graph of $y=(x+3)^3-2$.

1.

2. Sketch the graph of $y=\frac{1}{4}x^4-3$.

2.

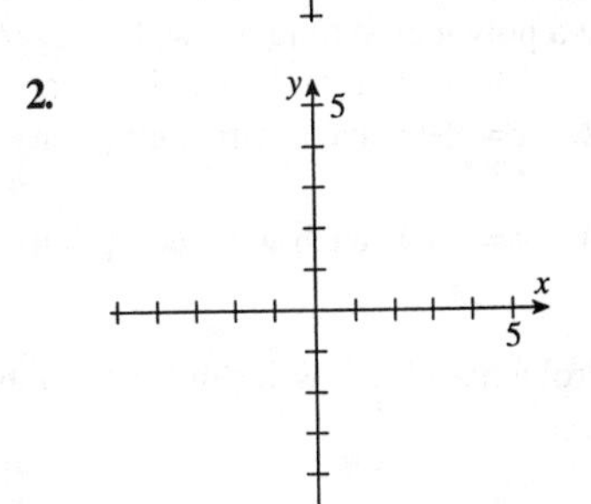

In Problems 3–4, decide whether the table fits a polynomial variation for the given value of n. If so, write an equation relating x and y.

3.

x	y $(n=3)$
1	5
2	40
4	320
5	625
9	3645

3. ____________________

4.

x	y $(n=4)$
1	0.25
2	4
4	20.25
8	64
16	156.25

4. ____________________

5. Write an equation of the function whose graph is that of $y=x^n$ shifted a units down $(a>0)$ and compressed vertically by a factor of $\frac{1}{2}$.

5. ____________________

6. Write an equation of the function whose graph is that of $y=x^n$ reflected in the x-axis and then shifted a units to the left $(a>0)$.

6. ____________________

7. Find all the zeros of the function $P(x)=x^4-5x^2+4$ by solving the equation $P(x)=0$.

7. ____________________

In Problems 8–9, identify the zeros of the function and state the multiplicity of each zero.

8. $f(x)=(x^2-1)(x^2+2x+1)$

8. ____________________

Chapter 7 Additional Exercises *(cont.)*

Name:____________________

9. $g(x) = (x^2 + 3x + 2)(x^2 - x - 2)$ **9.** ____________________

In Problems 10–11, find all zeros of the function to verify the *Fundamental Theorem of Algebra*.

10. $f(x) = x^3 + 3x^2 - 2x - 2$ **10.** ____________________

11. $h(x) = x^5 + x^3 - 2x$ **11.** ____________________

Find a polynomial function with integer coefficients having the given zeros with the indicated multiplicities.

12. $x = \pm\sqrt{3}$, each with multiplicity 2; $x = 1$, multiplicity 1 **12.** ____________________

13. $x = \pm\sqrt{7}$, each with multiplicity 1; $x = \frac{2}{3}$, multiplicity 1 **13.** ____________________

In Problems 14–15, write an equation in factored form for the sixth degree polynomial whose graph resembles the one shown.

14. **14.** ____________________

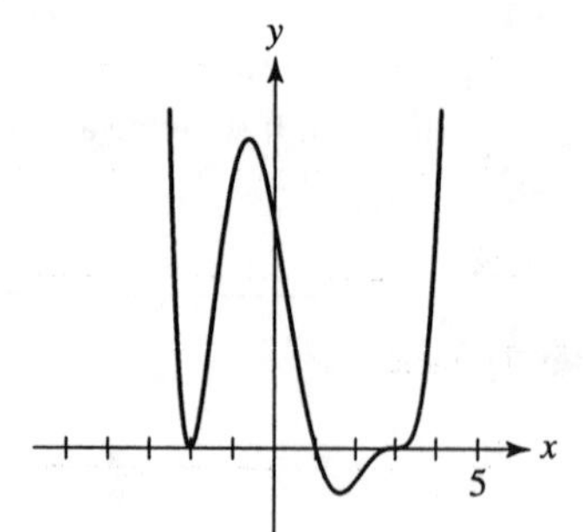

15. **15.** ____________________

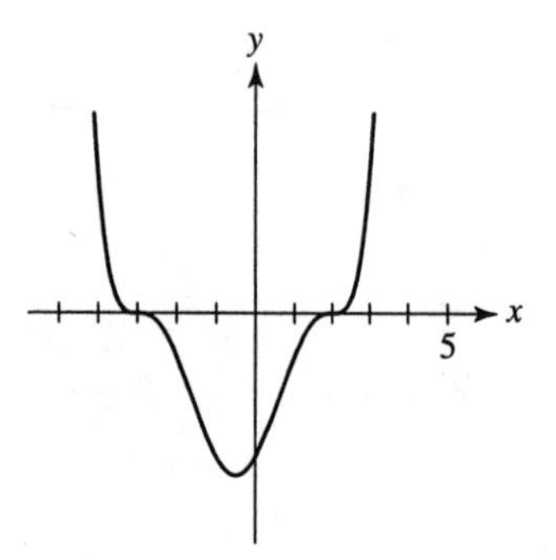

In Problems 16–17, use a graphing calculator to find all extreme values of the polynomial function, and identify any which are absolute. Round to two decimal places if necessary.

16. $f(x) = x^4 - 6x^2 + 3$ **16.** ____________________

17. $g(x) = x^3 - 6x + 3$ **17.** ____________________

In Problems 18–19, use a graphing calculator to identify the intervals where the function is increasing and decreasing. Round to two decimal places if necessary.

18. $f(x) = x^4 - 6x^2 + 3$ **18.** ____________________

Chapter 7 Additional Exercises *(cont.)*

Name:______________

19. $g(x) = x^3 - 6x + 3$

19. ______________

In Problems 20–21, use a graphing calculator to estimate its range. Round to two decimal places if necessary.

20. $f(x) = x^4 - 6x^2 + 3$

20. ______________

21. $g(x) = x^3 - 6x + 3$

21. ______________

In Problems 22–23, describe the end behavior of the graph of the function.

22. $f(x) = 10 - 2x^2 + x^3 - x^4$

22. ______________

23. $f(x) = 2x^{20} + x^{32} - x^{101}$

23. ______________

In Problems 24–26, decide whether the graph of $y = f(x)$ is symmetric about the y-axis, the origin, or neither by evaluating $f(-x)$.

24. $f(x) = -x^4 + x^2 - 99$

24. ______________

25. $f(x) = -x^4 + x^3$

25. ______________

26. $f(x) = (-x^3 + 4x)(x^2 + 1)$

26. ______________

In Problems 27–29, find the smallest degree of a polynomial function that fits the table.

27.

x	y
–5	539
–3	47
–1	–5
1	–1
3	59
5	559
7	2267

27. ______________

28.

x	y
–3	10
–2	1
–1	0
0	1
1	–2
2	–15
3	–44

28. ______________

In Problems 29–31, find a polynomial function of the smallest possible degree to fit the given ordered pairs.

29. (–3, 0), (–2, 3), (–1, 4), (0, 3), (1, 0)

29. ______________

30. (0, –4), (1, –11), (2, –20), (3, 5), (4, 124)

30. ______________

31. (–2, –45), (–1, –12), (0, –1), (1, –6), (2, –21)

31. ______________

Chapter 7 Additional Exercises *(cont.)*

Name:____________________

In Problems 32–34, use the bisection method to approximate the largest real zero of the function with an error of no more than 0.01.

32. $f(x) = x^3 + 8x^2 - 4$ **32.** ____________________

33. $g(x) = 3x^4 - 4x^3 + x^2 - 1$ **33.** ____________________

34. $h(x) = x^5 - 3x^4 + x^2 + 4x - 3$ **34.** ____________________

In Problems 35–37, solve the inequality graphically. Round answers to two decimal places.

35. $x^3 + 3x^2 - x - 4 > 0$ **35.** ____________________

36. $2x^4 - 4x^3 + x - 1 \le 0$ **36.** ____________________

37. $x^3 - x^2 + 2x + 1 < 0$ **37.** ____________________

In Problems 38–40, solve the inequality analytically, using either the test-value method or the scan method.

38. $9x^4 - x^2 < 0$ **38.** ____________________

39. $(2x-1)(x-1)^2(x-7)^3 \ge 0$ **39.** ____________________

40. $3x - 2x^2 - x^3 \le 0$ **40.** ____________________

Use the following scenario for Problems 41–46. A package is being shipped in a cylindrical carton with radius r. The combined length and girth of the carton is 144 inches.

41. Express the volume (V) as a function of the radius r. **41.** ____________________

42. What is the domain of the function you found in Problem 41? **42.** ____________________

43. Use a graphing calculator to find the dimensions that will hold the most. Round to two decimal places. **43.** ____________________

44. Use a graphing calculator to find the maximum volume. Round to two decimal places. **44.** ____________________

45. What is the range of the function you found in Problem 41? **45.** ____________________

46. Use a graphing calculator to find the values of r for which $V(r) > 24{,}000$. Round to two decimal places. **46.** ____________________

Chapter 7 Additional Exercises *(cont.)*

Name:________________

Use the following population table for Problems 47–48.

Year	Population (thousands)
1980	300
1985	275
1990	300
1995	525

47. Find a cubic function to fit the data. Let $x = 0$ for 1980.

47. ________________

48. Use your function to predict the population in the year 2000.

48. ________________

49. If $f(8) = 9$, what does the Factor Theorem tell you about $g(x) = f(x) + 9$?

49. ________________

50. $x = 12$ is a zero for the polynomial $f(x) = x^4 - 5x^3 - 68x^2 - 176x - 192$. Use a graphing calculator to graph the function. How many other zeros are there? What theorem did you use to determine this? Are they real numbers?

50. ________________

A Mathematical Looking Glass

Sonar

(to follow Section 7-3)

On July 17, 1996, a TWA airliner on a flight from New York to Paris crashed into the Atlantic Ocean off Long Island after an apparent midair explosion. A team of FBI agents was sent to the site to determine whether the explosion was caused by mechanical failure or by a bomb. Their investigation depended on their ability to locate and retrieve pieces of the wreckage, including two flight recorders, from the ocean floor.

A key piece of equipment in the FBI's search was a side-scanning sonar, which emits sound waves underwater and detects reflected waves from distant objects. To locate objects on the basis of sonar data, it is necessary to know how fast the sound waves travel. The speed of sound underwater depends on the depth, salinity, and temperature of the water. Specifically, at a depth of d meters, a salinity of s grams per liter of water, and a temperature of T degrees Celsius, the speed of sound in meters per second is

$$V = 0.0003T^3 - 0.055T^2 + (8.8 - 0.12s)T + (1400 + 0.017d)$$

The relationship is approximately valid for $0 \leq d \leq 1000$, $0 \leq s \leq 45$, and $0 \leq T \leq 35$.

Exercises

1. The Atlantic Ocean near Long Island has a salinity of about 35 grams per liter. On the same screen, in the window [0, 35] by [1400, 1500] graph V as a function of T for each value of d.

   a. $d = 0$ b. $d = 500$

   c. $d = 1000$

2. (Making Observations)

   a. How do the graphs in Exercise 1 relate to each other?

   b. What graphing transformation principle could you have used to predict that relationship?

3. (Making Observations)

   a. Sound always travels faster in warmer water. How do your graphs in Exercise 1 support this statement?

   b. In the window [0, 100] by [1400, 1500], graph V as a function of T with $d = 0$ and $s = 50$. How does your graph tell you that the equation given in **Sonar** does not model the speed of sound accurately for water with a salinity of 50 grams per liter?

A Mathematical Looking Glass

Just Do It 2

(to follow Section 7-4)

Colin Levkanich is a weight training instructor at Gold's Gym. Because he is friendly, people often ask him fitness questions in unusual settings. He was recently chagrined over his inability to answer a question, posed in the lobby of the Benedum Theatre, concerning a friend's ideal weight. One solution would be for Colin to carry a height-weight chart with him at all times, but that would be inconvenient. Here is another possible solution.

Ideal weight depends primarily on height, age, gender, and body type. Since Colin deals almost exclusively with women between the ages of 25 and 40, he can ignore age and gender as variables. Furthermore, if he knows the ideal weight for a medium-framed woman of a given height, he can add or subtract ten pounds to adjust for large- or small-framed women. For these reasons, he can regard ideal weight as a function of height alone. Weight is roughly proportional to volume, so it is reasonable for him to model ideal weight as a cubic function of height.

Since Colin can do mental arithmetic quickly, he could memorize a cubic function that fits the data on his chart. He could then calculate an ideal weight for any height.

Exercises

1. (Creating Models) Construct a cubic polynomial

$$W = aH^3 + bH^2 + cH + d$$

to fit the following four data points, taken from Colin's height-weight chart.

height (H, in cm)	ideal weight (W, in kg)
150	53.0
160	58.0
170	63.0
180	69.5

(*Hint*: Use the methods of Section 7-4 to construct a system of linear equations in a, b, c, and d. Use matrix methods on your calculator to solve the system.)

2. (Writing to Learn) Graph the cubic function from Exercise 1. For what heights do you believe the model gives reasonably good results? (1 inch = 2.54 cm, and 1 pound = 0.454 kg.) Explain your answer.

3. a. (Creating Models) The first three data points in Colin's chart are collinear. Write an equation for the linear function through these points.

   b. (Making Observations) Graph this function, and find an interval of heights for which the model gives reasonably good results.

   c. (Writing to Learn) Compare the linear and cubic models in terms of simplicity and accuracy. Which do you believe is better for Colin's purposes? Explain your answer.

Chapter 7

Section 7.1

1. The graphs all pass through (–1, –1), (0, 0), and (1, 1). The left tail of each graph points down, and the right tail points up. For larger powers, the graph is flatter near the origin, and steeper when $|x| > 1$.

3.

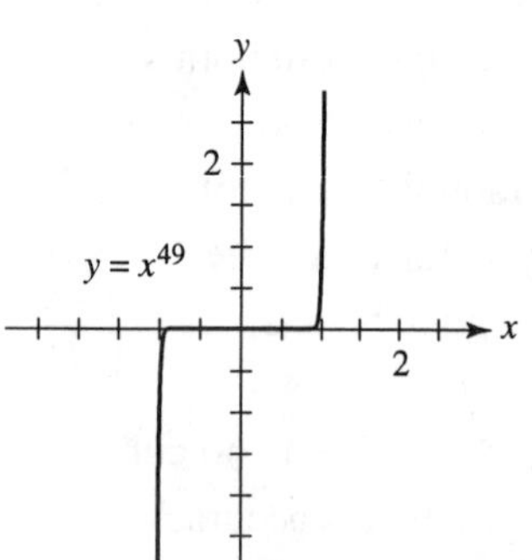

$y = x^{50}$

5.

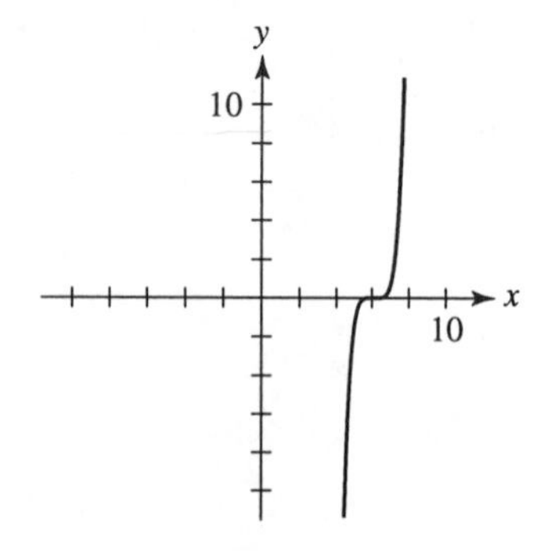

7.

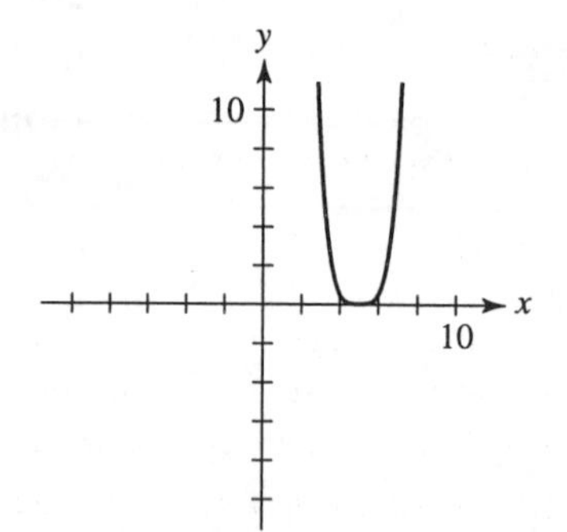

9. a.

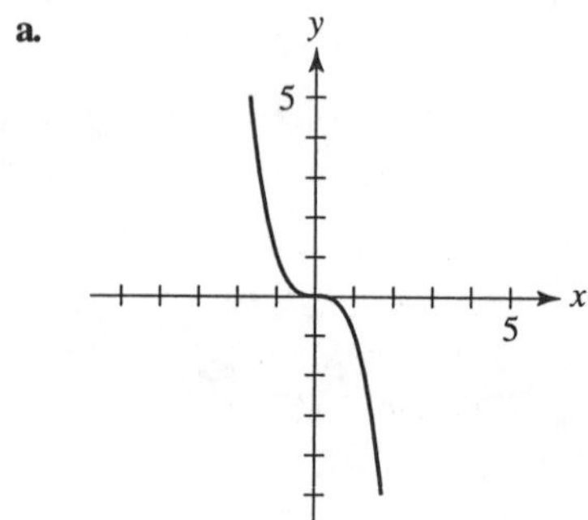

b. $y = (-x)^3$

11. a.

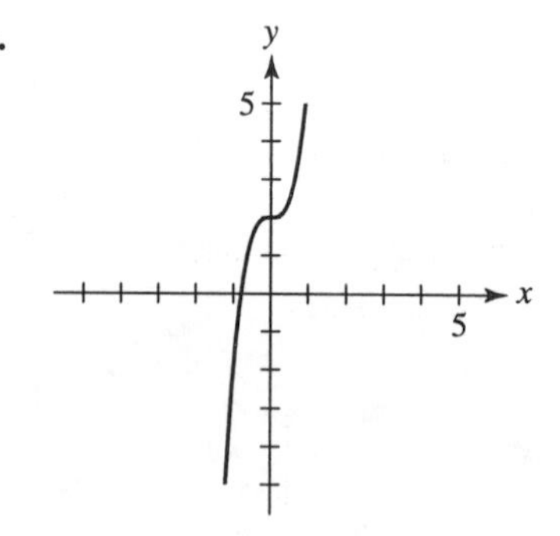

b. $y = 4x^3 + 2$

13. In Section 3-2 it is stated:
If $f(x) = mx$ for some real m, then whenever the input x is doubled, the output $f(x)$ is also doubled.

In Section 7-1 it is stated:
For nth power polynomial variations, doubling x multiplies y by 2^n. Replacing n by 1 in the second statement yields the first statement.

15. The ratio $\frac{y}{x^3}$ is 2 for all rows, so the table fits the function $y = 2x^3$.

17. The ratio $\frac{y}{x^4}$ is 0.5 for all rows, so the table fits the function $y = 0.5x^4$.

19. a. An animal's weight W is proportional to the cross-sectional area of its femur or humerus bone. If the circumference of a bone is C, its radius is $\frac{C}{2\pi}$, so its cross-sectional area is $\pi\left(\frac{C}{2\pi}\right)^2 = \frac{C^2}{4\pi}$. Thus $W = m\left(\frac{C^2}{4\pi}\right)$ for some constant m. It is more convenient to call the constant $\frac{m}{4\pi}$ by another name, say k. Then $W = kC^2$. To find the value of k, use the fact that for a hippopotamus the sum of the two circumferences is 400 mm. Thus $C = 200$ and $W = 5500$.

$$W = kC^2$$
$$5500 = k(200)^2$$
$$k = \frac{5500}{200^2} \cong 0.14$$
$$W \cong 0.14C^2$$

b. For *Brachiosaurus* the sum of the circumferences was 1400 mm, so $C = 700$. Thus $W = 0.14(700)^2 = 68,600$ pounds. This is approximately within the range of other estimates.

21. a.

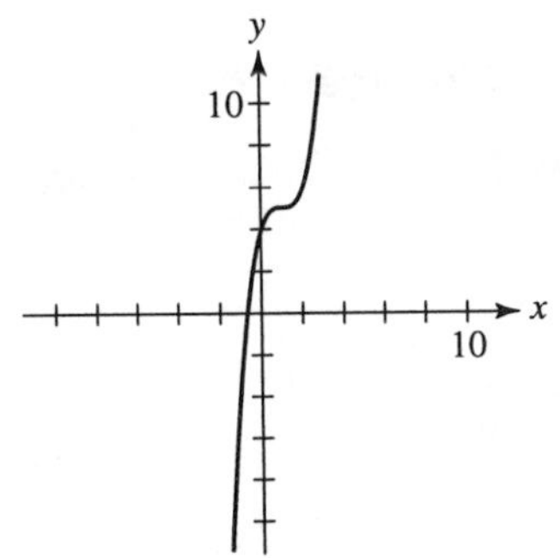

b. The domain is $(-\infty, \infty)$.
The range is $(-\infty, \infty)$.

23. a.

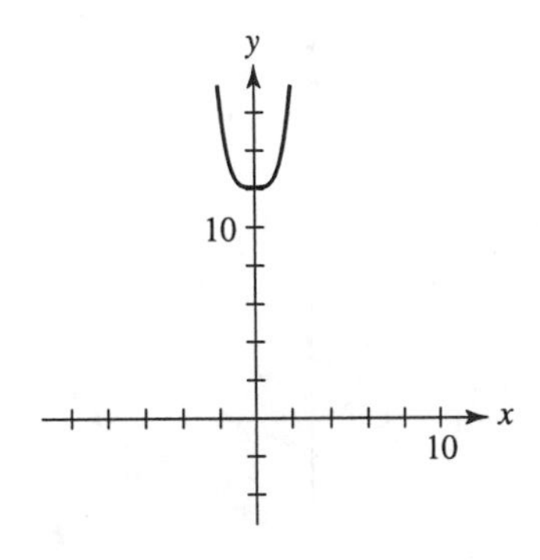

b. The domain is $(-\infty, \infty)$.
The range is $[12, \infty)$.

25. a. The domain and range are unchanged.

b. The domain is unchanged. If n is odd, the range is unchanged. If n is even, the range changes from $[0, \infty)$ to $(-\infty, 0]$.

c. The domain and range are unchanged.

d. The domain is unchanged. If n is odd, the range is unchanged. If n is even and the new equation is $y = x^n + k$, the range changes from $[0, \infty)$ to $[k, \infty)$.

27. a.

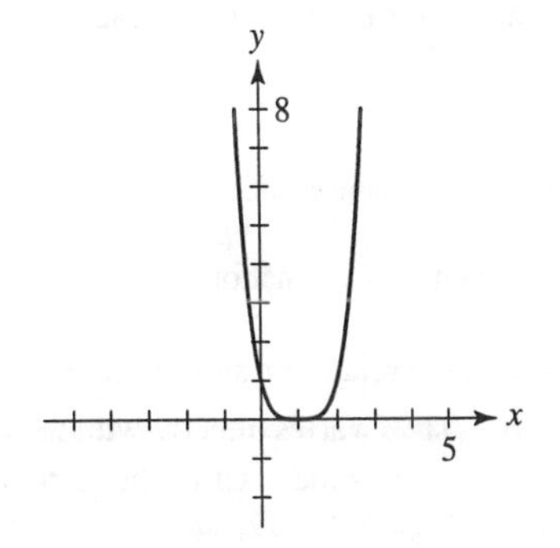

b. $y = [-(x-1)]^4$
$y = (x-1)^4$

29. a.

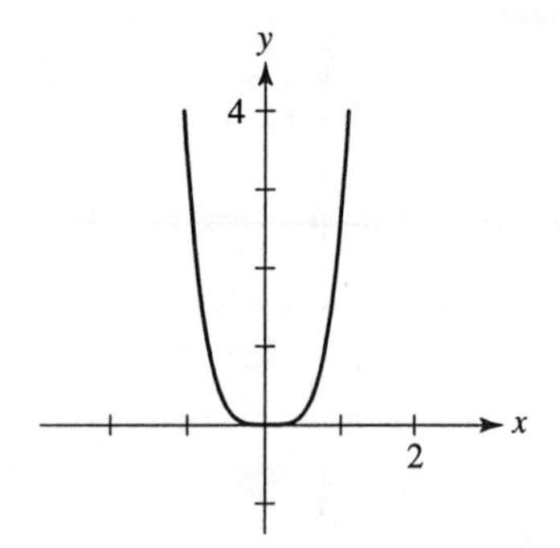

b. $y = 3x^4$

31. a.

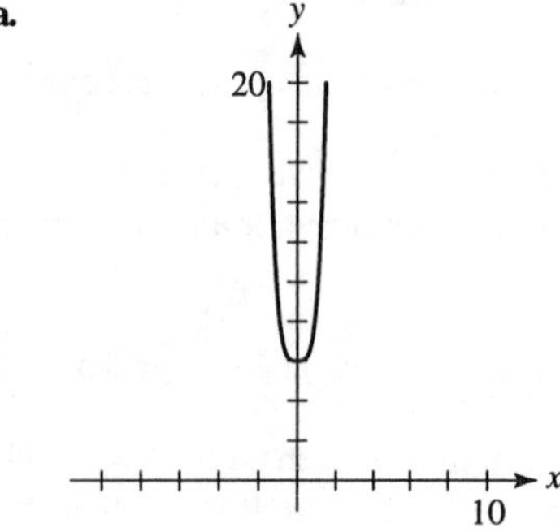

b. $y = 3(x^4 + 2)$

33. The ratio $\frac{y}{x}$ is equal to 15 for each row, so $y = 15x$.

35. The ratio $\frac{y}{x^3}$ is not the same for each row. The data does not fit a function $y = ax^3$.

37. The equation relating r and M has the form $M = ar^3$, so M varies directly with the cube of r. If M_0 represents the mass of the earth, then $M = M_0$ when $r = 6600$, so

$$M_0 = a(6600)^3$$

$$a = \frac{M_0}{6600^3}$$

This means that

$$M = \left(\frac{M_0}{6600^3}\right)r^3.$$

You need to find the value of r when $M = 0.11M_0$:

$$0.11M_0 = \left(\frac{M_0}{6600^3}\right)r^3$$

$$0.11 = \left(\frac{1}{6600^3}\right)r^3$$

$$r^3 = (6600)^3(0.11)$$

$$r = 6600\sqrt[3]{0.11} \cong 3162.34 \text{ kilometers}$$

Since the radius of Mars is 3398 kilometers, the satellite would need to be about 236 kilometers under the surface of Mars to have the required period of revolution. Tell your colleagues that their plans cannot be carried out without some alterations.

Section 7-2

1. a. $x^2 + x - 6 = 0$
$(x + 3)(x - 2) = 0$
$x = -3, 2$

b. $x^3 - 5x^2 + 6x = 0$
$x(x - 2)(x - 3) = 0$
$x = 0, 2, 3$

c. $2x^3 + 4x^2 - 2x - 4 = 0$
$x^3 + 2x^2 - x - 2 = 0$
$x^2(x + 2) - (x + 2) = 0$
$(x^2 - 1)(x + 2) = 0$
$(x + 1)(x - 1)(x + 2) = 0$
$x = \pm 1, -2$

d. $x^4 - 10x^2 + 9 = 0$
$(x^2 - 1)(x^2 - 9) = 0$
$(x + 1)(x - 1)(x + 3)(x - 3) = 0$
$x = \pm 1, \pm 3$

3. a. $x^2 - 8x + 16 = 0$
$(x - 4)^2 = 0$
$x = 4$

b. $x^5 - 9x^3 = 0$
$x^3(x + 3)(x - 3) = 0$
$x = 0, \pm 3$

c. $x^3 - x^2 + x - 1 = 0$
$x^2(x - 1) + (x - 1) = 0$
$(x^2 + 1)(x - 1) = 0$
$(x + i)(x - i)(x - 1) = 0$
$x = 1, \pm i$

d. $x^4 - 4x^2 + 4 = 0$
$(x^2 - 2)^2 = 0$
$\left(x+\sqrt{2}\right)^2\left(x-\sqrt{2}\right)^2 = 0$
$x = \pm\sqrt{2}$

5. The graph shows a zero at about (–1, 0). The zero occurs at exactly $x = -1$ since $f(-1) = (-1)^3 + 2(-1)^2 - 5(-1) - 6 = 0$. Therefore $x + 1$ is a factor of $f(x)$. Use synthetic division to divide $f(x)$ by $x + 1$.

–1	1	2	–5	–6
		–1	–1	6
	1	1	–6	0

$f(x) = (x+1)(x^2 + x - 6)$
$f(x) = (x+1)(x+3)(x-2)$
The other zeros are $x = -3$ and $x = 2$.

7. The graph shows a zero at about (–2, 0). The zero occurs at exactly $x = -2$ since $f(-2) = -(-2)^3 - (-2)^2 + 3(-2) + 2 = 0$. Therefore $x + 2$ is a factor of $f(x)$. Use synthetic division to divide $f(x)$ by $x + 2$.

–2	–1	–1	3	2
		2	–2	–2
	–1	1	1	0

$f(x) = (x+2)(-x^2 + x + 1)$
The other zeros are the solutions to
$-x^2 + x + 1 = 0$
$$x = \frac{-1 \pm \sqrt{1^2 - 4(-1)(1)}}{2(-1)} = \frac{-1 \pm \sqrt{5}}{-2}$$
$x \cong -0.62,\ 1.62$

9. a.

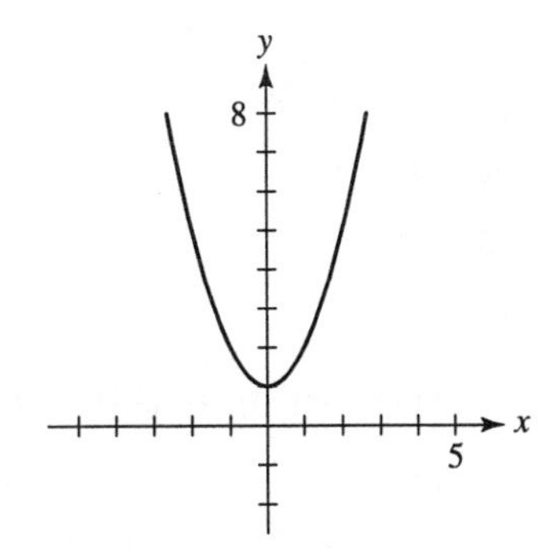

b. Each real zero must correspond to an x-intercept on the graph. Since the graph has no x-intercepts, the function has no real zeros.

c. $x^2 + 1 = 0$
$x^2 = -1$
$x = \pm i$

11. In Exercise 3a, the degree of $Q(x)$ is 2. There is one zero of multiplicity 2. If zeros are counted according to multiplicity, the number of zeros is equal to the degree of $Q(x)$.

In Exercise 3b, the degree of $Q(x)$ is 5. There is one zero of multiplicity 3 and two zeros of multiplicity 1 each. If zeros are counted according to multiplicity, the number of zeros is equal to the degree of $Q(x)$.

In Exercise 3c, the degree of $Q(x)$ is 3. There are three zeros of multiplicity 1 each. If zeros are counted according to multiplicity, the number of zeros is equal to the degree of $Q(x)$.

In Exercise 3d, the degree of $Q(x)$ is 4. There are two zeros of multiplicity 2 each. If zeros are counted according to multiplicity, the number of zeros is equal to the degree of $Q(x)$.

13. $t^4 + t^3 - 42t^2 = 0$
$t^2(t^2 + t - 42) = 0$
$t^2(t+7)(t-6) = 0$
$t = 0, -7, 6$

15. $v^3 - 3v^2 - 2v + 6 = 0$
$v^2(v-3) - 2(v-3) = 0$
$(v^2 - 2)(v-3) = 0$
$\left(v+\sqrt{2}\right)\left(v-\sqrt{2}\right)(v-3) = 0$
$v = 3, \pm\sqrt{2}$

17. a. $g(-1) = (-1)^7 + (-1)^5 - 2(-1)^3 = 0$, so -1 is a zero of g.

b. $g(x) = x^3(x^4 + x^2 - 2)$
$g(x) = x^3(x^2-1)(x^2+2)$
$g(x) = x^3(x+1)(x-1)(x^2+2)$
The zeros are $x = 0, \pm 1, \pm\sqrt{2}i$.

19. a. $g(-1) = 2(-1)^4 + 5(-1)^2 + 2 = 9$, so -1 is not a zero of g

b. $g(x) = (2x^2+1)(x^2+2)$, so the zeros are
$x = \pm\sqrt{\frac{1}{2}}i, \pm\sqrt{2}i.$

21. $b = 4$, multiplicity 3
$b = -\pi$, multiplicity 5

23. $h(r) = [(r+2)(r-4)]^3[(r+5)(r-3)]^2$
$h(r) = (r+2)^3(r-4)^3(r+5)^2(r-3)^2$
$r = -2$, multiplicity 3
$r = 4$, multiplicity 3
$r = -5$, multiplicity 2
$r = 3$, multiplicity 2

25. $z = x^4(x^2-2) - 4(x^2-2)$
$z = (x^4-4)(x^2-2)$
$z = (x^2+2)(x^2-2)(x^2-2)$
$z = (x^2+2)\left(x+\sqrt{2}\right)^2\left(x-\sqrt{2}\right)^2$
$z = \pm\sqrt{2}$, multiplicity 2 each
$z = \pm\sqrt{2}i$, multiplicity 1 each

27. The zeros are the solutions to
$2x^2 + 4x + 10 = 0$
$$x = \frac{-4 + \sqrt{4^2 - 4(2)(10)}}{2(2)} = -1 \pm 2i$$
There are two zeros of multiplicity 1 each, so the number of zeros is equal to the degree of the function.

29. The zeros are the solutions to
$r^5 + 6r^3 + 9r = 0$
$r(r^4 + 6r^2 + 9) = 0$
$r(r^2+3)^2 = 0$
$r = 0, \pm\sqrt{3}i$
There is one zero of multiplicity 1 and two zeros of multiplicity 2 each, so the number of zeros is equal to the degree of the function.

31. a. The degree of the polynomial is 3, and we have found only one solution. The number of solutions counted according to multiplicity must be 3.

b. Using synthetic division:

12	1	–34	408	–1728
		12	–264	1728
	1	–22	144	0

Therefore $P(x) = (x-12)(x^2 - 22x + 144)$. Any other real solutions must solve $x^2 - 22x + 144 = 0$, but the only solutions to this equation are $x = \frac{22 \pm \sqrt{92}i}{2}$, which are not real.

33. a. $P(x) = \left(x - \frac{1}{2}\right)^2(x^2-3)$
$P(x) = \left(x^2 - x + \frac{1}{4}\right)(x^2-3)$
$P(x) = x^4 - x^3 - \frac{11}{4}x^2 + 3x - \frac{3}{4}$

b.

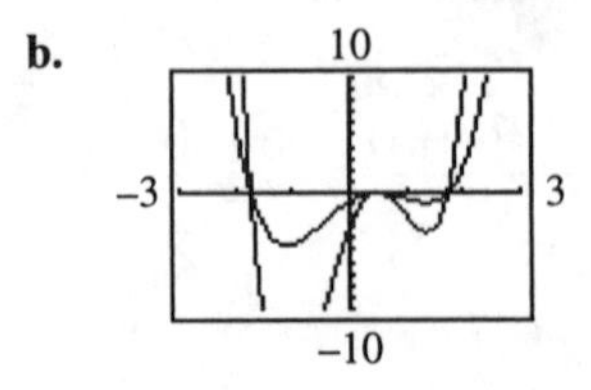

The two graphs have the same x-intercepts.

c. $P(x) = 0$ if and only if $4P(x) = 0$.

d. If the zeros of Q are the same as those of P with the same multiplicities, then Q also has factors of
$\left(x - \frac{1}{2}\right)^2$, $\left(x+\sqrt{3}\right)$, and $\left(x-\sqrt{3}\right)$. Since Q

has no other zeros, it can have no other linear factors. Thus Q must be a constant multiple of $\left(x-\frac{1}{2}\right)^2\left(x+\sqrt{3}\right)\left(x-\sqrt{3}\right)$.

35. A polynomial function having the given zeros with the given multiplicities is
$P(x)=(x-3)^2\left(x-\frac{1}{2}\right)$.
The function
$2P(x)=(x-3)^2\left[2\left(x-\frac{1}{2}\right)\right]=(x-3)^2(2x-1)$
has integer coefficients.

37. A polynomial function having the given zeros with the given multiplicities is
$P(x)=(x+5)\left(x+\frac{3}{2}\right)^2$.
The function
$4P(x)=(x+5)\left[2\left(x+\frac{3}{2}\right)\right]^2=(x+5)(2x+3)^2$
has integer coefficients.

39. A polynomial function having the given zeros with the given multiplicities is
$P(x)=\left(x+\sqrt{2}\right)^2\left(x-\sqrt{2}\right)^2$
$=\left[\left(x+\sqrt{2}\right)\left(x-\sqrt{2}\right)\right]^2=(x^2-2)^2$.

41. Possible rational zeros are $\pm 1, 2, 4, 5, 10, 20$. The graph indicates zeros at or near the possible rational zeros $x=-4$, 1, and 4. Synthetic division shows that 4 is a zero:

4	1	−1	−17	20
		4	12	−20
	1	3	−5	0

Thus, $P(x)=(x-4)(x^2+3x-5)$.
The other zeros are the solutions to $x^2+3x-5=0$, which are
$x=\dfrac{-3\pm\sqrt{29}}{2}\cong -4.19,\ 1.19$.

43. Possible rational zeros are
$x=\pm\frac{1}{6}, \frac{1}{3}, \frac{1}{2}, 1, \frac{7}{6}, \frac{3}{2}, \frac{7}{3}, 3, \frac{7}{2}, 7, \frac{21}{2}, 21$.
A graph indicates zeros at or near the possible rational zeros $x=-\frac{7}{3}, \frac{1}{2}$, and 3. Synthetic division shows that all three are zeros. Since the degree of s is 3, s can have no other zeros.

45. Possible rational zeros are $x=\pm 1, 2, 4, 16$. A graph shows a zero at or near the possible rational zero $x=4$. Synthetic division confirms that 4 is a zero:

4	1	−2	−12	16
		4	8	−16
	1	2	−4	0

Thus, $f(x)=(x-4)(x^2+2x-4)$. The remaining zeros are the solutions to $x^2+2x-4=0$, which are
$\dfrac{-2\pm\sqrt{20}}{2}\cong -3.24,\ 1.24$.

47. Possible rational zeros are $w=\pm 1, 2, 3, 4, 6, 8, 12, 24$. A graph shows a zero at or near the possible rational zero $x=-3$. Synthetic division confirms that −3 is a zero:

−3	1	2	0	1	−24
		−3	3	−9	24
	1	−1	3	−8	0

Thus, $P(w)=(w+3)(w^3-w^2+3w-8)$.
The remaining zeros are the solutions to $w^3-w^2+3w-8=0$, which are inconvenient to obtain analytically. A graphical solution shows the other real zero to be $w\cong 1.80$.

Section 7-3

1. a. The real zeros are
$x=-4$, multiplicity 1
$x=-1$, multiplicity 1
$x=3$, multiplicity 3
The x-intercepts are (−4, 0), (−1, 0), (3, 0).

b. $P(x)=x(x^2+4)$
The only real zero is $x=0$, multiplicity 1.
The only x-intercept is (0, 0).

c. The real zeros are
$x=0$, multiplicity 2

$x = \pm 2\sqrt{7}$, multiplicity 1 each
The x-intercepts are (0, 0), $\left(-2\sqrt{7},\ 0\right)$, $\left(2\sqrt{7},\ 0\right)$.

d. $P(x) = (x^2+4)(x+2)(x-2)$
The real zeros are $x = \pm 2$, multiplicity 1 each. The x-intercepts are (–2, 0), (2, 0).

3. a. $P(x) = (x+2)^2(x-1)^2(x-3)^2$
The x-intercepts are (–2, 0), (1, 0), (3, 0).

b. $P(x) = x\left(x+\sqrt{6}\right)\left(x-\sqrt{6}\right) = x^3 - 6x$
The x-intercepts are (0, 0), $\left(\pm\sqrt{6},\ 0\right)$.

c. $P(x) = \left(x-\frac{3}{2}\right)^2\left(x+\frac{1}{3}\right)\left(x-\frac{1}{3}\right)$ has the required zeros, but does not have integer coefficients.
$36P(x) = (2x-3)^2(3x+1)(3x-1)$
The x-intercepts are
$\left(\frac{3}{2},\ 0\right), \left(-\frac{1}{3},\ 0\right), \left(\frac{1}{3},\ 0\right)$.

d. $P(x)$
$= \left(x+\sqrt{2}\right)^2\left(x-\sqrt{2}\right)^2\left(x+\sqrt{3}\right)\left(x-\sqrt{3}\right)$
$= (x^2-2)^2(x^2-3)$
The x-intercepts are $\left(\pm\sqrt{2},\ 0\right)$, $\left(\pm\sqrt{3},\ 0\right)$.

5. $f(x) = (x+4)^3(x-1)^2(x-3)$

7. $f(x) = x^2(x+4)^2(x-4)^2$

9. The function has a local minimum at (–2, –16) and a local maximum at (2, 16). It is decreasing on (–∞, –2) and (2, ∞), and increasing on (–2, 2).

11. The function has a local maximum at about (–1.05, 13.03), and a local minimum at about (1.05, –1.03). It is increasing over (–∞, –1.05) and (1.05, ∞) and decreasing over (–1.05, 1.05).

13. a. From Exercise 12, the volume of the container is
$V(r) = \pi r^2(108 - 2\pi r) = -2\pi^2 r^3 + 108\pi r^2$
cubic inches. Since both the radius r and the length $108 - 2\pi r$ must be positive, we must have $r > 0$ and $r < \frac{108}{2\pi} \cong 17.19$. Thus, the domain of V is (0, 17.19), the interval between the r-intercepts. The graph indicates that V has a local maximum at about (11.46, 14851.07), so that the optimal container has a radius of about 11.46 inches, and a length of
$108 - 2\pi(11.46) \cong 36.00$ inches.

b. The amount of popcorn Alice can send is
$(14{,}851.07 \text{ cubic inches})\cdot$
$\left(\frac{1 \text{ cubic foot}}{12^3 \text{ cubic inches}}\right)\left(\frac{1.3 \text{ ounces}}{1 \text{ cubic foot}}\right)$
$\cong 11.17$ ounces.

15. If the end is a square x inches on a side, then the girth is $4x$ inches, so that the length is $108 - 4x$ inches. The volume is then
$V(x)$ = (area of cross section)(length)
$= x^2(108-4x) = -4x^3 + 108x^2$ cubic inches.
Since both the square side x and the length $108 - 4x$ must be positive, we must have $x > 0$ and $x < 27$. The domain of V is therefore (0, 27), the interval between the x-intercepts. The graph indicates that V has a local maximum at about (18.00, 11664.00). The amount of popcorn Alice can send is $(11{,}664)\left(\frac{1}{12^3}\right)(1.3) \cong 8.76$ ounces.
This is less than she could send in a cylindrical package.

17. When $|x|$ is large, the absolute value of $-100x^3$ is small in comparison with that of x^4, so the sign of the sum agrees with that of x^4.

19. The degree is odd, and the leading coefficient is positive, so the left tail of the graph points down, and the right tail points up.

21. The degree is odd, and the leading coefficient is negative, so the left tail of the graph points up, and the right tail points down.

23. The degree is even and the leading coefficient is negative, so both tails of the graph point down. The equation matches the graph in Figure 7-19d.

25. The degree is even and the leading coefficient is positive, so both tails of the graph point up. The equation matches the graph in Figure 7-19b.

27. $f(-x) = 4-(-x)^2 = 4-x^2 = f(x)$, so the graph is symmetric about the y-axis.

29. $f(-x) = 3(-x) = -3x = -f(x)$, so the graph is symmetric about the origin.

31. $f(-x) = 2(-x)^6 - (-x)^4 + 3$
$= 2x^6 - x^4 + 3 = f(x)$, so the graph is symmetric about the y-axis.

33. (Sample response) According to the test described in Section 6-3, a graph is symmetric about the y-axis if replacing x by $-x$ produces an equivalent equation. If y is a function of x, this amounts to saying that $f(-x) = f(x)$. The test in Section 6-3 is more general, because it can be applied even if y is not a function of x.

35. $y = (x-1)(x-4)$

37. $y = (x-0.1)(x-1)(x-10)(x-100)$

39. $y = (x+5)(x+3)(x+1)(x-2)(x-4)(x-6)$

41. $y = x^3(x+5)(x+2)(x-4)$

43. a. By the quadratic formula,

$$x^2 = \frac{6 \pm \sqrt{20}}{2} = 3 \pm \sqrt{5}\text{, so}$$

$$x = \pm\sqrt{3 \pm \sqrt{5}} \cong \pm 0.87,\ \pm 2.29$$

b. The function has a local maximum at (0, 4), and absolute minima at about (±1.73, –5.00).

c. The degree is even and the leading coefficient is positive, so both tails of the graph point up.

d. Answers will vary. The window [–5, 5] by [–10, 10] works well.

e. The function is decreasing over (–∞, –1.73) and (0, 1.73), and increasing over (–1.73, 0) and (1.73, ∞).

45. a. The graph indicates zeros at $x \cong 0.58, 4.75, 17.67$.

b. The function has a local minimum at about (2.52, –65.38), and a local maximum at about (12.81, 478.57).

c. The degree is odd and the leading coefficient is negative, so the left tail of the graph points up, and the right tail points down.

d. Answers will vary. The window [–10, 20] by [–500, 500] works well.

e. The function is decreasing over (–∞, 2.52) and (12.81, ∞), and increasing over (2.52, 12.81).

47. a. The graph indicates zeros at $x \cong -3.19, 2.19$.

b. The function has an absolute maximum at about (–1.85, 40.29).

c. The degree is even and the leading coefficient is negative, so both tails of the graph point down.

d. Answers will vary. The window [–5, 5] by [–50, 50] works well.

e. The function is increasing over (–∞, –1.85), and decreasing over (–1.85, ∞).

49. $U(-x) = (-x)^4 - 1 = x^4 - 1 = U(x)$, so U is even.

51. $W(-x) = (-x-1)^4 = (x+1)^4$, which is neither $W(x)$ nor $-W(x)$, so W is neither even nor odd.

53. a. Over [–2.5, –2], the average rate of change is $\frac{y(-2)-y(-2.5)}{(-2)-(-2.5)} \cong \frac{-0.67-(-2.29)}{0.5}$ $= 3.24$. Over [3, 3.5], the average rate of change is $\frac{y(3.5)-y(3)}{3.5-3} \cong \frac{3.79-0}{0.5} = 7.58$. The average rate of change is roughly twice as great over [3, 3.5].

b. The secant line over [3, 3.5] is roughly twice as steep as the secant line over [–2.5, –2].

c. Over [–1.5, –0.5], the average rate of change is $\frac{y(-0.5)-y(-1.5)}{-0.5-(-1.5)} \cong \frac{1.46-3.38}{1} = -1.92.$
Over [–0.5, 0.5], the average rate of change is $\frac{y(0.5)-y(-0.5)}{0.5-(-0.5)} \cong \frac{-1.46-1.46}{1} = -2.92.$
The average rate of change has a larger absolute value over [–0.5, 0.5].

d. The secant line over [–0.5, 0.5] is steeper.

55. The domain is (–∞, ∞). Since the degree is odd, the range is also (–∞, ∞).

57. The domain is (–∞, ∞). The graph indicates a range of about (–∞, 0].

59. The domain is (–∞, ∞). The graph indicates a range of about [64.00, ∞).

61. (Sample response) In this extremely large viewing window (that is, *from a distance*), details such as turning points are too small to see, and the two graphs appear identical. In particular, their end behavior is the same.

Section 7-4

1. a.

x	$f(x)$	first differences	second differences	third differences
1	3			
		7.125		
1.5	10.125		6.75	
		13.875		2.25
2	24		9	
		22.875		2.25
2.5	46.875		11.25	
		34.125		2.25
3	81		13.5	
		47.625		
3.5	128.625			

b.

x	$f(x)$	first differences	second differences	third differences
–1	$-a$			
		a		
0	0		0	
		a		$6a$
1	a		$6a$	
		$7a$		$6a$
2	$8a$		$12a$	
		$19a$		$6a$
3	$27a$		$18a$	
		$37a$		
4	$64a$			

x	$f(x)$	first differences	second differences	third differences
x_0	$ax_0{}^3$			
		$3ax_0{}^2h+3x_0h^2+h^3$		
x_0+h	$ax_0{}^3+3ax_0{}^2h+3x_0h^2+h^3$		$6ax_0h^2$ $+6h^3$	
		$3ax_0{}^2h+9x_0h^2+7h^3$		$6h^3$
x_0+2h	$ax_0{}^3+6ax_0{}^2h+12x_0h^2+8h^3$		$6ax_0h^2$ $+12h^3$	
		$3ax_0{}^2h+15x_0h^2+19h^3$		$6h^3$
x_0+3h	$ax_0{}^3+9ax_0{}^2h+27x_0h^2+27h^3$		$6ax_0h^2$ $+18h^3$	
		$3ax_0{}^2h+21x_0h^2+37h^3$		$6h^3$
x_0+4h	$ax_0{}^3+12ax_0{}^2h+48x_0h^2+64h^3$		$6ax_0h^2$ $+24h^3$	
		$3ax_0{}^2h+27x_0h^2+61h^3$		
x_0+5h	$ax_0{}^3+15ax_0{}^2h+75x_0h^2+125h^3$			

3.

x	y	first differences	second differences	third differences
2	0			
		15		
3	15		18	
		33		6
4	48		24	
		57		6
5	105		30	
		87		6
6	192		36	
		123		6
7	315		42	
		165		
8	480			

The table fits a third-degree polynomial.

5.

x	y	1st diff.	2nd diff.
–3	–1		
		–2	
–1	–3		2
		0	
1	–3		2
		2	
3	–1		2
		4	
5	3		2
		6	
7	9		2
		8	
9	17		

The table fits a second-degree polynomial.

7. (Sample response) A table containing four data points with equally spaced x-values has only one third difference. Thus the third differences are constant by default, so the table must fit a cubic polynomial function.

9. Any ordered pairs $(x_0, f(x_0))$ will work. Some possibilities are (4, 26), (5, 61), (6, 120), (7, 209), (8, 334), and (9, 501).

11. The data fits a function
$f(x) = ax^3 + bx^2 + cx + d$.
$f(0) = 1$, so $a(0)^3 + b(0)^2 + c(0) + d = 1$
$f(1) = 3$, so $a(1)^3 + b(1)^2 + c(1) + d = 3$
$f(2) = 1$, so $a(2)^3 + b(2)^2 + c(2) + d = 1$
$f(3) = 3$, so $a(3)^3 + b(3)^2 + c(3) + d = 3$

$$\begin{aligned} d &= 1 \\ a + b + c + d &= 3 \\ 8a + 4b + 2c + d &= 1 \\ 27a + 9b + 3c + d &= 3 \end{aligned}$$

Substitute 1 for d in the last three equations.

$$\begin{aligned} a + b + c &= 2 \\ 8a + 4b + 2c &= 0 \\ 27a + 9b + 3c &= 2 \end{aligned}$$

Divide the second equation by 2 and subtract the first equation from it.

$$\begin{array}{r} 4a + 2b + c = 0 \\ \underline{a + b + c = 2} \\ 3a + b \quad = -2 \end{array}$$

Multiply the first equation by 3 and subtract it from the third equation.

$$\begin{array}{r} 27a + 9b + 3c = 2 \\ \underline{3a + 3b + 3c = 6} \\ 24a + 6b \quad = -4 \end{array}$$

In the resulting 2×2 system, multiply the first equation by 6 and subtract it from the second.

$$\begin{array}{r} 24a + 6b = -4 \\ \underline{18a + 6b = -12} \\ 6a \quad = 8 \end{array}$$

$a = \frac{4}{3}$

Substitute $\frac{4}{3}$ for a in the equation $3a + b = -2$ to obtain $b = -6$. Substitute $\frac{4}{3}$ for a and -6 for b in the equation $a + b + c = 2$ to obtain $c = \frac{20}{3}$. The function is $f(x) = \frac{4}{3}x^3 - 6x^2 + \frac{20}{3}x + 1$.

13. The data fits a function
$f(x) = ax^4 + bx^3 + cx^2 + dx + e$.
$f(0) = 1$, so $a(0)^4 + b(0)^3 + c(0)^2 + d(0) + e = 1$
$f(1) = -3$, so $a(1)^4 + b(1)^3 + c(1)^2 + d(1) + e = -3$
$f(2) = -5$, so
$a(2)^4 + b(2)^3 + c(2)^2 + d(2) + e = -5$
$f(3) = 19$, so
$a(3)^4 + b(3)^3 + c(3)^2 + d(3) + e = 19$
$f(4) = 117$, so
$a(4)^4 + b(4)^3 + c(4)^2 + d(4) + e = 117$

$$\begin{aligned} e &= 1 \\ a + b + c + d + e &= -3 \\ 16a + 8b + 4c + 2d + e &= -5 \\ 81a + 27b + 9c + 3d + e &= 19 \\ 256a + 64b + 16c + 4d + e &= 117 \end{aligned}$$

Subtract the first equation from the second to obtain $a + b + c + d = -4$
Subtract the second equation from the third to obtain $15a + 7b + 3c + d = -2$.
Subtract the third equation from the fourth to obtain $65a + 19b + 5c + d = 24$.
Subtract the fourth equation from the fifth to obtain $175a + 37b + 7c + d = 98$.

In the resulting 4×4 system:
Subtract the first equation from the second to obtain $14a + 6b + 2c = 2$
Subtract the second equation from the third to obtain $50a + 12b + 2c = 26$.
Subtract the third equation from the fourth to obtain $110a + 18b + 2c = 74$.

In the resulting 3×3 system:
Subtract the first equation from the second to obtain $36a + 6b = 24$.
Subtract the second equation from the third to obtain $60a + 6b = 48$.

In the resulting 2×2 system:
Subtract the first equation from the second to obtain $a = 1$.

Substitute 1 for a in the equation $36a + 6b = 24$ to obtain $b = -2$. Substitute 1 for a and -2 for b in the equation $14a + 6b + 2c = 2$ to obtain $c = 0$. Substitute 1 for a, -2 for b, and 0 for c in the equation $a + b + c + d = -4$ to obtain $d = -3$.

The function is $f(x) = x^4 - 2x^3 - 3x + 1$.

15. A graph shows three real zeros, of which the largest is in (1, 2). To find it:

interval	width of interval	value of P at left endpoint	value of P at right endpoint	midpoint	value of P at midpoint
(1, 2)	1	–1	10	1.5	1.13
(1, 1.5)	0.5	–1	1.13	1.25	–0.64
(1.25, 1.5)	0.25	–0.64	1.13	1.375	0.05
(1.25, 1.375)	0.125	–0.64	0.05	1.3125	–0.34
(1.3125, 1.375)	0.0625	–0.34	0.05	1.34375	–0.16
(1.34375, 1.375)	0.03125	–0.16	0.05	1.359375	–0.06
(1.359375, 1.375)	0.015625	–0.06	0.05	1.3671875	–0.01
(1.3671875, 1.375)	0.0078125				

The largest zero is $x \cong 1.37$, with an error of less than 0.01.

17. A graph shows three real zeros, of which the largest is in (1, 2). To find it:

interval	width of interval	value of P at left endpoint	value of P at right endpoint	midpoint	value of P at midpoint
(1, 2)	1	–1	39	1.5	–2.20
(1.5, 2)	0.5	–2.20	39	1.75	6.15
(1.5, 1.75)	0.25	–2.20	6.15	1.625	0.21
(1.5, 1.625)	0.125	–2.20	0.21	1.5625	–1.32
(1.5625, 1.625)	0.0625	–1.32	0.21	1.59375	–0.65
(1.59375, 1.625)	0.03125	–0.65	0.21	1.609375	–0.25
(1.609375, 1.625)	0.015625	–0.25	0.21	1.6171875	–0.02
(1.6172875, 1.625)	0.0078125				

The largest zero is $x \cong 1.62$, with an error of less than 0.01.

19. A graph shows two zeros near $t = 1$ and $t = 3$. Synthetic division confirms that both are zeros.

$$\begin{array}{r|rrrrr} 1 & 1 & 0 & 0 & -40 & 39 \\ & & 1 & 1 & 1 & -39 \\ \hline & 1 & 1 & 1 & -39 & 0 \end{array}$$

$$\begin{array}{r|rrrr} 3 & 1 & 1 & 1 & -39 \\ & & 3 & 12 & 39 \\ \hline & 1 & 4 & 13 & 0 \end{array}$$

Thus $P(t) = (t-1)(t-3)(t^2 + 4t + 13)$. The remaining zeros are the solutions of $t^2 + 4t + 13 = 0$, which are $t = -2 \pm 3i$, and are therefore not real.

21. A graph shows three real zeros. The only possible rational zeros are $x = \pm 1$, and the graph makes it clear that neither is a zero. Since the polynomial cannot be easily factored, we must settle for estimates of the zeros. Using either the graph or the bisection method, $x \cong -1.84,\ 1.20,\ 1.52$.

23.

x	y	1st diff.	2nd diff.	3rd diff.	4th diff.	5th diff.	6th diff.
1	0						
		9					
2	9		–5				
		4		2			
3	13		–3		0		
		1		2		0	
4	14		–1		0		–1
		0		2		–1	
5	14		1		–1		
		1		1			
6	15		2				
		3					
7	18						

The table fits a sixth-degree polynomial.

25.

x	y	1st diff.	2nd diff.
–0.9	–3.12		
		0.84	
–0.8	–2.28		–0.04
		0.80	
–0.7	–1.48		–0.04
		0.76	
–0.6	–0.72		–0.04
		0.72	
–0.5	0		–0.04
		0.68	
–0.4	0.68		–0.04
		0.64	
–0.3	1.32		–0.04
		0.60	
–0.2	1.92		

The table fits a second-degree polynomial.

27. The data fits a function

$f(x) = ax^3 + bx^2 + cx + d.$

$f(0) = 0$, so $a(0)^3 + b(0)^2 + c(0) + d = 0$

$f(1) = 10$, so $a(1)^3 + b(1)^2 + c(1) + d = 10$

$f(2) = 4$, so $a(2)^3 + b(2)^2 + c(2) + d = 4$

$f(3) = 7$, so $a(3)^3 + b(3)^2 + c(3) + d = 7$

$$\begin{aligned} d &= 0 \\ a + b + c + d &= 10 \\ 8a + 4b + 2c + d &= 4 \\ 27a + 9b + 3c + d &= 7 \end{aligned}$$

Subtract the first equation from the second to obtain $a + b + c = 10$.

Subtract the second equation from the third to obtain $7a + 3b + c = -6$.

Subtract the third equation from the fourth to obtain $19a + 5b + c = 3$.

In the resulting 3×3 system, subtract the first equation from the second to obtain $6a + 2b = -16$.

Subtract the second equation from the third to obtain $12a + 2b = 9$.

In the resulting 2×2 system, subtract the first equation from the second to obtain $a = \frac{25}{6}$.

Substitute $\frac{25}{6}$ for a in the equation $6a + 2b = -16$ to obtain $b = -\frac{41}{2}$.

Substitute $\frac{25}{6}$ for a and $-\frac{41}{2}$ for b in the equation $7a + 3b + c = -6$ to obtain $c = \frac{79}{3}$.

The equation is $f(x) = \frac{25}{6}x^3 - \frac{41}{2}x^2 + \frac{79}{3}x$.

29. The data fits a function

$f(x) = ax^4 + bx^3 + cx^2 + dx + e.$

$f(-1) = 25$, so

$a(-1)^4 + b(-1)^3 + c(-1)^2 + d(-1) + e = 25$

$f(0) = 18$, so

$a(0)^4 + b(0)^3 + c(0)^2 + d(0) + e = 18$

$f(1) = 12$, so

$a(1)^4 + b(1)^3 + c(1)^2 + d(1) + e = 12$

$f(2) = 7$, so

$a(2)^4 + b(2)^3 + c(2)^2 + d(2) + e = 7$

$f(3) = 3$, so

$a(3)^4 + b(3)^3 + c(3)^2 + d(3) + e = 3$

$$\begin{aligned} a - b + c - d + e &= 25 \\ e &= 18 \\ a + b + c + d + e &= 12 \\ 16a + 8b + 4c + 2d + e &= 7 \\ 81a + 27b + 9c + 3d + e &= 3 \end{aligned}$$

Subtract the first equation from the second to obtain $-a + b - c + d = -7$.

Subtract the second equation from the third to obtain $a + b + c + d = -6$.

Subtract the third equation from the fourth to obtain $15a + 7b + 3c + d = -5$.

Subtract the fourth equation from the fifth to obtain $65a + 19b + 5c + d = -4$.

In the resulting 4×4 system:

Subtract the first equation from the second to obtain $2a + 2c = 1$.

Subtract the second equation from the third to obtain $14a + 6b + 2c = 1$.

Subtract the third equation from the fourth to obtain $50a + 12b + 2c = 1$.

In the resulting 3×3 system:
Subtract the first equation from the second to obtain $12a + 6b = 0$.
Subtract the second equation from the third to obtain $36a + 6b = 0$.

In the resulting 2×2 system:
Subtract the first equation from the second to obtain $a = 0$.
Substitute 0 for a in the equation $12a + 6b = 0$ to obtain $b = 0$.
Substitute 0 for a in the equation $2a + 2c = 1$ to obtain $c = \frac{1}{2}$.
Substitute 0 for a, 0 for b, and $\frac{1}{2}$ for c in the equation $a + b + c + d = -6$ to obtain $d = -\frac{13}{2}$.
The function is $f(x) = \frac{1}{2}x^2 - \frac{13}{2}x + 18$.

31. A graph shows four real zeros, of which the largest is in (2, 3). To find it:

interval	width of interval	value of P at left endpoint	value of P at right endpoint	midpoint	value of P at midpoint
(2, 3)	1	–4	31	2.5	5.56
(2, 2.5)	0.5	–4	5.56	2.25	–0.75
(2.25, 2.5)	0.25	–0.75	5.56	2.375	1.97
(2.25, 2.375)	0.125	–0.75	1.97	2.3125	0.51
(2.25, 2.3125)	0.0625	–0.75	0.51	2.28125	–0.14
(2.28125, 2.3125)	0.03125	–0.14	0.51	2.296875	0.18
(2.28125, 2.296875)	0.015625	–0.14	0.18	2.2890625	0.02
(2.28125, 2.2890625)	0.0078125				

The largest zero is $x \cong 2.28$, with an error of less than 0.01.

33. A graph shows one real zero, in (4, 5). To find it:

interval	width of interval	value of P at left endpoint	value of P at right endpoint	midpoint	value of P at midpoint
(4, 5)	1	4.56	–12.75	4.5	–1.88
(4, 4.5)	0.5	4.56	–1.88	4.25	1.80
(4.25, 4.5)	0.25	1.80	–1.88	4.375	0.09
(4.375, 4.5)	0.125	0.09	–1.88	4.4375	–0.86
(4.375, 4.4375)	0.0625	0.09	–0.86	4.40625	–0.38
(4.375, 4.40625)	0.03125	0.09	–0.38	4.390625	–0.14
(4.375, 4.390625)	0.015625	0.09	–0.14	4.3828125	–0.03
(4.375, 4.3828125)	0.0078125				

The zero is $x \cong 4.38$, with an error of less than 0.01.

35. A graph shows only one real zero, in [1, 2]. Using either the graph or the bisection method, $x \cong 1.34$.

37. A graph shows two real zeros, one of which is near 2. Synthetic division confirms that 2 is a zero:

2	3	–4	0	0	–16
		6	4	8	16
	3	2	4	8	0

Thus $P(x) = (x-2)(3x^3 + 2x^2 + 4x + 8)$. The quotient $3x^3 + 2x^2 + 4x + 8$ has only one real zero. Using either the graph or the bisection method, $x \cong -1.27$.

Section 7-5

1. A graph shows three *x*-intercepts, at about (1.20, 0), (2.55, 0), and (4.25, 0). The inequality is true when the graph is above the *x*-axis, that is, in $(1.20,\ 2.55) \cup (4.25,\ \infty)$.

3. A graph shows two *x*-intercepts, at (–1, 0) and (3, 0). The inequality is true when the graph is on or below the *x*-axis. The graph is never below the *x*-axis, and is on it only at the two *x*-intercepts. The solution is $\{-1, 3\}$.

5. (Sample response) According to the new cost function, $C(x)$ represents the cost of producing 100*x* shirts. According to the former cost function, the cost of producing 100*x* shirts is $C(100x) = 3.50(100x) + 612.50 = 350x + 612.50$. Thus the new cost function should be $C(x) = 350x + 612.50$

Similarly, according to the former revenue function, the revenue produced by the sale of 100*x* shirts is

$R(100x) = -0.005(100x)^2 + 16(100x)$
$= -50x^2 + 1600x$

7. **a.** $P(x) = R(x) - C(x)$

$P(x) = (-50x^2 + 1600x)$
$\quad - (0.08x^3 - 6x^2 + 350x + 612.50)$

$P(x) = -0.08x^3 - 44x^2 + 1250x - 612.50$

b. The *x*-intercepts on the graph occur at about (0.50, 0) and (26.60, 0). The graph is above the *x*-axis between the intercepts. The solution is [0.50, 26.60], representing sales levels between 50 and 2660 shirts. Since $P(0.50) > 0$ and $P(26.60) < 0$, your profitable sales levels are from 50 through 2659 shirts.

c. The third *x*-intercept is at about (–577.10, 0) and the graph is above the *x*-axis to the left of this intercept. The solution is (–∞, –577.10] ∪ [0.50, 26.60]. This differs from the solution in part (b) because the negative values of *x* do not make sense in the physical context.

9. First, solve the equation $8x^2 - x^4 = 0$:

$x^2(8 - x^2) = 0$

$x = 0,\ \pm 2\sqrt{2} \cong \pm 2.83$

Next indicate the solutions on a number line:

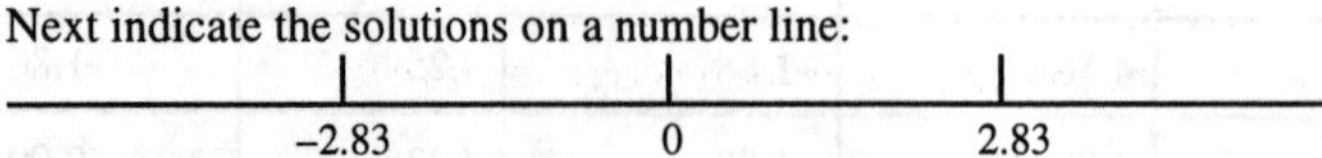

Then choose a test value in each of the intervals (–∞, –2.83), (–2.83, 0), (0, 2.83), (2.83, ∞), and determine whether the inequality is true at each test value:

$f(-3) = -9 < 0$ $\qquad f(1) = 7 > 0$

$f(-1) = 7 > 0$ $\qquad f(3) = -9 < 0$

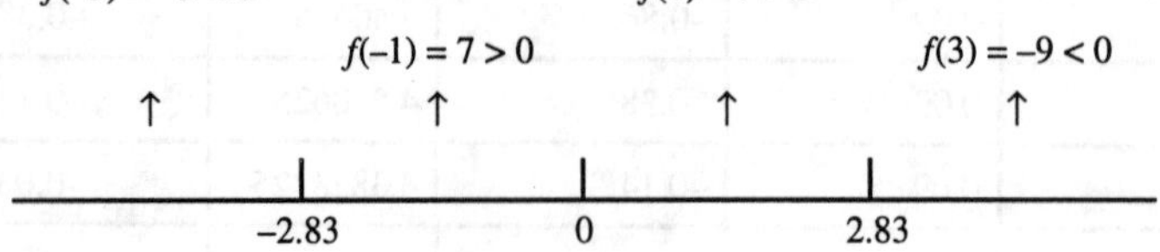

The inequality is true in (–∞, –2.83] ∪ {0} ∪ [2.83, ∞).

11. First, solve the equation $x^5 + x^4 + 2x^3 = 0$:

$x^3(x^2 + x + 2) = 0$

$x^3 = 0$ or $x^2 + x + 2 = 0$

$x = 0$ or $x = \dfrac{-1 \pm \sqrt{7}i}{2}$

The only real solution is $x = 0$. Indicate it on a number line:

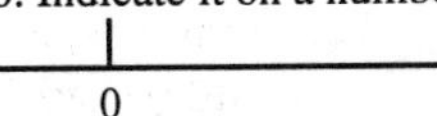

Then choose a test value in each of the intervals $(-\infty, 0)$ and $(0, \infty)$, and determine whether the inequality is true at each test value:

$f(-1) = -2 < 0$ ↑ $\qquad$ $f(1) = 4 > 0$ ↑

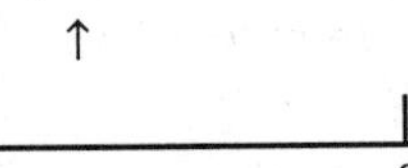
0

The inequality is true in $[0, \infty)$.

13. First, solve $(x + 7)(3x + 4) = 0$, obtaining
$x = -7,\ -\frac{4}{3}$. Next, indicate the solutions on a number line:

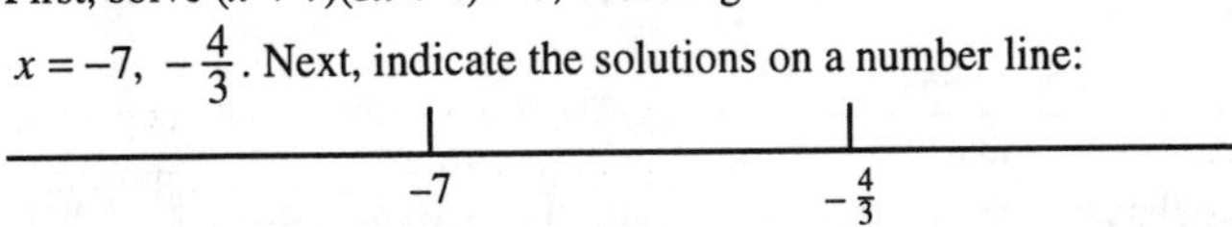

The leading term in the expanded form of $f(x)$ is $3x^2$, so the left tail of its graph points up, and $f(x) > 0$ on $(-\infty, -7)$.
The zeros of each have multiplicity 1, so the sign of $f(x)$ changes at both. The pattern of signs for $f(x)$ is shown below:

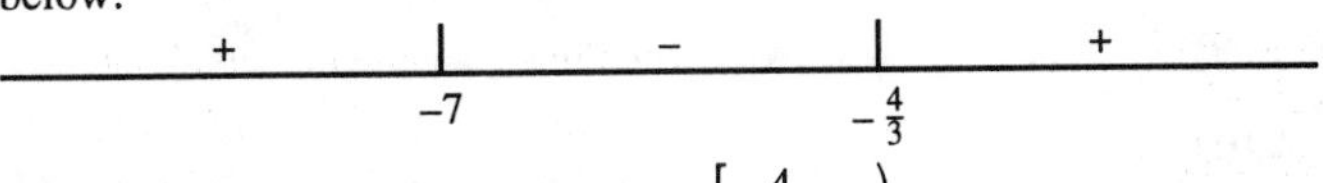

The inequality is true on $(-\infty, -7] \cup \left[-\frac{4}{3}, \infty\right)$.

15. First, solve $(1 - x)(7 - 2x)^2(5 - x)^2 = 0$, obtaining $x = 1,\ \frac{7}{2},\ 5$. Next, indicate the solutions on a number line:

1 $\quad \frac{7}{2} \quad$ 5

The leading term in the expanded form of $f(x)$ is $(-x)(-2x)^2(-x)^2 = -4x^5$, so the left tail of the graph points up, and $f(x) > 0$ on $(-\infty, 1)$.

The zero at $x = 1$ has multiplicity 1, so the sign of $f(x)$ changes there. The other two zeros each have multiplicity 2, so the sign of $f(x)$ does not change at either. The pattern of signs for $f(x)$ is shown below:

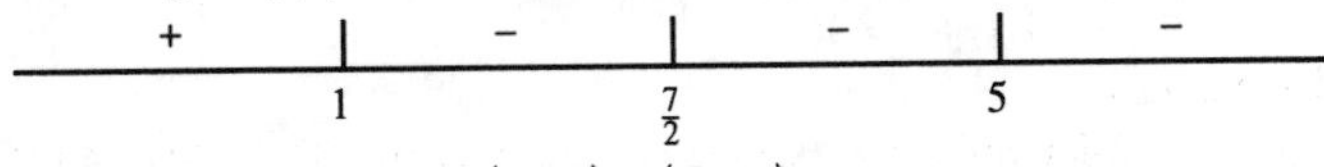

The inequality is true in $\left(1, \frac{7}{2}\right) \cup \left(\frac{7}{2}, 5\right) \cup (5, \infty)$.

17. From Exercise 13 of Section 7-3, the volume of Alice's cylindrical package is $108\pi r^2 - 2\pi^2 r^3$ cubic inches. The amount of popcorn it will hold is

$$\frac{108\pi r^2 - 2\pi^2 r^3 \text{ cubic inches}}{1} \cdot \frac{1 \text{ cubic foot}}{1728 \text{ cubic inches}} \cdot \frac{1.3 \text{ ounces}}{1 \text{ cubic foot}}$$

$\cong 0.26r^2 - 0.015r^3$ ounces. The acceptable radii must satisfy the inequality

$0.26r^2 - 0.015r^3 \geq 10$, or equivalently,

$0.015r^3 - 0.26r^2 + 10 \leq 0$.

First, solve $0.015r^3 - 0.26r^2 + 10 = 0$, using either the graph or the bisection method to obtain $x \cong -5.41,\ 8.88,\ 13.87$. In the context of the problem, we are concerned with only positive values of r, so indicate the positive solutions on a number line, along with the boundary $r = 0$:

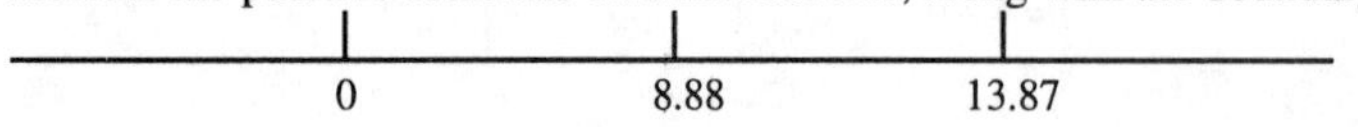

0 8.88 13.87

Choose a test value in each of the intervals

(0, 8.88), (8.88, 13.87), (13.87, ∞) and determine whether the inequality is true at each test value:

$f(1) \cong 9.76 > 0$ $f(14) = 0.2 > 0$

$f(10) = -1 < 0$

↑ ↑ ↑

0 8.88 13.87

The inequality is true in [8.88, 13.87], so the package should have a radius of at least 8.88 inches, but no more than 13.87 inches.

19. A graph shows three x-intercepts, at (–5, 0), (0, 0), and (3, 0). The inequality is true when the graph is below the x-axis, that is, in $(-\infty, -5) \cup (0, 3)$.

21. A graph shows two x-intercepts, at (–2, 0) and (2, 0). The inequality is true when the graph is on or below the x-axis, that is, in [–2, 2].

23. First, solve $x^3 + 2x^2 - 15x = 0$:

$x(x + 5)(x - 3) = 0$

$x = -5, 0, 3$

Indicate these solutions on a number line:

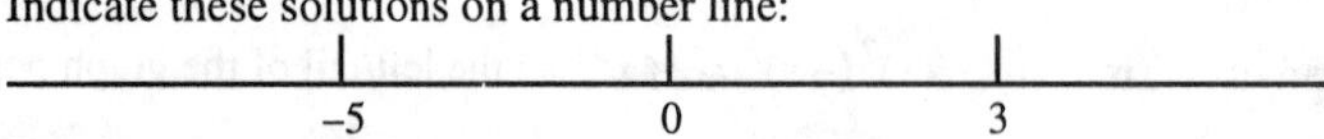

–5 0 3

Choose a test value in each of the intervals

(–∞, –5), (–5, 0), (0, 3), (3, ∞), and determine whether the inequality is true at each test value:

$f(-6) = -54 < 0$ $f(1) = -12 < 0$

$f(-1) = 16 > 0$ $f(4) = 36 > 0$

↑ ↑ ↑ ↑

–5 0 3

The inequality is true in $(-\infty, -5) \cup (0, 3)$.

25. First, solve $x^6 - x^4 - 16x^2 + 16 = 0$:

$x^4(x^2-1) - 16(x^2-1) = 0$

$(x^2-1)(x^4-16) = 0$

$(x-1)(x+1)(x-2)(x+2)(x^2+4) = 0$

$x = \pm 1, 2$

Indicate these solutions on a number line:

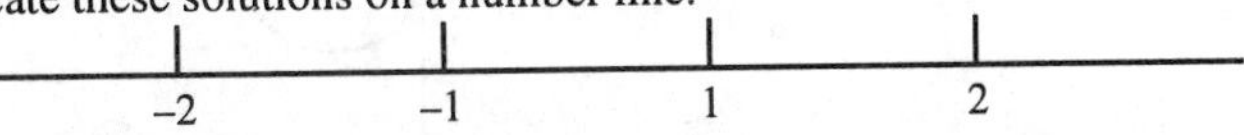

Choose a test value in each of the intervals
$(-\infty, -2)$, $(-2, -1)$, $(-1, 1)$, $(1, 2)$, $(2, \infty)$, and determine whether the inequality is true at each test value:

$f(-3) = 520 > 0$ $f(0) = 16 > 0$ $f(3) = 520 > 0$

$f(-1.5) \cong -13.67 < 0$ $f(1.5) = -13.67 < 0$

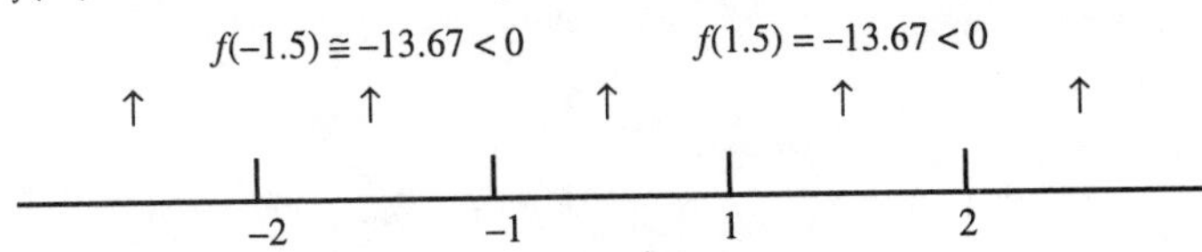

The inequality is true in $[-2, -1] \cup [1, 2]$.

27. First, solve $(2x + 5)(3x - 1)(x - 7) = 0$,
obtaining $x = -\frac{5}{2}, \frac{1}{3}, 7$.

Indicate these solutions on a number line:

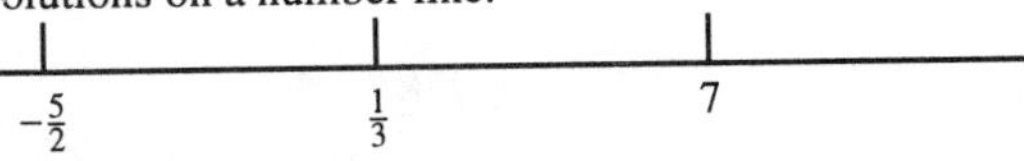

The leading term in the expanded form of $f(x)$ is
$(2x)(3x)(x) = 6x^3$, so the left tail of the graph points down, and $f(x) < 0$ on $\left(-\infty, -\frac{5}{2}\right)$. Each zero has multiplicity 1, so the sign of $f(x)$ changes at each. The pattern of signs for $f(x)$ is shown below:

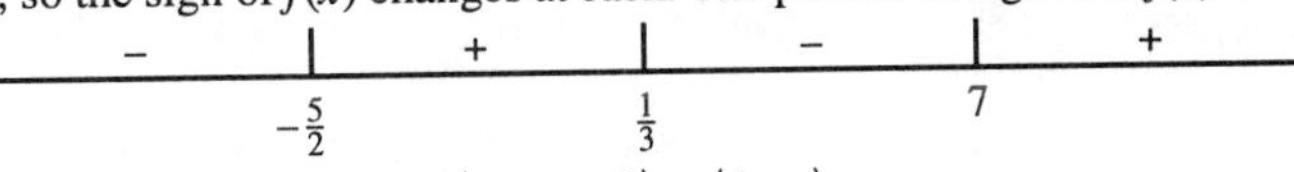

The inequality is true in $\left(-\infty, -\frac{5}{2}\right) \cup \left(\frac{1}{3}, 7\right)$.

29. First, solve $(x+4)^3(x-12)^3 = 0$, obtaining
$x = -4, 12$. Indicate these solutions on a number line:

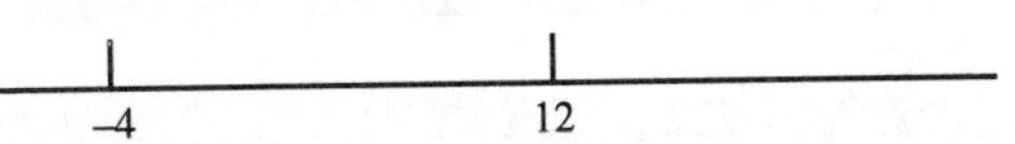

The leading term in the expanded form of $f(x)$ is $(x^3)(x^3) = x^6$, so the left tail of the graph points up, and $f(x) > 0$ on $(-\infty, -4)$. Each zero has multiplicity 3, so the sign of $f(x)$ changes at each. The pattern of signs for $f(x)$ is shown below:

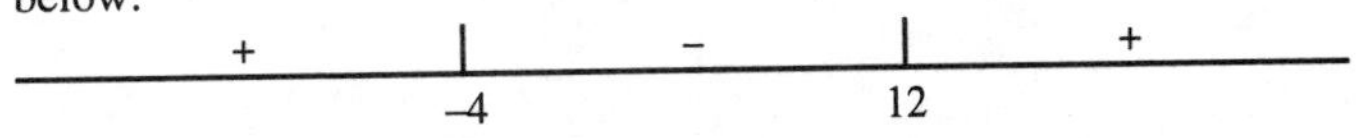

The inequality is true in $(-\infty, -4) \cup (12, \infty)$.

31.

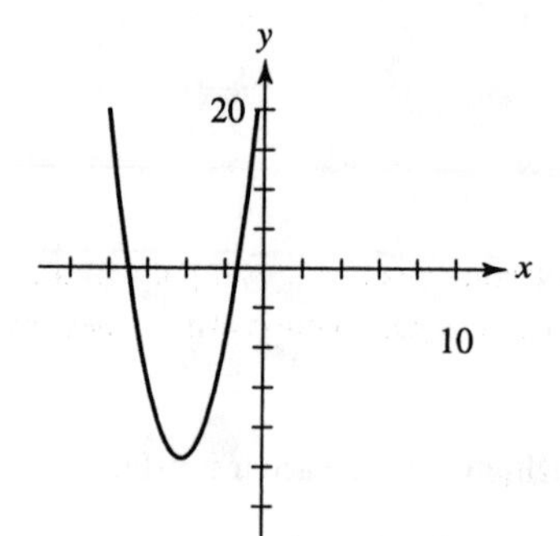

33.

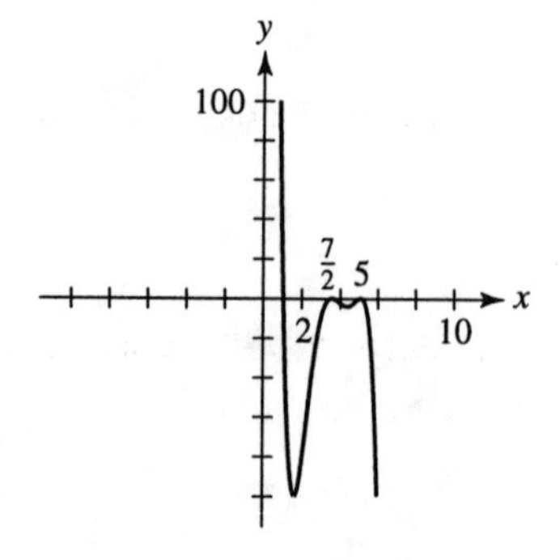

35. (Sample response) The sign pattern in Exercise 15 indicates that the graph is above the x-axis when $x < 1$, and below it when $x > 1$ except at the x-intercepts $\left(\frac{7}{2}, 0\right)$ and $(5, 0)$. Conversely, the graph in Exercise 33 indicates that $y < 0$ in $\left(1, \frac{7}{2}\right) \cup \left(\frac{7}{2}, 5\right) \cup (5, \infty)$.

37.

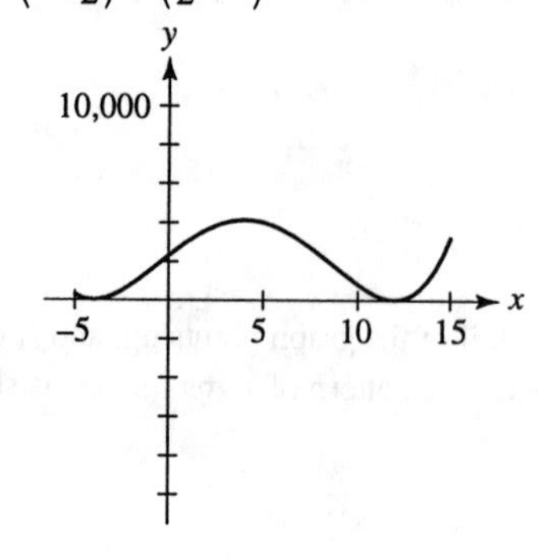

39.

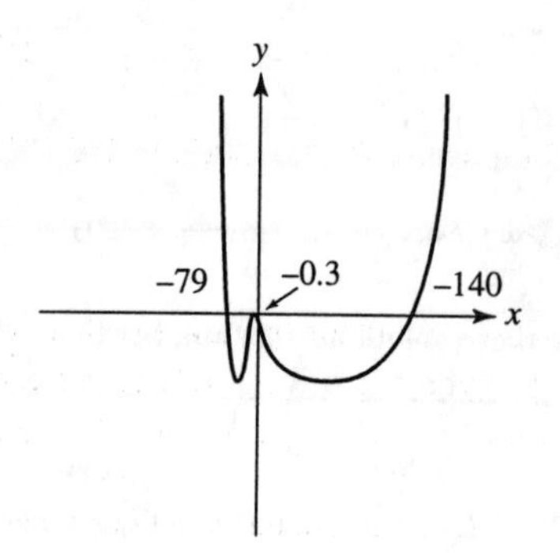

NUTSHELLS

Nutshell for Chapter 8 - RATIONAL FUNCTIONS

What's familiar?

- We can now expand the framework we created in Chapter 7, since rational functions include all polynomials. Once again, you begin your study of this category by looking at the simplest rational functions first. They are the *inverse power functions* $y = 1\backslash x$, $y = 1\backslash x^2$, and so on.
- Section 8-1 looks at rational functions numerically, analytically and graphically.
- Section 8-1 introduces you to the idea of *inverse variation*. This is similar to the idea of polynomial variation in Chapter 7.
- Section 8-2 discusses how the Highest Degree Theorem from Chapter 7 can be modified for use with rational functions.
- Section 8-2 comes back to the idea of continuity, which you saw in Section 3-4.
- Section 8-3 shows that the methods used to solve rational equations and inequalities build on those that you used for polynomial equations and inequalities in Chapter 7.

What's new?

- A striking feature of every inverse power graph is its vertical asymptote. The vertical asymptote separates the graph into *two* distinct pieces. This separation breaks the graph and causes it to be *discontinuous.*
- Another striking feature of every inverse power graph is its horizontal asymptote. This represents a type of end behavior not observed in polynomial graphs.
- Section 8-2 discusses how you can tell whether a more general rational function is discontinuous. You will learn how to tell whether a discontinuity appears graphically as a vertical asymptote or just a missing point on the graph.

List of hints for exercises in each chapter.

Chapter 8

- Before beginning Section 8-1, review the principles of graphical transformations. You may want to make a table of values to help locate designated points as you transform an inverse power graph.
- Pay particular attention to vertical asymptotes and discontinuities. The exercises will ask you to recognize these analytically, numerically, and graphically.
- Also, pay attention to the relationship between end behavior and horizontal asymptotes. The ideas in Section 7-3 for the end behavior of polynomial functions is expanded for rational functions.
- If you graph a function with a vertical asymptote on your calculator, be sure it is in "dot mode". This will give a more accurate graph.

Name:______________________

Chapter 8 Additional Exercises

Date:______________________

Decide whether the table fits an inverse variation $y = \frac{a}{x^n}$ for the given value of n. If it does, write an equation relating x and y.

1.

x	y ($n = 1$)
2	165
3	110
5	66
6	55
11	30

1. ______________________

2.

x	y ($n = 2$)
2	225
3	100
5	36
10	9
15	4

2. ______________________

3.

x	y ($n = 3$)
2	460
3	310
4	286
7	194
11	181

3. ______________________

4. If r varies inversely with s, and $r = 85$ when $s = 7$, what is the value of r when $s = 35$?

4. ______________________

5. If g varies inversely with m^2, and $g = 64$ when $m = 3$, what is the value of g when $m = 4$?

5. ______________________

6. If y varies inversely with x^4 and $y = 81$ when $x = \sqrt{5}$, what is the value of y when $x = 3$?

6. ______________________

Chapter 8 Additional Exercises *(cont.)*

Name:____________________

7. Sketch the graph of $y = \frac{6}{x+2}$ by transforming the graph of $y = \frac{1}{x}$. Identify the points obtained by moving (1, 1) and (–1, –1).

7.

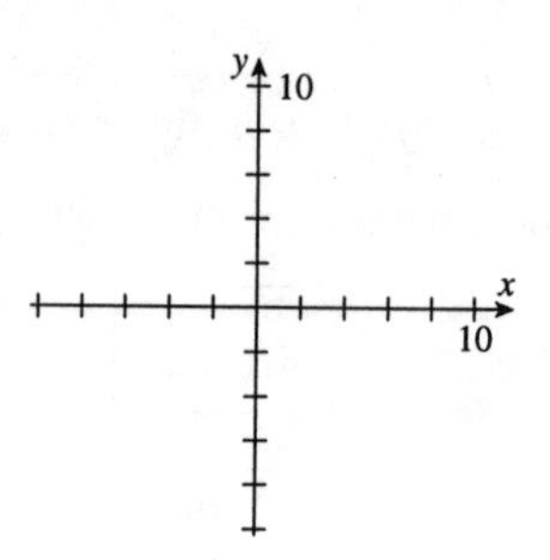

8. Sketch the graph of $f(x) = 3 - \frac{1}{x-4}$ by transforming the graph of $y = \frac{1}{x}$. Identify the vertical and horizontal asymptotes.

8.

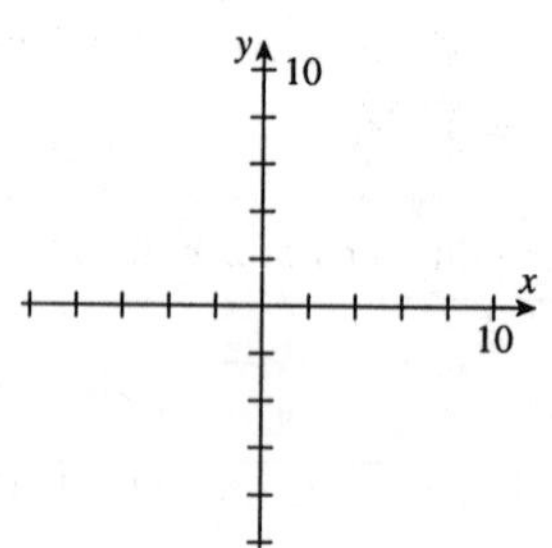

9. Sketch the graph of $L(x) = \frac{1}{(x-3)^2}$ by transforming the graph of $y = \frac{1}{x^2}$. Identify the points obtained by moving (1, 1) and (–1, 1).

9.

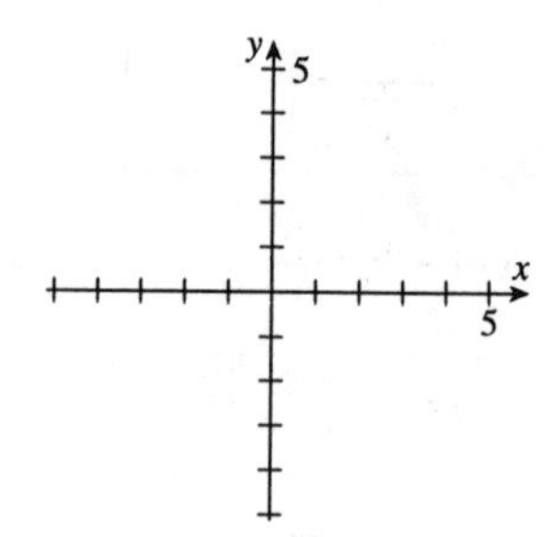

10. Sketch the graph of $M(x) = 2 + \frac{1}{x^3}$ by transforming the graph of $y = \frac{1}{x^3}$. Identify the vertical and horizontal asymptotes.

10.

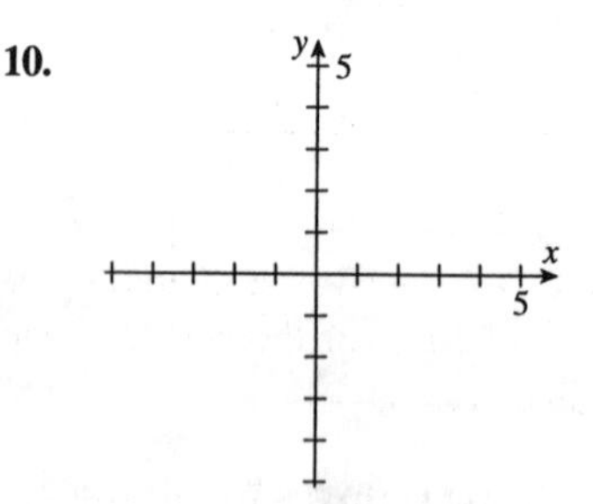

11. Sketch the graph of $y = \frac{2x-6}{x+2}$. Identify the intercepts and asymptotes.

11.

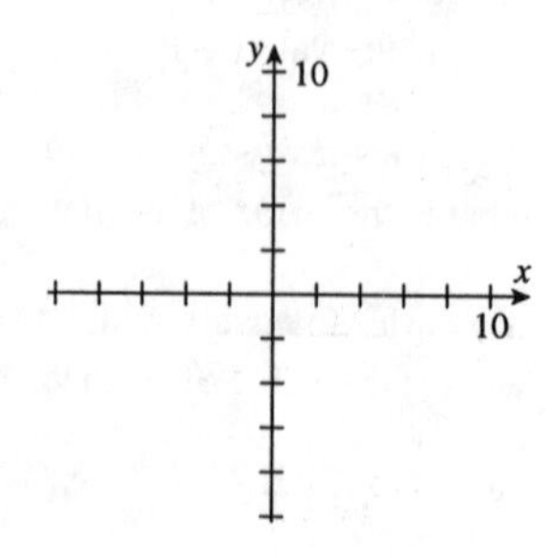

Chapter 8 Additional Exercises *(cont.)*

Name:________________________

12. Sketch the graph of $y = \frac{4x-6}{2x-1}$. Identify the intercepts and asymptotes.

12.

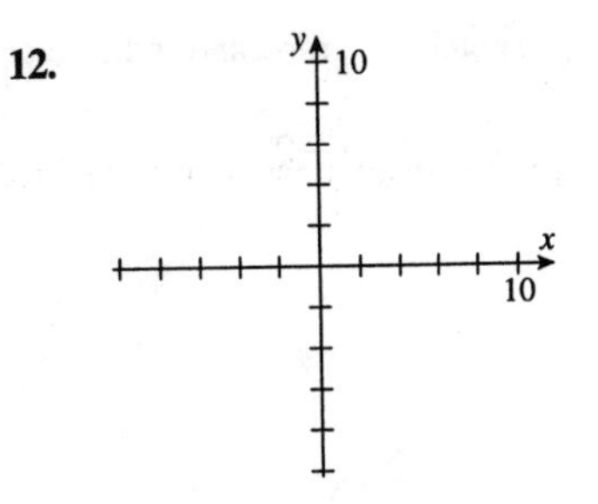

13. Identify the horizontal and vertical asymptotes of the graph of $y = \frac{8x+3}{5x-2}$.

13. ________________________

14. Identify the horizontal and vertical asymptotes of the graph of $y = \frac{6x-7}{3x+2}$.

14. ________________________

For 15–19, list all discontinuities (if any) of the function along with the type (vertical asymptote or missing point.)

15. $f(x) = \frac{4x-7}{x^2+6}$

15. ________________________

16. $g(x) = \frac{x^2-16}{x+4}$

16. ________________________

17. $h(x) = \frac{x^2-x-12}{x+4}$

17. ________________________

18. $f(x) = \frac{x-1}{x^2+2x-3}$

18. ________________________

19. $H(x) = \frac{4x-8}{x^2+4x+6}$

19. ________________________

In 20–23, use the Highest Degree Theorem to determine the end behavior of the graph (horizontal asymptote or whether each tail goes up or down.)

20. $f(x) = \frac{18x^2-7}{6x^2-5x+1}$

20. ________________________

21. $g(x) = \frac{2x^2-5}{7x^3+3x^2-4}$

21. ________________________

22. $H(x) = \frac{6x^3-3x+4}{2x^2-7}$

22. ________________________

23. $R(x) = \frac{6-x^3}{3x-4}$

23. ________________________

Chapter 8 Additional Exercises *(cont.)*

Name:____________________

In 24–26 sketch the graph and indicate any discontinuities.

24. $f(x)=\dfrac{x^2-x-2}{x-2}$

24.

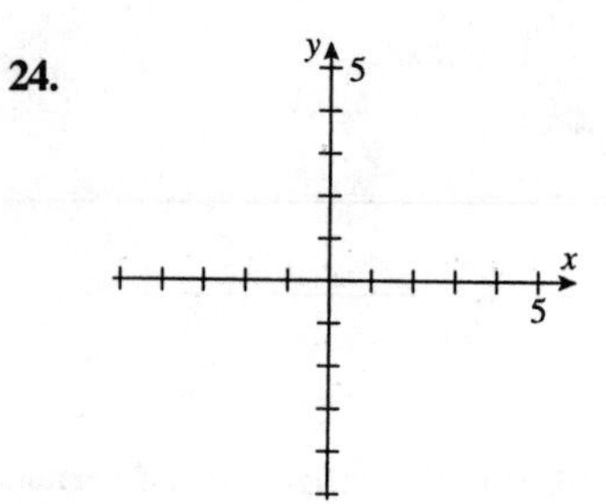

25. $g(x)=\dfrac{2x-3}{x+1}$

25.

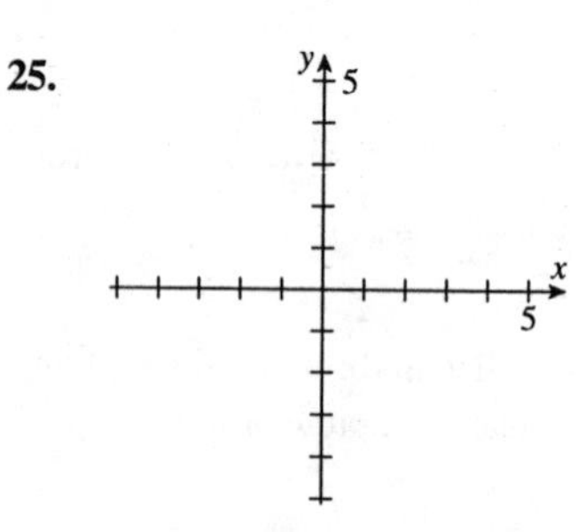

26. $R(x)=\dfrac{x+3}{x^2+2x-3}$

26.

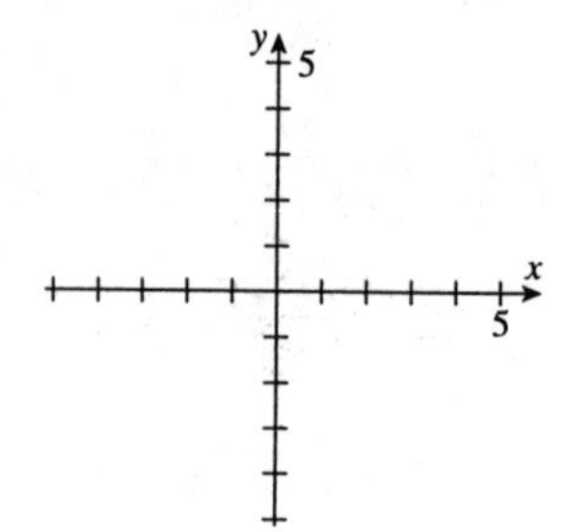

In 27–29, solve the inequality graphically.

27. $\dfrac{x+1}{x^2-4}\geq 0$

27. ____________________

Chapter 8 Additional Exercises *(cont.)*

Name:________________

28. $\dfrac{(x-1)(x+3)}{x-2} \le 0$

28. ________________

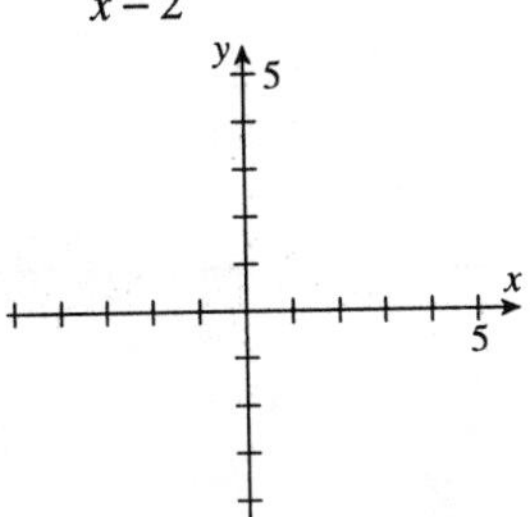

29. $\dfrac{1}{(x-2)^2} > 1$

29. ________________

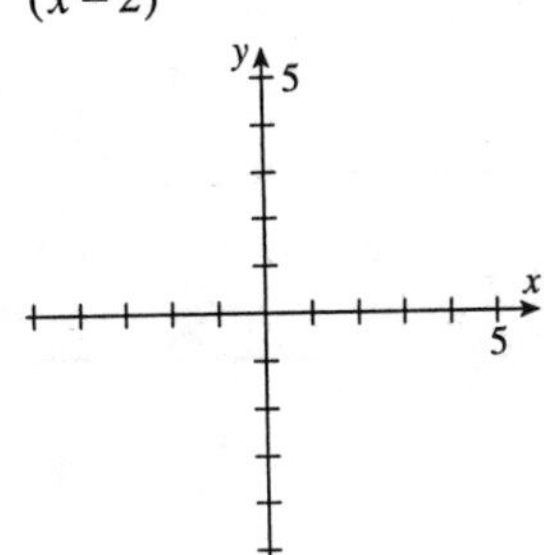

In 30–38, solve the inequality with any method.

30. $\dfrac{x^2-3x}{x+7} \ge 0$

30. ________________

31. $\dfrac{x}{x+3} < 0$

31. ________________

32. $\dfrac{6}{x} \ge 3$

32. ________________

33. $\dfrac{(x+19)^4(x-4)}{(x+3)(x-23)^2} < 0$

33. ________________

34. $\dfrac{5-x}{x^2} \le 0$

34. ________________

35. $\dfrac{x^2+3x-15}{x} \ge 1$

35. ________________

36. $\dfrac{-3}{x^2} \le 0$

36. ________________

37. $\dfrac{3x}{x^2+4} \ge 0$

37. ________________

38. $\dfrac{2x^2+4x+4}{x^2} > 1$

38. ________________

39. Explain why transforming the graph of $y = \dfrac{1}{x^2}$ into the graph of $y = \dfrac{5}{(x-2)^2} - 9$ moves the point (1, 1) to (3, –4).

39. ________________

Name:________________________

40. If the temperature of a given amount of gas is kept constant, the volume that it occupies varies inversely with the pressure. If the volume of the gas is 8 quarts when the pressure is 6 atmospheres, what is the volume when the pressure is 12 atmospheres?

40. ____________________

41. When a person walks, the pressure on the sole of each shoe varies inversely with the area of the sole. If the pressure is 6 pounds per square inch when the area is 24 square inches, what is the pressure when the area is 16 square inches?

41. ____________________

42. The illumination from a light source varies inversely with the square of the distance from the light source. If the illumination is 9 candles per square inch when the light is 10 inches away, what is the illumination when the same light is 15 inches away?

42. ____________________

43. If y varies inversely with x^2 and x is multiplied by a factor of 5, what is the change in y?

43. ____________________

44. As $|x|$ becomes large, explain why $f(x) = x + 3 + \frac{5}{x-1}$ approaches the graph of $y = x + 3$.

44. ____________________

Use for 45–47. A box with a square bottom needs to have a volume of 1.728 m^3.

45. Express the height of the box as a function $h(l)$ of the length (l) of a side of the square bottom. Explain the physical meaning of any discontinuities in $h(l)$.

45. ____________________

46. Express the surface area of the box as a function $S(l)$ and explain the physical meaning of any discontinuities in S.

46. ____________________

47. What physical information does the end behavior of the graph of S provide?

47. ____________________

48. Explain what intervals you would use to solve $\frac{x-2}{x^2-2x-15} \geq 0$ with the test-value method, and why you would use them.

48. ____________________

Use for 49–50. The intensity of light from a fixed source at a distance d from the source and with a constant reference intensity of 1 is $\frac{1}{d^2}$. A street light is 40 feet high.

49. Write a function I for the intensity of light x feet from the base of the street light.

49. ____________________

$h = 40'$

x

50. Find the distance from the base of the street light where the intensity of light is less than 0.0004.

50. ____________________

A Mathematical Looking Glass
Long's Peak
(to follow Section 8-1)

Long's Peak is located in Rocky Mountain National Park in Colorado. From about mid-July until early September each year, the 14,255-foot summit can be reached by a trail from the Long's Peak Ranger Station near Estes Park.

Because you are in excellent condition, you believe that you can start from the trailhead at 8:00 am, spend 2 hours at the top, and arrive back at the trailhead by 6:00 pm. Since part of the return trip will be easy downhill hiking, you can afford to go more slowly on the way up. However, it is dangerous to be stranded on the trail after dark, so you need to monitor your uphill speed very carefully.

Exercises

1. (Creating Models) Let U and D represent your average uphill and downhill speeds, in miles per hour.

   a. Express the number of hours you will need for the uphill leg in terms of U.

   b. Express the number of hours you will need for the downhill leg in terms of D.

   c. If you are to complete your trip by 6:00 pm, explain why you need to choose U and D so that $\frac{8}{U} + \frac{8}{D} = 8$.

   d. Verify that the equation in part (c) is equivalent to $D = \frac{U}{U - 1}$.

2. a. Divide U by $U - 1$, and verify that you obtain

$$f(U) = 1 + \frac{1}{U - 1}$$

b. Use transformations to obtain the graph of $f(U)$ from the basic graph $y = \frac{1}{U}$. Write the equations of the horizontal and vertical asymptotes.

c. What is the domain of $f(U)$ in the abstract? What is its domain in the context of the situation?

d. (Interpreting Mathematics) No matter how fast you plan to go downhill, your average uphill speed should not be too slow. What uphill speeds are feasible?

3. (Writing to Learn) If you were making this trip, what combination of uphill and downhill speeds would you use? Explain your choice.

A Mathematical Looking Glass
The Z-Computer
(to follow Section 8-1)

Figure S-12 is a reproduction of a gas mileage computer (which we will call a z-computer) from a *Rand McNally Road Atlas.*

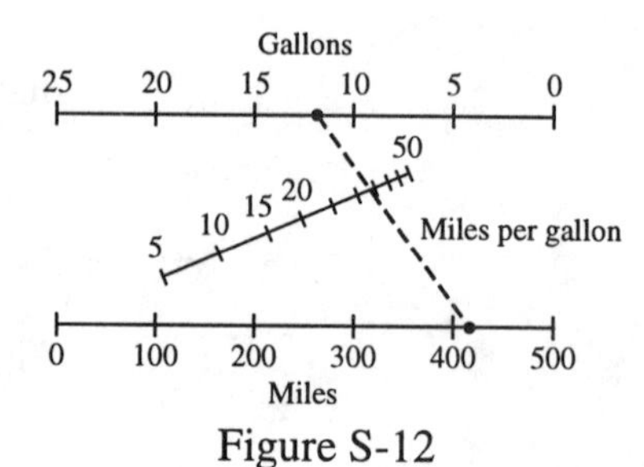

Figure S-12

In the sample computation in Figure S-12, a driver has gone 420 miles between gas stops, and has filled the tank by buying 12 gallons of gas. To compute his gas mileage, he joins the 420-mile mark on the bottom horizontal scale to the 12-gallon mark on the top horizontal scale. This line intersects the diagonal scale at a point indicating gas mileage of 35 miles per gallon.

For the *z*-computer to work, the lines joining 35 miles to 1 gallon, 70 miles to 2 gallons, and so on, should all intersect the diagonal scale at the point indicating 35 miles per gallon. More generally, the line joining M miles to G gallons should intersect the diagonal scale at a point P depending only on the value of $\frac{M}{G}$ (the miles per gallon), and not on the values of M and G separately. In Exercise 1, you will show that this is the case.

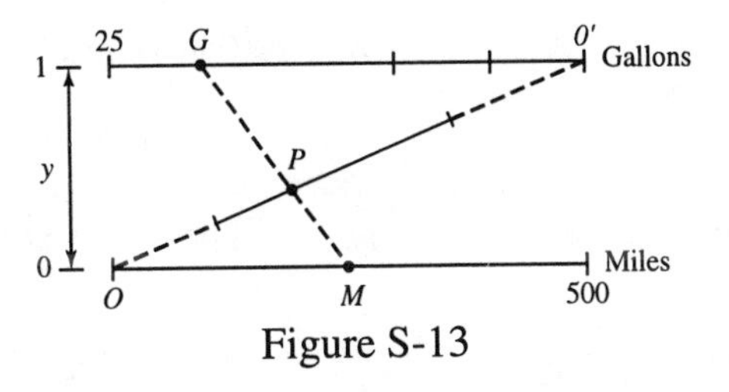

Figure S-13

To show how the *z*-computer should be calibrated, we have placed a *y*-axis next to the computer in Figure S-13. In Exercise 1, you will also show that if $\frac{M}{G} = r$, then the point P will have a *y*-coordinate of $\frac{r}{r + 20}$. You will thus demonstrate that the *z*-computer works, and you will simultaneously show how to calibrate the diagonal scale.

Exercises

1. (Writing to Learn)

   a. Explain why the triangles *PMO* and *PGO'* in Figure S-13 are similar.

   b. Explain why the ratio $\frac{OP}{PO'} = \frac{\frac{m}{500}}{\frac{g}{25}}$.

   c. Explain why the ratio $\frac{OP}{OO'} = \frac{\frac{m}{500}}{\frac{m}{500} + \frac{g}{25}}$.

   d. Explain why your result from part (c) leads to the conclusion that the *y*-coordinate at P is $\frac{r}{r + 20}$.

2. (Interpreting Mathematics) Sketch the graph of the function $y = \frac{r}{r + 20}$. What physical information, if any, is provided by:

   a. the vertical asymptote?

   b. the horizontal asymptote?

A Mathematical Looking Glass

Tommy

(to follow Section 8-3)

When attending a musical performance, where should you sit to hear the best possible sound? A little mathematical modeling can help you answer this question.

Suppose that a local theater company is performing the Who's rock opera, *Tommy*. The actors' voices will be amplified and broadcast through two speakers located at the ends of the proscenium arch, 60 feet apart, as indicated in Figure S-14. The band will be in the orchestra pit, directly in front of the center of the stage. For simplicity, we have indicated the orchestra pit as a single point in Figure S-14.

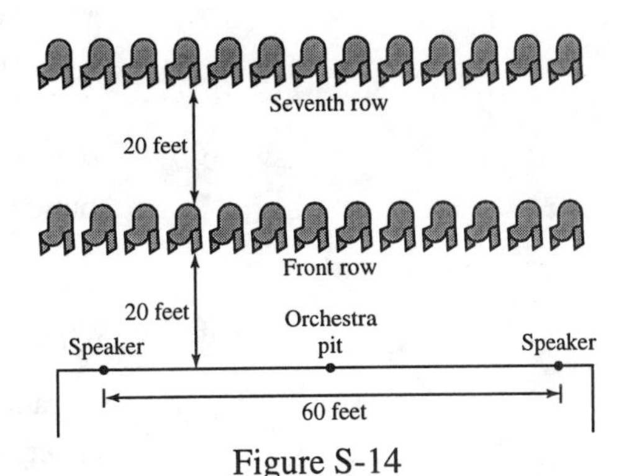

Figure S-14

To hear a balanced sound, you would like to sit where the sound intensity from the singers is between 1 and 2 times that from the band. In Exercises 1-3, you will find the suitable locations in the front row (20 feet from the stage) and in the seventh row (40 feet from the stage).

Exercises

1. (Problem-Solving) At a distance of D feet from the band or one of the speakers, the intensity of its sound is inversely proportional to D^2. For simplicity assume that the intensity from each is $\frac{1}{D^2}$. Suppose you are sitting in the front row, x feet from the center.

   a. Write three expressions in x to represent your distances from the band and each of the two speakers.

   b. Write an expression $B(x)$ for the intensity of the sound from the band.

   c. Write an expression $S(x)$ for the combined sound intensity from both speakers.

   d. Write a function $R(x) = \frac{S(x)}{B(x)}$ to represent the ratio of the sound intensity from the singers to that of the band.

   e. Explain why the suitable locations in the front row are described by the inequalities $R(x) \geq 1$ and $R(x) \leq 2$.

   f. Solve the inequalities in part (d).

2. (Interpreting Mathematics) Repeat Exercise 1, assuming that you are sitting in the seventh row, x feet from the center.

3. (Making Observations) In which of the two rows are there more acceptable seats?

Chapter 8

Section 8.1

1. The time required to drive 450 miles at a rate of s miles per hour is
$(450 \text{ miles})\left(\frac{1 \text{ hour}}{s \text{ miles}}\right) = \frac{450}{s}$ hours.
Thus $g(s) = \frac{450}{s}$.

3. a. $g(80) - g(65) = \frac{450}{80} - \frac{450}{65} \cong -1.30$ hours, or about 1 hour and 18 minutes.

   b. The net savings in time is about 58 minutes. The cost of that time is about $\frac{150}{58} \cong \$2.59$ per minute.

5. a.

x	$f(x)$
1	1
10	0.1
100	0.01
1000	0.001

b.

x	$f(x)$
–1	–1
–10	–0.1
–100	–0.01
–1000	–0.001

c.

x	$f(x)$
1	1
0.1	10
0.01	100
0.001	1000

d.

x	$f(x)$
–1	–1
–0.1	–10
–0.01	–100
–0.001	–1000

7. The tables illustrate that when $|x|$ is large, both $|f(x)|$ and $|g(x)|$ are small. They also illustrate that when $|x|$ is small, both $|f(x)|$ and $|g(x)|$ are large.

9. a. $g(600) = \frac{450}{600} = 0.75$ hours, or 45 minutes

   b. $g(10{,}000) = \frac{450}{10{,}000} = 0.045$ hours, or 2.7 minutes

   c. $\left(\frac{186{,}000 \text{ miles}}{1 \text{ second}}\right)\left(\frac{3600 \text{ seconds}}{1 \text{ hour}}\right)$
   $= 669{,}600{,}000$ miles per hour
   $g(669{,}600{,}000) = \frac{450}{669{,}600{,}000}$
   $\cong 0.00000067$ hours, or about 0.0024 seconds.

11. a. $g(2.5) = \frac{450}{2.5} = 180$ hours, or 7.5 days

    b. $g(0.01) = \frac{450}{0.01} = 45{,}000$ hours, or about 5.14 years

    c. $\left(\frac{0.01 \text{ miles}}{1 \text{ year}}\right)\left(\frac{1 \text{ year}}{365 \text{ days}}\right)\left(\frac{1 \text{ day}}{24 \text{ hours}}\right)$
    $= \frac{1}{876{,}000}$ miles per hour
    $g\left(\frac{1}{876{,}000}\right) =$
    $(450)(876{,}000) = 394{,}200{,}000$ hours, or 45,000 years

13. The product $xy = 84$ for each row, so the table fits the equation $y = \frac{84}{x}$.

15. The product $x^2y = 900$ for each row, so the table fits the equation $y = \frac{900}{x^2}$.

17. The equation relating V and P is $P = \frac{a}{V}$ for some real number a, and $20 = \frac{a}{6}$, so $a = 120$. Thus $P = \frac{120}{V}$. When $V = 10$, $P = \frac{120}{10} = 12$.

19. The equation relating r and s is $s = \frac{a}{r^3}$ for some real number a, and $250 = \frac{a}{1^3}$, so $a = 250$. Thus $s = \frac{250}{r^3}$. When $r = 5$, $s = \frac{250}{5^3} = 2$.

21. The graph of $y = \frac{1}{x}$ has a vertical asymptote at $x = 0$ (on the y-axis). When the graph is shifted 1 unit to the left, the asymptote shifts 1 unit to the left, to the line $x = -1$. Similarly, when the graph is shifted 2 units up, the vertical asymptote shifts from $y = 0$ to $y = 2$.

23. The graph is obtained by the following transformations:

$$y = \frac{1}{x} \to y = \frac{1}{x+3} \to y = \frac{1}{x+3} - 2$$

The graph is shifted 3 units to the left and 2 units down. The vertical asymptote is the line $x = -3$, and the horizontal asymptote is the line $y = -2$. The domain is $(-\infty, -3) \cup (-3, \infty)$ and the range is $(-\infty, -2) \cup (-2, \infty)$.

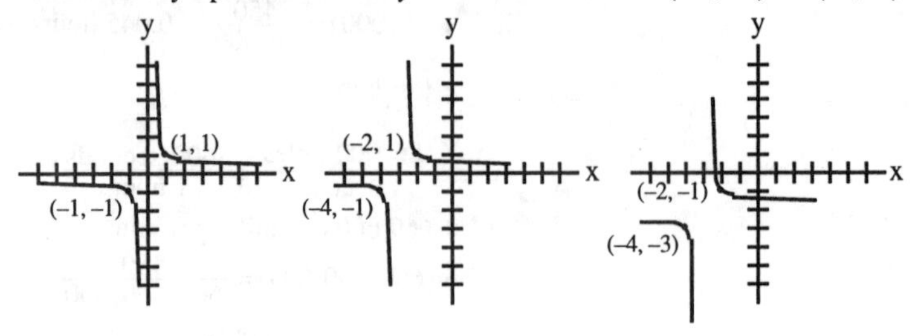

25. The graph is obtained by the following transformations:

$$y = \frac{1}{x} \to y = -\frac{1}{x} \to y = -\frac{3}{x} \to y = -\frac{3}{x} + 7$$

The graph is reflected in the x-axis, stretched vertically by a factor of 3, then shifted 7 units up. The vertical asymptote is the y-axis, and the horizontal asymptote is the line $y = 7$. The domain is $(-\infty, 0) \cup (0, \infty)$ and the range is $(-\infty, 7) \cup (7, \infty)$.

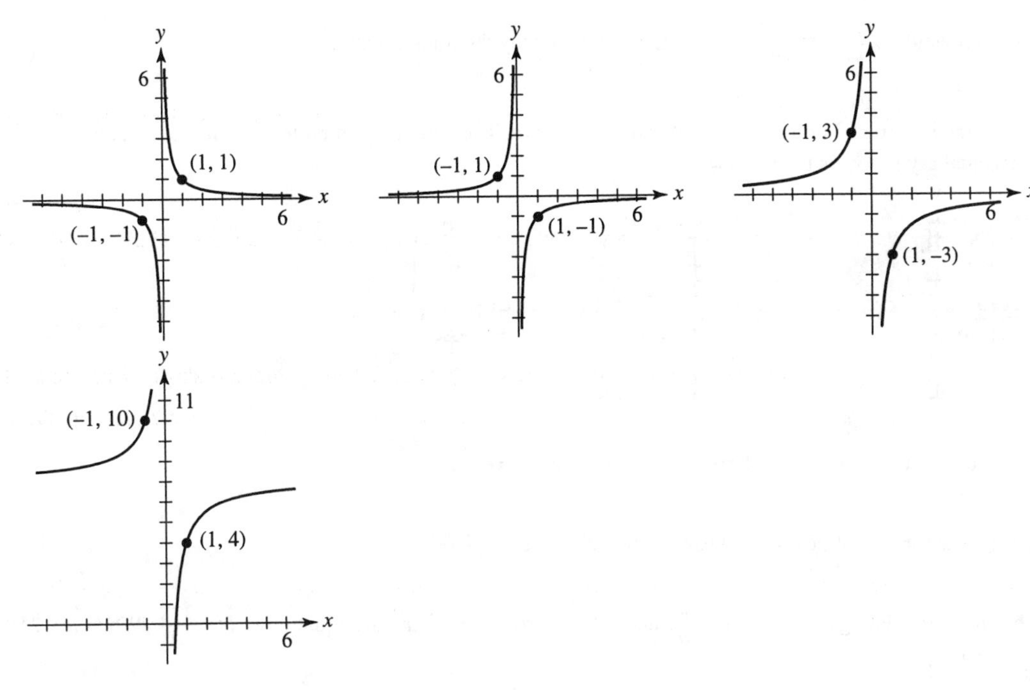

27. Division yields $y = 3 + \frac{4}{x+1}$. The graph is obtained by the transformations

$y = \frac{1}{x} \to y = \frac{4}{x} \to y = \frac{4}{x+1} \to y = \frac{4}{x+1} + 3$

The graph is stretched vertically by a factor of 4, then shifted 1 unit to the left and 3 units up. The vertical asymptote is the line $x = -1$, and the horizontal asymptote is the line $y = 3$.

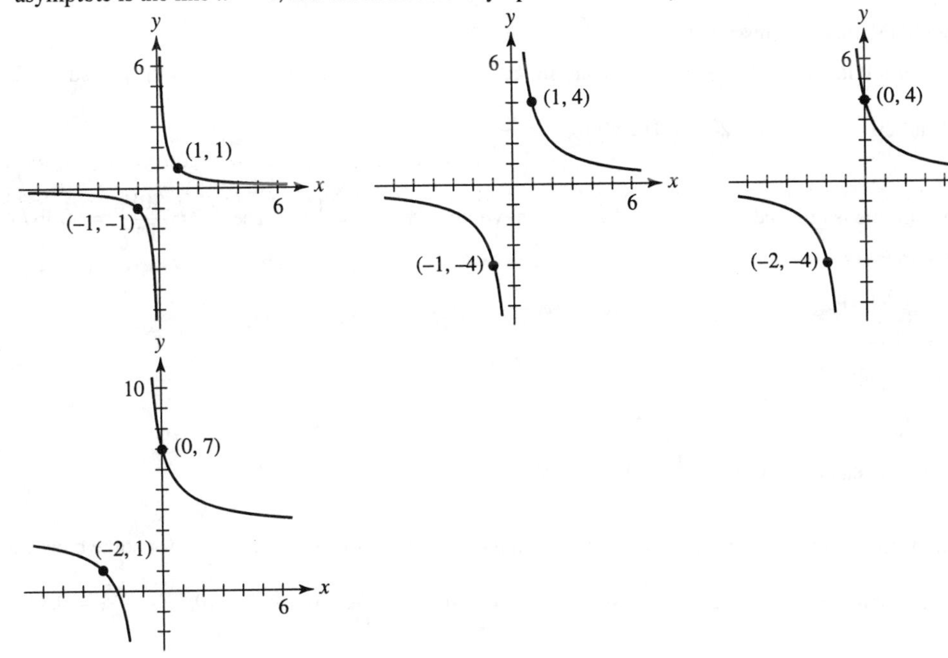

29. Division yields $y=-1+\frac{1}{x-2}$. The graph is obtained by the transformations

$y=\frac{1}{x} \to y=\frac{1}{x-2} \to y=\frac{1}{x-2}-1$

The graph is shifted 2 units to the right and 1 unit down. The vertical asymptote is the line $x = 2$, and the horizontal asymptote is the line $y = -1$.

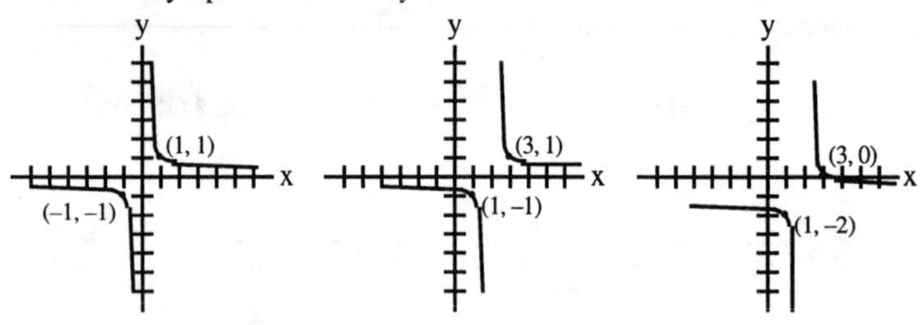

31. The product $x^2y = 0.1$ for each row, so the table fits $y=\frac{0.1}{x^2}$.

33. The product $x^2y = 4096$ for each row, so the table fits $y=\frac{4096}{x^2}$.

35. The equation relating a and b is $b=\frac{k}{a}$, and $54=\frac{k}{7}$, so $k = 378$, and the equation is $b=\frac{378}{a}$. When $a = 21$, $b=\frac{378}{21}=18$.

37. The equation relating t and v is $v=\frac{a}{t^3}$, and $4=\frac{a}{3^3}$, so $a = 108$, and the equation is $V=\frac{108}{t^3}$. When $t = 6$, $V=\frac{108}{6^3}=\frac{1}{2}$.

39. The equation relating the pressure P in lb/in^2
and the volume V in in^3 is $P=\frac{k}{V}$ for some constant k. Let V_0 represent the volume when the pressure is 20 lb/in^2. Then $20=\frac{k}{V_0}$, so $k=20V_0$. Therefore $P=\frac{20V_0}{V}$.

a. If the volume is reduced to one third of its present volume, then $V=\frac{1}{3}V_0$. Then $P=\frac{20V_0}{\frac{1}{3}V_0}=60 \text{ lb / in}^2$.

b. If $P = 100$, then

$100=\frac{20V_0}{V}$

$100V=20V_0$

$V=\frac{1}{5}V_0$

The volume can be reduced to $\frac{1}{5}$ of the present volume.

41. The relationship between the distance D in meters and the force F in newtons is $F=\frac{k}{D^2}$ for some constant k. Since $F = 5$ when $D = 3$, we have $5=\frac{k}{3^2}$, so $k = 45$. Therefore $F=\frac{45}{D^2}$. When $D = 10$, $F=\frac{45}{10^2}=0.45$ newtons.

43. The graph is obtained by the transformation

$$y = \frac{1}{x} \to y = \frac{10}{x} \to y = \frac{10}{x-2}$$

The graph is stretched vertically by a factor of 10, then shifted 2 units to the right. The vertical asymptote is the line $x = 2$, and the horizontal asymptote is the x-axis. The domain is $(-\infty, 2) \cup (2, \infty)$. The range is $(-\infty, 0) \cup (0, \infty)$.

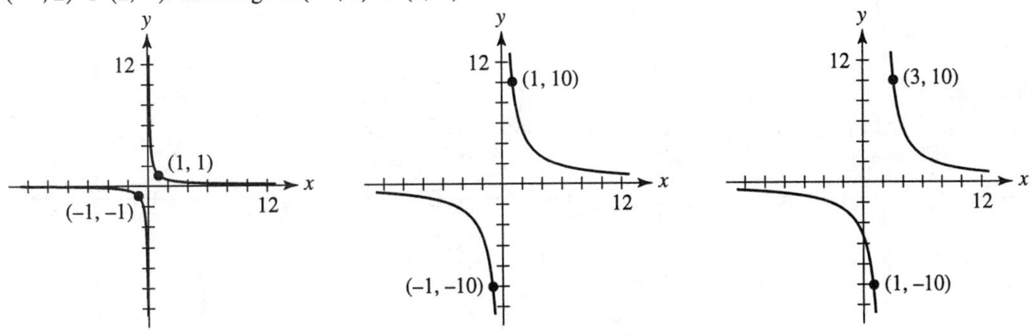

45. The graph is obtained by the transformations

$$y = \frac{1}{x} \to y = \frac{2}{x} \to y = \frac{2}{x+7}$$

The graph is stretched vertically by a factor of 2, then shifted 7 units to the left. The vertical asymptote is the line $x = -7$, and the horizontal asymptote is the x-axis. The domain is $(-\infty, -7) \cup (-7, \infty)$. The range is $(-\infty, 0) \cup (0, \infty)$.

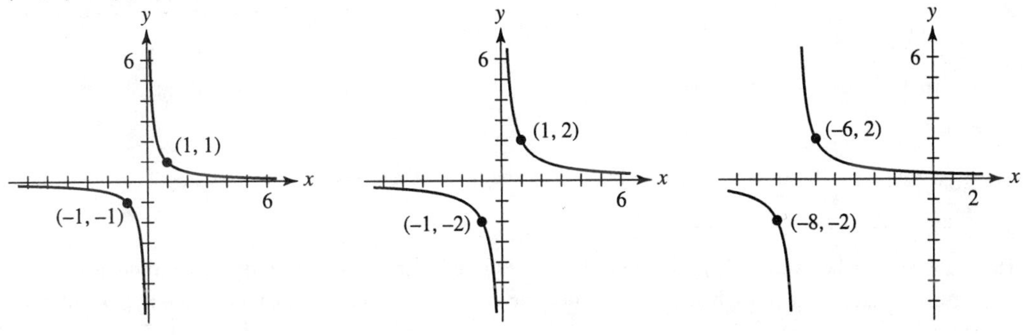

47. Division yields $y = \frac{2}{x+5} + 1$. The graph is obtained by the transformations

$y = \frac{1}{x} \to y = \frac{2}{x} \to y = \frac{2}{x+5} \to y = \frac{2}{x+5} + 1$

The graph is stretched vertically by a factor of 2, then shifted 5 units to the left and 1 unit up. The vertical asymptote is the line $x = -5$, and the horizontal asymptote is the line $y = 1$. The domain is $(-\infty, -5) \cup (-5, \infty)$. The range is $(-\infty, 1) \cup (1, \infty)$.

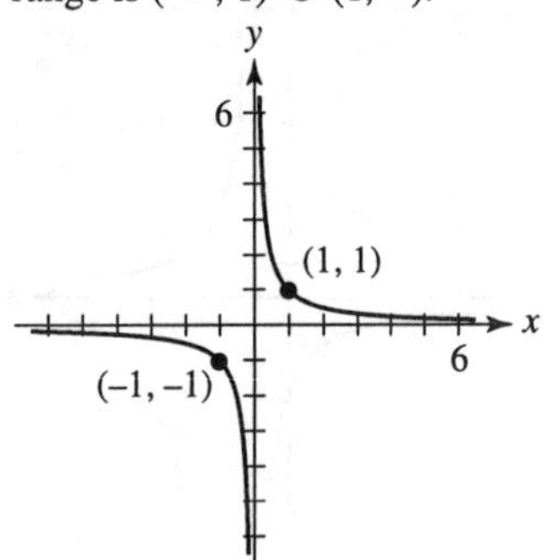

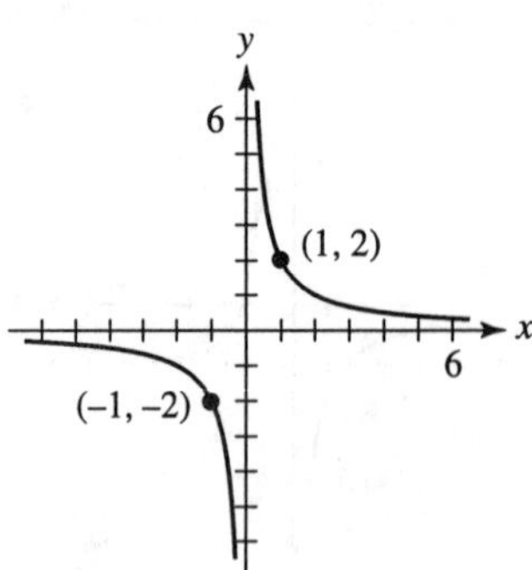

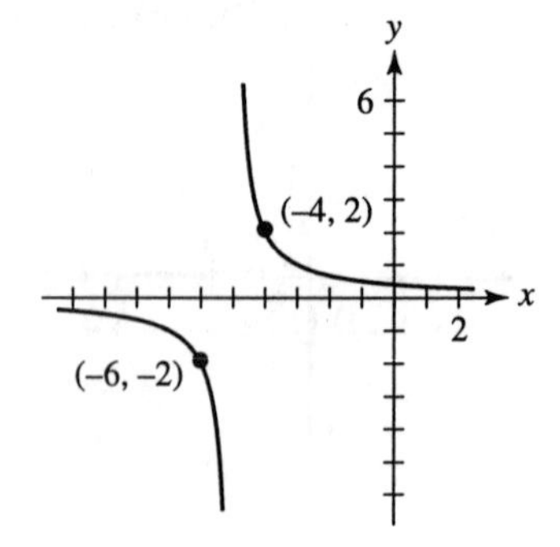

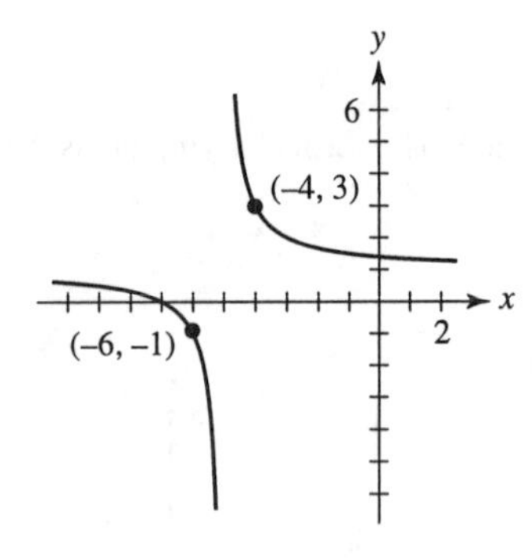

49. Division yields $y = \frac{4}{x} + 2$. The graph is obtained by the transformations

$y = \frac{1}{x} \to y = \frac{4}{x} \to y = \frac{4}{x} + 2$

The graph is stretched vertically by a factor of 4, then shifted 2 units up. The vertical asymptote is the y-axis, and the horizontal asymptote is the line $y = 2$. The domain is $(-\infty, 0) \cup (0, \infty)$. The range is $(-\infty, 2) \cup (2, \infty)$.

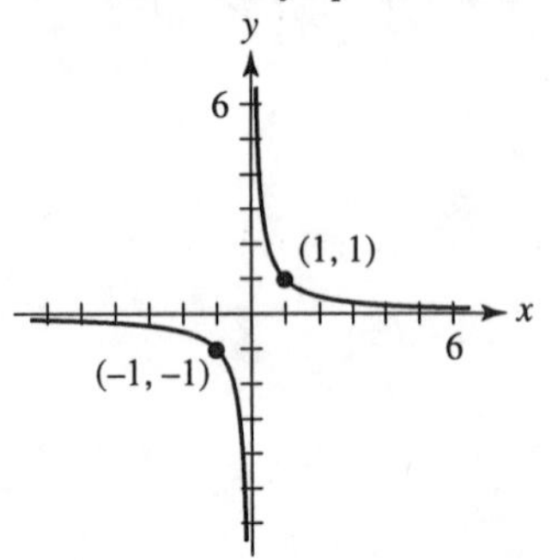

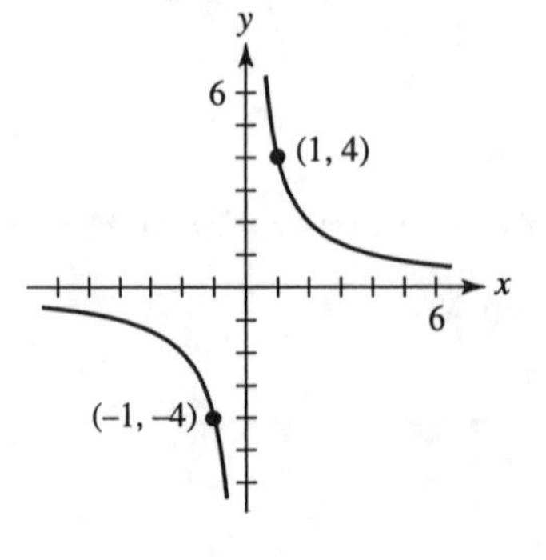

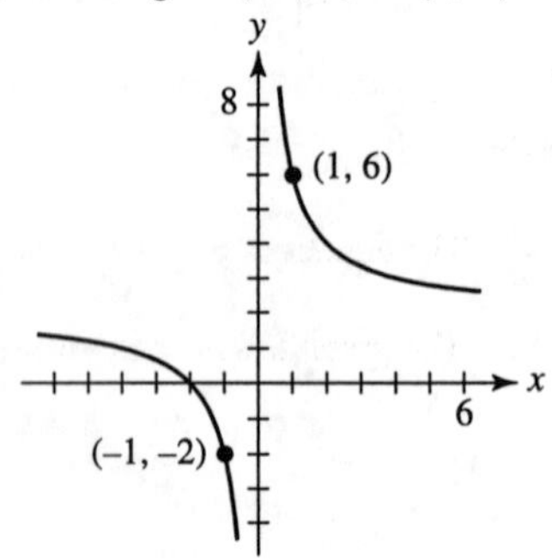

51. The graph is obtained from that of $y = \frac{1}{x^3}$ by the transformation

$y = \frac{1}{x^3} \rightarrow y = \frac{6}{x^3}$

The graph is stretched vertically by a factor of 6. The vertical asymptote is the y-axis, and the horizontal asymptote is the x-axis. The domain is
$(-\infty, 0) \cup (0, \infty)$. The range is $(-\infty, 0) \cup (0, \infty)$.

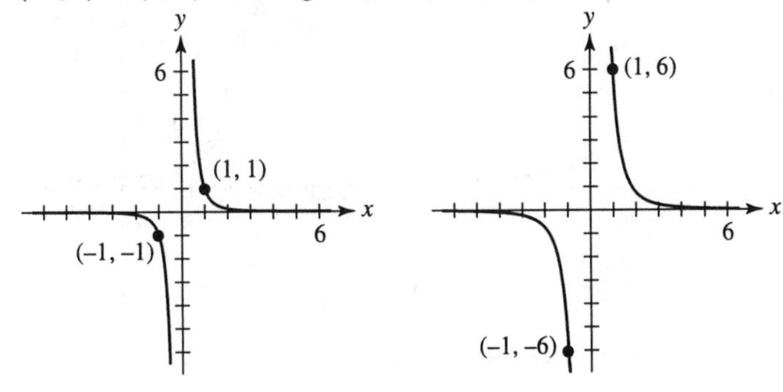

53. The graph is obtained from that of $u = \frac{1}{x^2}$ by the transformations

$u = \frac{1}{x^2} \rightarrow u = -\frac{1}{x^2} \rightarrow u = -\frac{1}{x^2} + 3$

The graph is reflected in the x-axis, then shifted 3 units up. The vertical asymptote is the u-axis, and the horizontal asymptote is the line $u = 3$. The domain is $(-\infty, 0) \cup (0, \infty)$. The range is
$(-\infty, 3)$.

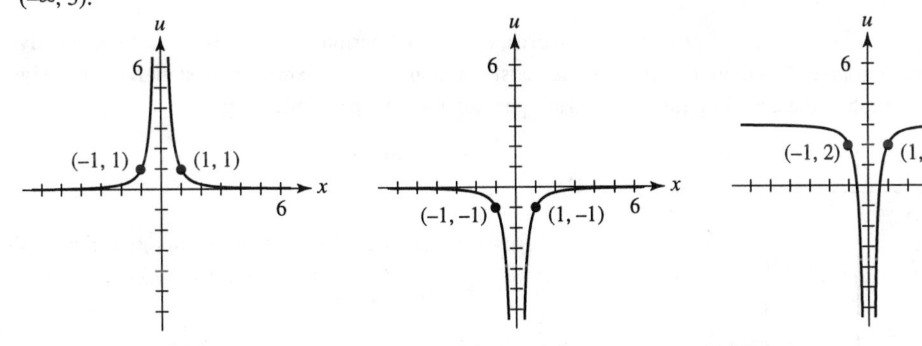

Section 8-2

1. a. The equation $x^2 + 1 = 0$ has only the solutions $x = \pm i$, so $x^2 + 1$ is not 0 for any real value of x.

b. The graph has no x-intercepts.

3. a.

x	$h(x)$
0	–1
0.9	–10
0.99	–100
0.999	–1000
1	undefined
1.001	1000
1.01	100
1.1	10
2	1

b. As the values of x become close to 1, $|h(x)|$ becomes arbitrarily large.

5. (Sample response) If x is near 1, then the value of $x - 1$ is near 0. The reciprocal of a number near 0 has a large absolute value, so $|h(x)| = \left|\frac{1}{x-1}\right|$ is large.

7. The function is discontinuous when $x = -4$. The graph suggests that the discontinuity is a vertical asymptote.

9. The function is discontinuous when $x = 5$. The graph suggests that the discontinuity is a missing point.

11. (Sample response) As the level of production increases, the given description indicates that the cost of advertising increases slowly at first, then more rapidly. Thus the cost function does not exhibit a constant rate of change. As x increases, the graph of the cost function becomes steeper, suggesting a parabolic shape.

13. a.

x	y
100	11.50
10	73.60
1	703.51
0.1	7003.50

Each row shows the average cost per shirt at a given level of production.

b. When the production level is low, the average cost per shirt is very high. On the graph of $y = A(x)$, x-values near 0 correspond to large y-values. This suggests that the graph has a vertical asymptote at $x = 0$.

15. a. $f(s)=\dfrac{0.01s^2+3.50s-1950}{s-300}$
$=\dfrac{0.01(s^2+350s-195\,000)}{s-300}$
$=\dfrac{0.01(s-300)(s+650)}{s-300}$
$=0.01(s+650)$ is $s\neq 300$.
When x is near 300, $f(s)$ is near $0.01(300+650)=9.50$.

b. If you increase production, each additional shirt will cost about \$9.50 to produce. Since you are selling shirts for only \$9 each, you will be losing money on the additional shirts. Therefore you should not increase production.

17. a.
$$\begin{array}{r} 1 \\ x-1\overline{)\,x-6} \\ \underline{x-1} \\ -5 \end{array}$$

$\dfrac{x-6}{x-1}=1-\dfrac{5}{x-1}$

b. The graph can be obtained from the basic graph $y=\dfrac{1}{x}$ by the sequence of transformations
$y=\dfrac{1}{x}\rightarrow y=-\dfrac{1}{x}\rightarrow y=-\dfrac{5}{x}\rightarrow$
$y=-\dfrac{5}{x-1}\rightarrow y=-\dfrac{5}{x-1}+1$
The transformations reflect the basic graph in the x-axis, stretch it vertically by a factor of 5, then shift it 1 unit to the right and 1 unit up.

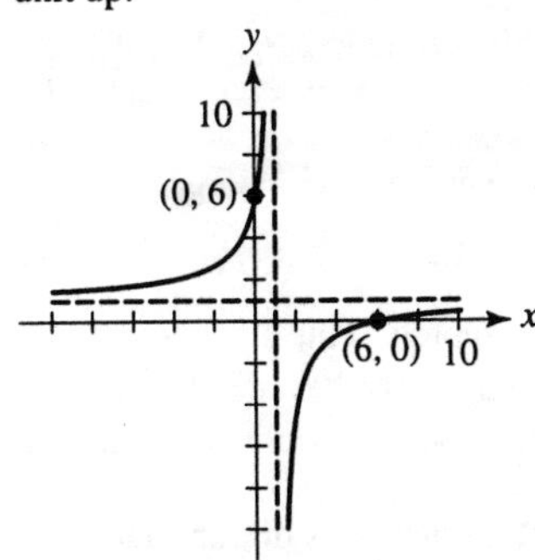

The vertical shift moves the horizontal asymptote 1 unit up, so that it is on the line $y=1$.

19. a. (Sample response) When $|x|$ is large, then $|x-1|$ is also large. Since the reciprocal of a large number is near 0, any expression $\dfrac{\text{constant}}{x-1}$ is small when $|x|$ is large.

b. (Sample response) Since $h(x)$ is the sum of the expressions $x+2$ and $-\dfrac{4}{x-1}$, and since the second expression is near 0 when $|x|$ is large, $h(x)$ must be near $x+2$ when $|x|$ is large.

21.

x	$g(x)$ (approximate)
10	0.44
100	0.95
1000	0.995
10000	0.9995

The table suggests that if $|x|$ is large, $g(x)$ is near 1.

23. The end behavior of $f(x)$ agrees with that of $y=\dfrac{x^2}{3x^2}=\dfrac{1}{3}$, so the graph has a horizontal asymptote at $y=\dfrac{1}{3}$.

25. The end behavior of $A(t)$ agrees with that of $y=-\dfrac{t^5}{6t^3}=-\dfrac{1}{6}t^2$. Thus the graph has no horizontal asymptote, and both tails point down.

27. a. He produces 2000 open-edition pieces each year, so the number produced in t years is $P(t)=2000t$.

b. He produced 500 original and limited-edition pieces prior to 1990, and has produced 50 per year since then. The number produced in t years since 1990 is therefore $Q(t)=500+50t$.

c. $f(t)=\dfrac{2000t}{500+50t}$
This represents the ratio of open-edition pieces to original and limited-edition pieces in Robert Arnold's work.

29. **a.** The function is discontinuous when $x = 4$.

b. When x is near 4, the numerator is near 12 and the denominator is small, so the quotient is large. The discontinuity is a vertical asymptote.

c. The end behavior agrees with that of $y = \frac{x}{x} = 1$, so the graph has a horizontal asymptote at $y = 1$.

31. **a.** The function is discontinuous when $x = 0$.

b. The equation can be simplified:
$y = \frac{x(x-4)}{x} = x - 4$ if $x \neq 0$.
When x is near 0, y is near -4, so the discontinuity is a missing point.

c. The end behavior agrees with that of $y = x - 4$, so the graph has no horizontal asymptote. The left tail points down and the right tail points up.

33. **a.** The function is discontinuous when $x + 3 = 0$, that is, when $x = -3$.

b. When x is near -3, the numerator of the fractional term is near 12 and the denominator is small, so the quotient is large. The discontinuity is a vertical asymptote.

c. The function can be rewritten
$$y = \frac{6-2x}{x+3} + \frac{x+3}{x+3}$$
$$y = \frac{(6-2x)+(x+3)}{x+3}$$
$$y = \frac{9-x}{x+3}$$
The end behavior agrees with that of $y = -\frac{x}{x} = -1$, so the graph has a horizontal asymptote at $y = -1$.

35. **a.** The function is discontinuous when $x^2 - 3 = 0$, that is, when $x = \pm\sqrt{3}$.

b. When x is near $\pm\sqrt{3}$, the numerator is 1 and the denominator is small, so the quotient is large. Both discontinuities are vertical asymptotes.

c. The end behavior agrees with that of $y = \frac{1}{x^2}$, which has a horizontal asymptote at $y = 0$. Thus the given function also has a horizontal asymptote at $y = 0$.

37. **a.** The function is discontinuous when $x^2 - 8x + 7 = 0$, that is, when $x = 1, 7$.

b. The equation can be simplified to $y = \frac{x-1}{(x-1)(x-7)} = \frac{1}{x-7}$ if $x \neq 1$. When x is near 1, y is near $-\frac{1}{6}$, so the discontinuity at $x = 1$ is a missing point. When x is near 7, the numerator is 1 and the denominator is small, so the quotient is large. The discontinuity at $x = 7$ is a vertical asymptote.

c. The end behavior agrees with that of $y = \frac{x}{x^2} = \frac{1}{x}$, which has a horizontal asymptote at $y = 0$. Thus the given function also has a horizontal asymptote at $y = 0$.

39. **a.** The function is discontinuous when $x + 2 = 0$, that is, when $x = -2$.

b. The equation can be simplified:
$y = \frac{(x+2)^2}{x+2} = x + 2$ if $x \neq -2$.
When x is near -2, y is near 0, so the discontinuity is a missing point.

c. The end behavior agrees with that of $y = \frac{x^2}{x} = x$, which has no horizontal asymptote. Thus the given function also has no horizontal asymptote. The left tail points down and the right tail points up.

41. **a.** The function is discontinuous when $2 - x^4 = 0$, that is, when $x = \pm\sqrt[4]{2}$.

b. When x is near $\sqrt[4]{2}$, the numerator is near $2 - \sqrt[4]{2^3} \cong 0.32$ and the denominator is small, so the quotient is large. When x is near $-\sqrt[4]{2}$, the numerator is near $2+\sqrt[4]{2^3} \cong 3.68$ and the denominator is small, so the quotient is large. Both discontinuities are vertical asymptotes.

c. The end behavior of $f(x)$ agrees with that of $y=-\frac{x^4}{x^4}=-1$, so the graph has a horizontal asymptote at $y=-1$.

43. a-b. Since $x^4+16>0$ for all real values of x, $R(x)$ is never discontinuous.

c. The end behavior of $R(x)$ agrees with that of $y=\frac{10x^2}{x^4}=\frac{10}{x^2}$, which has a horizontal asymptote at $y=0$. Thus $R(x)$ also has a horizontal asymptote at $y=0$.

45. The discontinuity occurs when $50t+500=0$, that is, when $t=-10$. The discontinuity may be interpreted as follows. Robert Arnold had produced 500 original and limited-edition pieces prior to 1990. If he had produced them at his current rate of 50 per year, he would have begun in 1980, at $t=-10$.

47. a. Since the volume of the can must be 21.7 cubic inches, $\pi r^2 h=21.7$, so $h=\frac{21.7}{\pi r^2}$.

b. The surface area is

$$S(r)=2\pi r^2+2\pi r\left(\frac{21.7}{\pi r^2}\right)\cong 6.28r^2+\frac{43.4}{r}.$$

c. S is discontinuous when $r=0$, and its graph has a vertical asymptote there. This says that if the radius of the can is near 0, the surface area is extremely large (because the height must be large to produce a volume of 21.7 in^3).

d. The end behavior of S agrees with that of $S=r^2$, so the right tail of the graph points up. This says that if the radius of the can is large, the surface area is also large.

49. a. The function $f(x)$ is the sum of the expressions x^2+2 and $\frac{3x-2}{2x^2+x-4}$. Exercise 48 shows that the second expression is near 0 if $|x|$ is large, so that $f(x)$ is near x^2+2.

b. The function $f(x)$ is the sum of $P(x)$ and the remainder from the division. The remainder is a rational function in which the denominator has a larger degree than the numerator. By the Highest Degree Theorem, the remainder is near 0 when $|x|$ is large, so $f(x)$ is near $P(x)$.

51. Division yields $y=x+\frac{1}{x}$. Thus the slant asymptote is $y=x$.

53. Division yields $y=3x+4.5+\frac{13.5}{2x-3}$. Thus the slant asymptote is $y=3x+4.5$.

55. Division yields $y=2x+2-\frac{2}{x^2-x+1}$. Thus the slant asymptote is $y=2x+2$.

57. Division yields $y=0.5x+\frac{-6x^2+1}{2x^3+8x}$. Thus the slant asymptote is $y=0.5x$.

Section 8-3

1. Multiply both sides by $x-1$:
$2x+7=5(x-1)$
$2x+7=5x-5$
$12=3x$
$x=4$
Since $x-1\neq 0$ when $x=4$, the solution is valid.

3. Multiply both sides by $x-5$:
$2x-1-3(x-5)=9$
$2x-1-3x+15=9$
$-x=-5$
$x=5$
Since $x-5=0$ when $x=5$, the solution is extraneous, so the equation has no solution.

5. a. Multiply both sides by x:

$700+3.50x+0.01x^2=9x$

$0.01x^2-5.5x+700=0$

$x=\dfrac{5.5\pm\sqrt{5.5^2-4(0.01)(700)}}{2(0.01)}$

$=200,\ 350.$

Since the denominator $x\neq 0$ when $x=200$ or 350, both solutions are valid.

b. If your production level is either 200 or 350 shirts per month, the average cost per shirt is \$9.

7. The inequality is true when the graph of $y=\dfrac{(x+3)(x-7)}{x-2}$ is on or above the x-axis. The graph is above the x-axis in $(-3, 2)\cup(7,\infty)$. It is on the x-axis when the numerator is 0, that is, when $x=-3, 7$. The solution to the inequality is $[-3, 2)\cup[7,\infty)$.

9. Rewrite the inequality:

$\dfrac{1}{x}-\dfrac{1}{x^2}>0$

$\dfrac{x-1}{x^2}>0$

The inequality is true when the graph of $y=\dfrac{x-1}{x^2}$ is above the x-axis, that is, in $(1,\infty)$.

11. a. Define $f(x)=\dfrac{x+1}{x^2-9}$, and solve $f(x)=0$:

$x+1=0$

$x=-1$

Since the denominator $x^2-9\neq 0$ when $x=-1$, the solution is valid. Also, observe that $f(x)$ is discontinuous when $x=\pm 3$. Indicate the zeros and discontinuities on a number line.

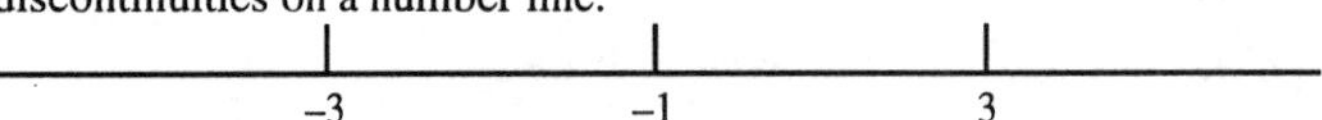

Choose a test value in each of the intervals $(-\infty,-3), (-3,-1), (-1,3), (3,\infty)$.

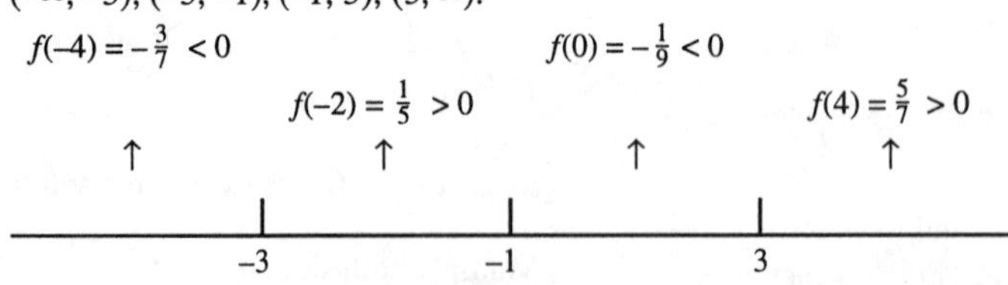

The inequality is also true when $x=-1$. The solution is $(-\infty,-3)\cup[-1, 3)$.

b. Define $f(x)=\dfrac{(x+3)(x-7)}{x-2}$, and solve

$f(x)=0$:

$(x+3)(x-7)=0$

$x=-3, 7$

Since the denominator $x-2\neq 0$ when $x=-3$ or $x=7$, the solution is valid. Also observe that $f(x)$ is discontinuous when $x=2$. Indicate the zeros and discontinuities on a number line.

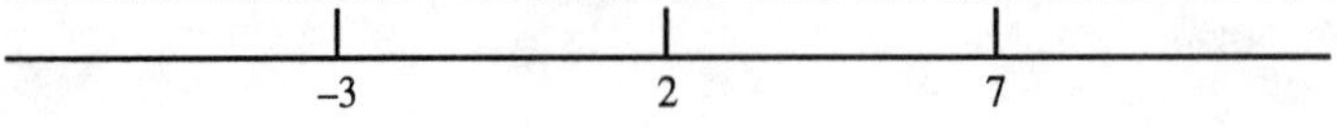

Choose a test value in each of the intervals
$(-\infty, -3)$, $(-3, 2)$, $(2, 7)$, $(7, \infty)$.

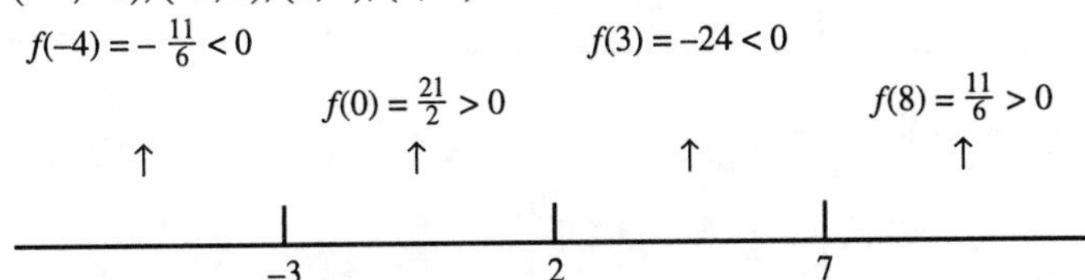

The inequality is also true when $x = -3$ or $x = 7$. The solution is $[-3, 2) \cup [7, \infty)$.

c. Define $f(x) = \dfrac{1}{(x+4)^2} - 1$

$$f(x) = \frac{1}{(x+4)^2} - \frac{(x+4)^2}{(x+4)^2}$$

$$f(x) = \frac{1-(x+4)^2}{(x+4)^2}$$

$$f(x) = \frac{1-(x^2+8x+16)}{(x+4)^2}$$

$$f(x) = \frac{-x^2-8x-15}{(x+4)^2}$$

Solve $f(x) = 0$:

$-x^2 - 8x - 15 = 0$

$-(x+5)(x+3) = 0$

$x = -5, -3$

Since the denominator $(x+4)^2 \neq 0$ when $x = -5$ or $x = -3$, the solution is valid. Also, observe that $f(x)$ is discontinuous when $x = -4$. Indicate the zeros and discontinuities on a number line.

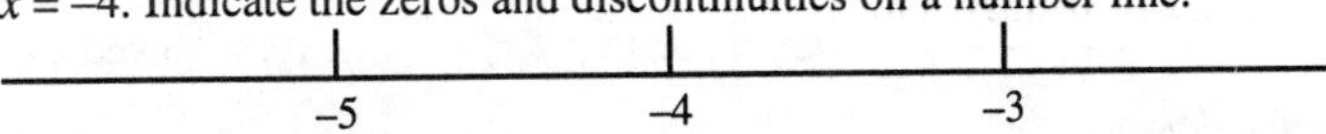

Choose a test value in each of the intervals
$(-\infty, -5)$, $(-5, -4)$, $(-4, -3)$, $(-3, \infty)$.

$f(-6) = -\frac{3}{4} < 0$ $\qquad$ $f(-3.5) = 3 > 0$

$f(-4.5) = 3 > 0$ $\qquad$ $f(-2) = -\frac{3}{4} < 0$

↑ ↑ ↑ ↑

−5 −4 −3

The solution is $(-\infty, -5) \cup (-3, \infty)$.

d. Define $f(x) = \dfrac{1}{x} - \dfrac{1}{x^2}$

$$f(x) = \frac{x}{x^2} - \frac{1}{x^2}$$

$$f(x) = \frac{x-1}{x^2}$$

Solve $f(x) = 0$:

$x - 1 = 0$

$x = 1$

Since the denominator $x^2 \neq 0$ when $x = 1$, the solution is valid. Also, observe that $f(x)$ is discontinuous when $x = 0$. Indicate the zeros and discontinuities on a number line.

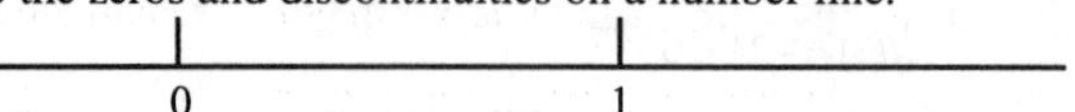

Choose a test value in each of the intervals $(-\infty, 0)$, $(0, 1)$, $(1, \infty)$.

$f(-1) = -2 < 0$ $f(2) = 1 > 0$

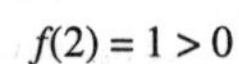

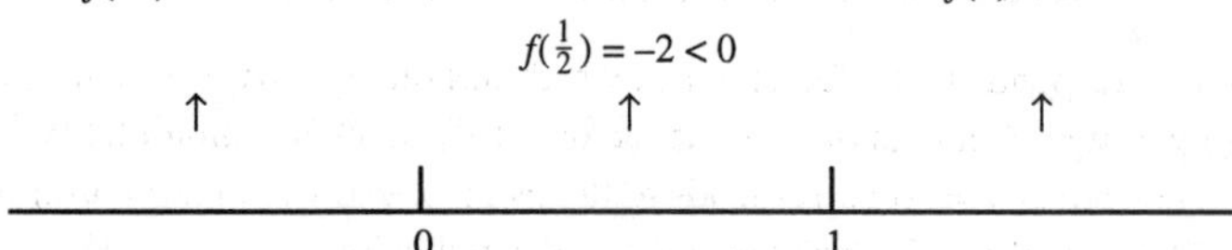

The solution is $(1, \infty)$.

13. a. As in Exercise 11a, define $f(x) = \frac{x+1}{x^2 - 9} = \frac{x+1}{(x+3)(x-3)}$, and determine that $f(x) = 0$ when $x = -1$ and is discontinuous when $x = \pm 3$. Indicate these x-values on a number line.

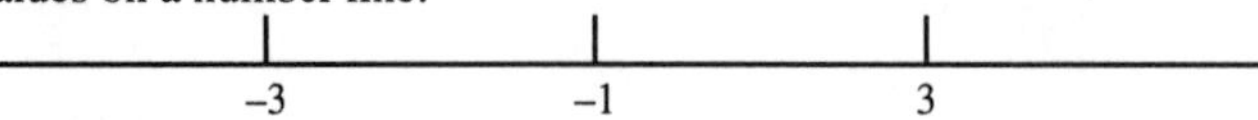

The end behavior of $f(x)$ agrees with that of $y = \frac{x}{x^2} = \frac{1}{x}$, so $f(x) < 0$ on $(-\infty, -3)$. The discontinuity at $x = -3$ corresponds to the factor $x + 3$, which has multiplicity 1, so the sign of $f(x)$ changes there. The zero at $x = -1$ corresponds to the factor $x + 1$, which has multiplicity 1, so the sign of $f(x)$ changes there. The discontinuity at $x = 3$ corresponds to the factor $x - 3$, which has multiplicity 1, so the sign of $f(x)$ changes there. The sign of $f(x)$ in each interval is indicated below.

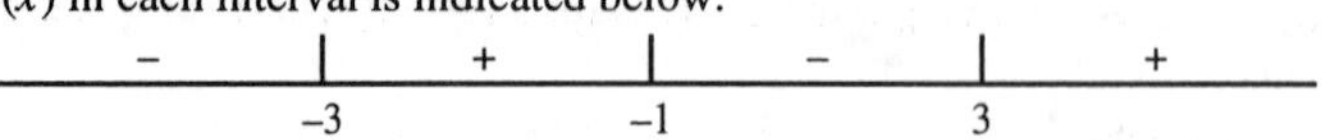

The inequality is also true when $x = -1$. The solution is $(-\infty, -3) \cup [-1, 3)$.

b. As in Exercise 11b, define $f(x) = \frac{(x+3)(x-7)}{x-2}$, and determine that $f(x) = 0$ when $x = -3$ and $x = 7$ and is discontinuous when $x = 2$. Indicate these x-values on a number line.

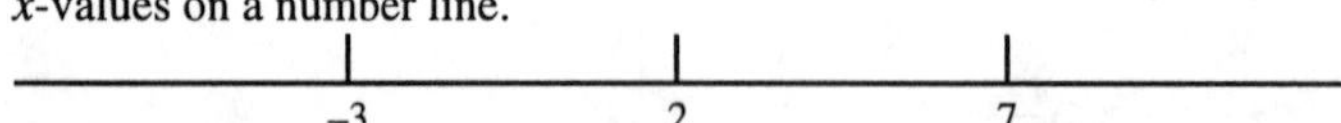

If the numerator of $f(x)$ is multiplied out, the term of highest degree is x^2. The end behavior of $f(x)$ agrees with that of $y = \frac{x^2}{x} = x$, so $f(x) < 0$ on $(-\infty, -3)$. The zero at $x = -3$ corresponds to the factor $x + 3$, which has multiplicity 1, so the sign of $f(x)$ changes there. The discontinuity at $x = 2$ corresponds to the factor $x - 2$, which has multiplicity 1, so the sign of $f(x)$ changes there. The zero at $x = 7$ corresponds to the factor $x - 7$, which has multiplicity 1, so the sign of $f(x)$ changes there. The sign of $f(x)$ in each interval is indicated below.

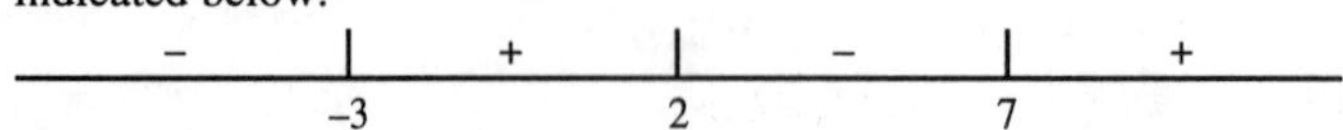

The inequality is also true when $x = -3$ or $x = 7$. The solution is $[-3, 2) \cup [7, \infty)$.

c. As in Exercise 11c, define $f(x)=\dfrac{-x^2-8x-15}{(x+4)^2}=\dfrac{-(x+5)(x+3)}{(x+4)^2}$, and determine that $f(x)=0$ when $x=-5$ and $x=-3$ and is discontinuous when $x=-4$. Indicate these x-values on a number line.

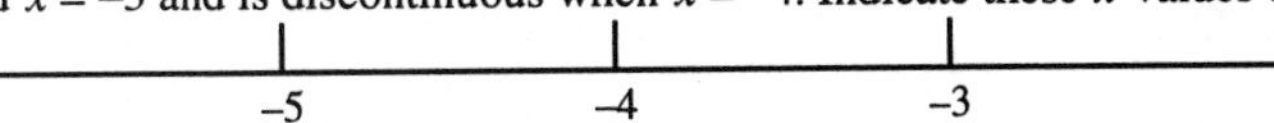

If the numerator of $f(x)$ is multiplied out, the term of highest degree is $-x^2$. If the denominator is multiplied out, the term of highest degree is x^2. The end behavior of $f(x)$ agrees with that of $y=\dfrac{-x^2}{x^2}=-1$, so $f(x)<0$ on $(-\infty,-5)$. The zero at $x=-5$ corresponds to the factor $x+5$, which has multiplicity 1, so the sign of $f(x)$ changes there. The discontinuity at $x=-4$ corresponds to the factor $x+4$, which has multiplicity 2, so the sign of $f(x)$ does not change there. The zero at $x=-3$ corresponds to the factor $x+3$, which has multiplicity 1, so the sign of $f(x)$ changes there. The sign of $f(x)$ in each interval is indicated below.

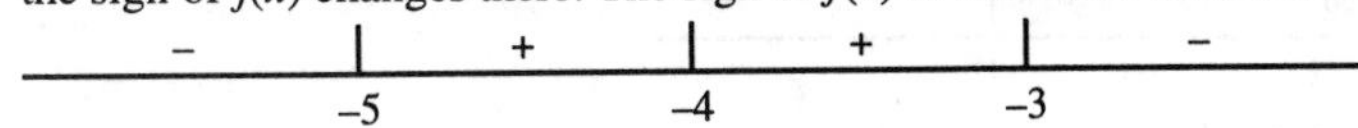

The solution is $(-\infty,-5)\cup(-3,\infty)$.

d. As in Exercise 11d, define $f(x)=\dfrac{x-1}{x^2}$, and determine that $f(x)=0$ when $x=1$ and is discontinuous when $x=0$. Indicate these x-values on a number line.

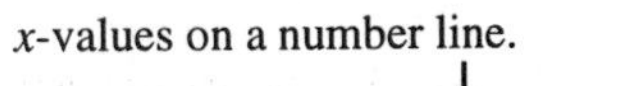

The end behavior of $f(x)$ agrees with that of $y=\dfrac{x}{x^2}=\dfrac{1}{x}$, so $f(x)<0$ on $(-\infty,0)$. The discontinuity at $x=0$ corresponds to the factor x^2, which has multiplicity 2, so the sign of $f(x)$ does not change there. The zero at $x=1$ corresponds to the factor $x-1$, which has multiplicity 1, so the sign of $f(x)$ changes there. The sign of $f(x)$ in each interval is indicated below.

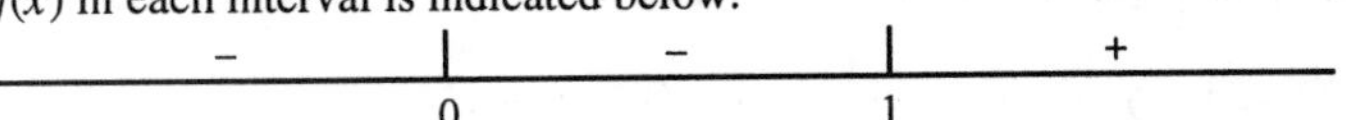

The solution is $(1,\infty)$.

15. Define $f(x)=\dfrac{x^2+3x}{x-4}$, and solve $f(x)=0$:

$x^2+3x=0$
$x(x+3)=0$
$x=0,-3$

Since the denominator $x-4\neq 0$ when $x=0$ or -3 both solutions are valid. Also observe $f(x)$ is discontinuous when $x=4$. Indicate the zeros and discontinuities on a number line:

Choose a test value in each of the intervals $(-\infty,-3)$, $(-3,0)$, $(0,4)$, $(4,\infty)$.

$f(-4)=-\frac{1}{2}<0$ ↑ (on $(-\infty,-3)$)

$f(-1)=\frac{2}{5}>0$ ↑ (on $(-3,0)$)

$f(1)=-\frac{4}{3}<0$ ↑ (on $(0,4)$)

$f(5)=40>0$ ↑ (on $(4,\infty)$)

Number line: -3, 0, 4

The inequality is true in $(-3,0)\cup(4,\infty)$.

17. Define $f(x) = \frac{x^2+3x-1}{x+2} - 1$ and solve $f(x) = 0$:

$\frac{x^2+3x-1}{x+2} - 1 = 0$

$(x^2+3x-1)-(x+2) = 0$

$x^2+2x-3 = 0$

$(x+3)(x-1) = 0$

$x = -3, 1$

Since the denominator $x + 2 \neq 0$ when $x = -3$ or 1, both solutions are valid. Also, observe that $f(x)$ is discontinuous when $x = -2$. Indicate the zeros and discontinuities on a number line:

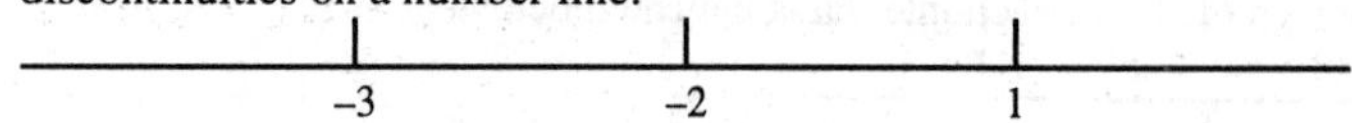

Choose a test value in each of the intervals $(-\infty, -3), (-3, -2), (-2, 1), (1, \infty)$.

$f(-4) = -2.5 < 0$ $f(0) = -1.5 < 0$

$f(-2.5) = 3.5 > 0$ $f(2) = 1.25 > 0$

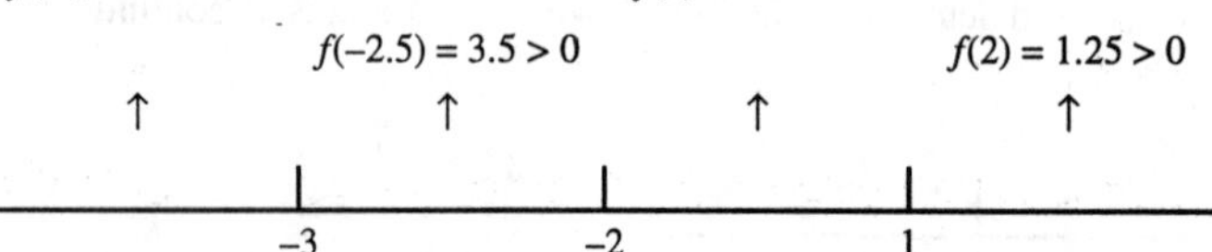

The inequality is true in $[-3, -2) \cup [1, \infty)$.

19. Define $f(x) = \frac{x}{x-1}$. The solution to $f(x) = 0$ is $x = 0$, and $f(x)$ is discontinuous when $x = 1$.

Choose a test value in $(-\infty, 0)$: $f(-1) = \frac{1}{2} > 0$

Choose a test value in $(0, 1)$: $f\left(\frac{1}{2}\right) = -1 < 0$

Choose a test value in $(1, \infty)$: $f(2) = 2 > 0$

The inequality is true in $(0, 1)$.

21. Define $f(x) = \frac{x^2+x-6}{x} - 1$, and solve $f(x) = 0$:

$\frac{x^2+x-6}{x} - 1 = 0$

$(x^2+x-6) - x = 0$

$x^2 - 6 = 0$

$x = \pm\sqrt{6}$

Also, observe that $f(x)$ is discontinuous when $x = 0$.

Choose a test value in $(-\infty, -\sqrt{6})$:

$f(-3) = -1 < 0$

Choose a test value in $(-\sqrt{6},\ 0)$:
$f(-1) = 5 > 0$
Choose a test value in $(0,\ \sqrt{6})$:
$f(1) = -5 < 0$
Choose a test value in $(\sqrt{6},\ \infty)$:
$f(3) = 1 > 0$
The inequality is true in $[-\sqrt{6},\ 0) \cup [\sqrt{6},\ \infty)$.

23. Define $f(x) = \dfrac{(x+20)^2(x-10)^3}{(x+10)(x-30)^4}$. The solutions to $f(x) = 0$ are $x = -20$, 10, and $f(x)$ is discontinuous when $x =$ -10, 30.
Choose a test value in $(-\infty, -20)$:
$$f(-30) = \frac{(-10)^2(-40)^3}{(-20)(-60)^4} > 0$$
Choose a test value in $(-20, -10)$:
$$f(-15) = \frac{(5)^2(-25)^3}{(-5)(-45)^4} > 0$$
Choose a test value in $(-10, 10)$:
$$f(0) = \frac{(20)^2(-10)^3}{(10)(-30)^4} < 0$$
Choose a test value in $(10, 30)$:
$$f(20) = \frac{(40)^2(10)^3}{(30)(-10)^4} > 0$$
Choose a test value in $(30, \infty)$:
$$f(40) = \frac{(60)^2(30)^3}{(50)(10)^4} > 0$$
The inequality is true in $(-\infty, -20) \cup (-20, -10) \cup (10, 30) \cup (30, \infty)$.

25. Define $f(x) = \dfrac{x}{x^2+1}$. The solution to $f(x) = 0$ is $x = 0$, and $f(x)$ is never discontinuous.
Choose a test value in $(-\infty, 0)$: $f(-1) = -\frac{1}{2} < 0$
Choose a test value in $(0, \infty)$: $f(1) = \frac{1}{2} > 0$
The inequality is true in $[0, \infty)$.

27. Define $f(x) = \dfrac{x^3+2x-1}{x^3} - 1$, and solve
$f(x) = 0$:
$$\frac{x^3+2x-1}{x^3} - 1 = 0$$
$$x^3 + 2x - 1 - x^3 = 0$$
$$2x - 1 = 0$$
$$x = \frac{1}{2}$$
Also, observe that $f(x)$ is discontinuous when $x = 0$.
Choose a test value in $(-\infty, 0)$: $f(-1) = 3 > 0$

Choose a test value in $\left(0, \frac{1}{2}\right)$: $f(0.1) = -800 < 0$

Choose a test value in $\left(\frac{1}{2}, \infty\right)$: $f(1) = 1 > 0$

The inequality is true in $(-\infty, 0) \cup \left[\frac{1}{2}, \infty\right)$.

29. Define $f(x) = \dfrac{(x-1)(x-7)}{(x-3)^4}$. Solve $f(x) = 0$ to obtain $x = 1, 7$. Observe that $f(x)$ is discontinuous when $x = 3$. Indicate these values on a number line.

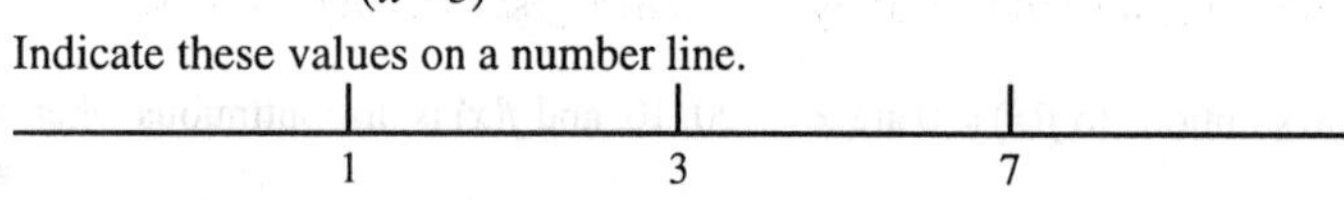

If $f(x)$ is multiplied out, the terms of highest degree in its numerator and denominator are x^2 and x^4. Thus the end behavior of $f(x)$ agrees with that of $y = \dfrac{x^2}{x^4} = \dfrac{1}{x^2}$, so $f(x) > 0$ on $(-\infty, 1)$. The zero at $x = 1$ has odd multiplicity, so the sign of f changes there. The discontinuity at $x = 3$ corresponds to a factor of even multiplicity in the denominator, so the sign of f does not change there. The zero at $x = 7$ has odd multiplicity, so the sign of f changes there. The sign pattern for f is shown below.

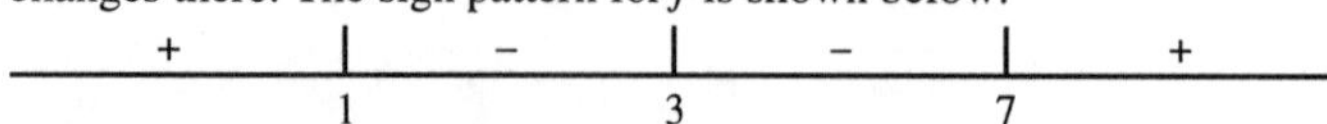

The inequality is true in $(-\infty, 1) \cup (7, \infty)$.

31. Define $f(x) = \dfrac{x^2(x+5)^2}{x+2}$. Solve $f(x) = 0$ to obtain $x = -5, 0$. Observe that $f(x)$ is discontinuous when $x = -2$. Indicate these values on a number line.

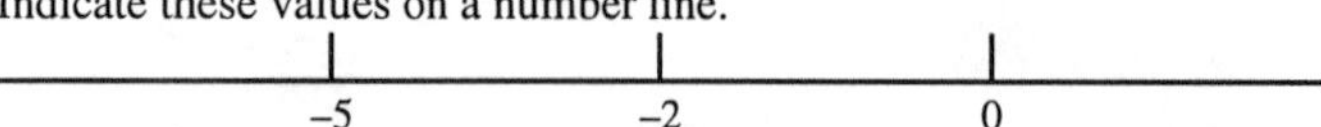

If $f(x)$ is multiplied out, the terms of highest degree in its numerator and denominator are x^4 and x. Thus the end behavior of $f(x)$ agrees with that of $y = \dfrac{x^4}{x} = x^3$, so $f(x) < 0$ on $(-\infty, -5)$. The zero at $x = -5$ has even multiplicity, so the sign of f does not change there. The discontinuity at $x = -2$ corresponds to a factor of odd multiplicity in the denominator, so the sign of f changes there. The zero at $x = 0$ has even multiplicity, so the sign of f does not change there. The sign pattern for f is shown below:

– | – | + | +
–5 –2 0

The inequality is true in $(-\infty, -2) \cup \{0\}$.

33. a. The values are the solutions to

$$2\pi r^2 + \frac{43.4}{r} < 100$$

Rewrite the inequality

$$2\pi r^2 + \frac{43.4}{r} - 100 < 0$$

$$\frac{2\pi r^3 + 43.4 - 100r}{r} < 0$$

An analytical solution is not convenient. The graph of $y = \dfrac{2\pi r^3 + 43.4 - 100r}{r}$ indicates r-intercepts at about $(-4.19, 0)$, $(0.44, 0)$, and $(3.75, 0)$. The graph is below the r-axis in $(-4.19, 0) \cup (0.44, 3.75)$. In the context of the problem, the solution is $(0.44, 3.75)$, indicating that the radius of the can should be between 0.44 and 3.75 inches.

b. The graph of $y = S(r)$ indicates that the minimum value occurs at about (1.51, 43.06). The can has a radius of 1.51 inches, and a height of $\frac{21.7}{\pi(1.51)^2} \cong 3.02$ inches.

c. (Sample response) A can of the optimal shape is difficult to hold and drink from. Furthermore, cans of a nonstandard shape would be hard to recognize in a supermarket, and would not fit most vending machines. Finally, the cost of retooling a canning plant would probably outweigh any savings in materials.

NUTSHELLS

Nutshell for Chapter 9 - EXPONENTIAL AND LOGARITHMIC FUNCTIONS

What's familiar?

- In this chapter, you study *two* categories of functions numerically, analytically, and graphically.
- In Section 9-1, you learn to fit an *exponential* function to a table and sketch the graphs of many exponential functions by transforming the basic graphs $y = b^x$.
- Section 9-1 also leads you to discover some properties of *geometric sequences*, that is, sequences that are exponential functions.
- In Section 9-4, you learn to sketch the graphs of many *logarithmic* functions by transforming the basic graphs $y = \log_b x$.

What's new?

- The category of exponential and logarithmic functions *does* not include that of rational functions. In fact the two categories are entirely separate. Exponential and logarithmic functions exhibit properties quite different from those of rational functions, but their properties are closely related to each other.
- Section 9-2 discusses a special number called *e*. This number is especially useful for performing the operations of calculus on exponential and logarithmic functions.
- Section 9-3 develops the idea of the *inverse function*, which is the key to the relationship between exponential and logarithmic functions.
- Section 9-4 develops some properties of logarithms that correspond to the familiar laws of exponents.
- Section 9-5 shows you how to find a "line of best fit" for a nonlinear collection of data points. Using the properties of logarithms you can also apply this method to find "logarithmic, exponential, and power curves of best fit". These methods are especially useful for creating mathematical models from the approximate or inaccurate data that often come with actual problems.

List of hints for exercises in each chapter.

Chapter 9

- Remember that e is a real number not a variable.
- Use your calculator quite a bit when looking at inverse functions. The visual nature of graphs will add to your understanding.
- Be sure that you have learned to do curve-fitting on your calculator before beginning "Curve-Fitting on Your Calculator" on page 383.
- When working with logarithms, keep in mind, that they are "disguised exponents"!
- Make sure you understand the proofs of the properties of logarithms by writing them in your own words. it will really help you understand what they are and why they work the way they do.

Name:__________________

Date:__________________

Chapter 9 Additional Exercises

In 1–3, decide whether the table fits an exponential function. If it does, find its equation.

1.

x	y
1	15
2	45
3	135
4	405

1. __________________

2.

x	y
1	0.01
2	0.05
3	0.25
4	1.25

2. __________________

3.

x	y
1	0.1
2	0.5
3	0.005
4	0.025

3. 

In 4–7, sketch the graph of each function by transforming an appropriate graph $y = b^x$. In each case, state the domain and range of the function and decide whether the function is increasing or decreasing.

4. $f(x) = 3 \cdot \left(\frac{1}{2}\right)^x$

4. __________________

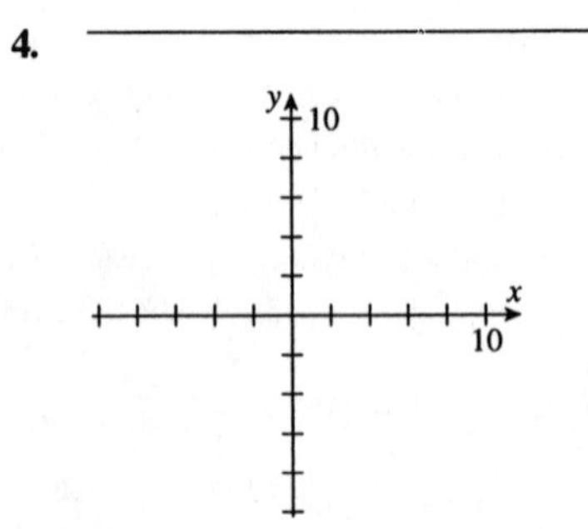

Chapter 9 Additional Exercises *(cont.)*

Name:________________

5. $f(x) = -\frac{1}{4} \cdot 3^{x-2}$

5. ________________

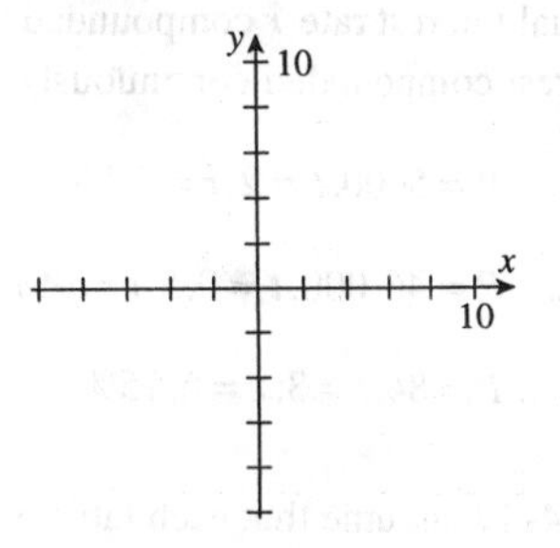

6. $f(x) = 4^{x-1} - 3$

6. ________________

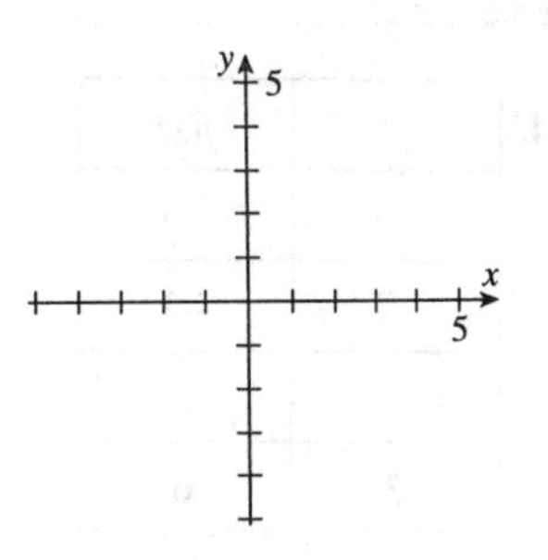

7. $f(x) = 4 - \left(\frac{1}{2}\right)^{3x}$

7. ________________

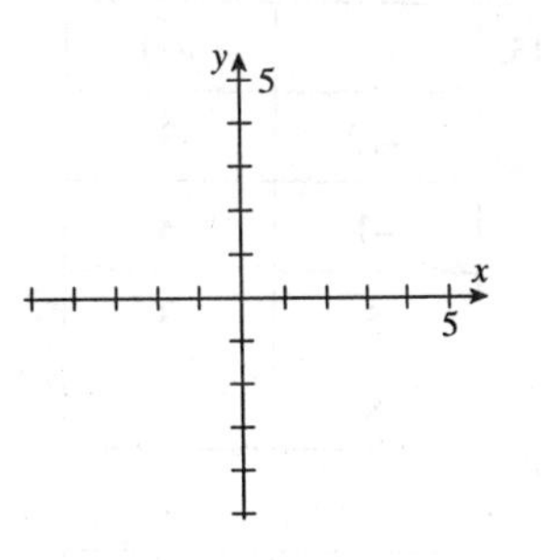

In 8–10,

a. Decide whether the sequence is geometric,

b. If the sequence is geometric, find a formula for a_n, and use your formula to find a_8.

c. If the sequence is geometric, decide whether the infinite series $a_1 + a_2 + a_3 + \ldots$ has a sum. If the sum exists, find it.

8. $a = \{1, 6, 36, 216, \ldots\}$

8. ________________

9. $a = \{1, 0.7, 0.49, 0.343, \ldots\}$

9. ________________

10. $a = \{48, 12, 3, \frac{3}{4} \ldots\}$

10. ________________

Chapter 9 Additional Exercises *(cont.)*

Name:________________

In 11–13, find the amount obtained when $\$P$ is invested for t years at an annual interest rate r compounded monthly. Repeat the calculation for interest compounded continuously.

11. $P = 5000, t = 2, r = 5.25\%$

11. ________________

12. $P = 46,000, t = 0.5, r = 9\%$

12. ________________

13. $P = 84, t = 3; r = 6.75\%$

13. ________________

In 14–17, assume that each table is complete, and decide whether it describes a one-to-one function. If it does, write a table to describe its inverse.

14.

x	$f(x)$
–3	–6
–1	–2
1	2
3	6

14. ________________

15.

x	$f(x)$
–12	3
–1	3
5	3
8	3

15. ________________

16.

x	$f(x)$
–2	$\frac{1}{9}$
–1	$\frac{1}{3}$
0	1
2	9

16. ________________

Chapter 9 Additional Exercises *(cont.)*

Name:________________________

In 17–20, graph the function and decide whether it is one-to-one. If it is, sketch the graph of its inverse.

17. $f(x) = 3 - x^2$

17. ____________________

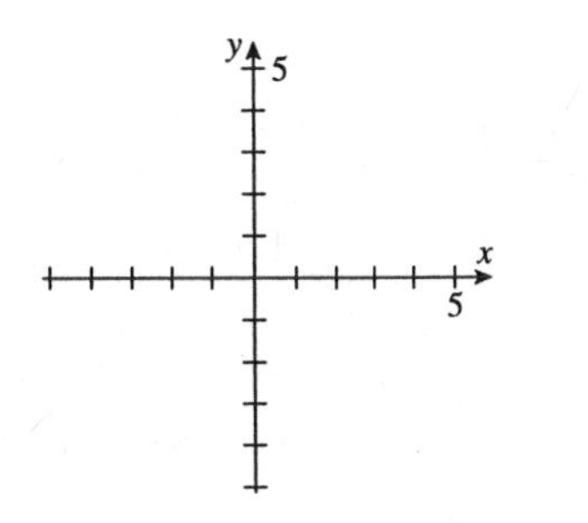

18. $g(x) = \sqrt{x+2}$

18. ____________________

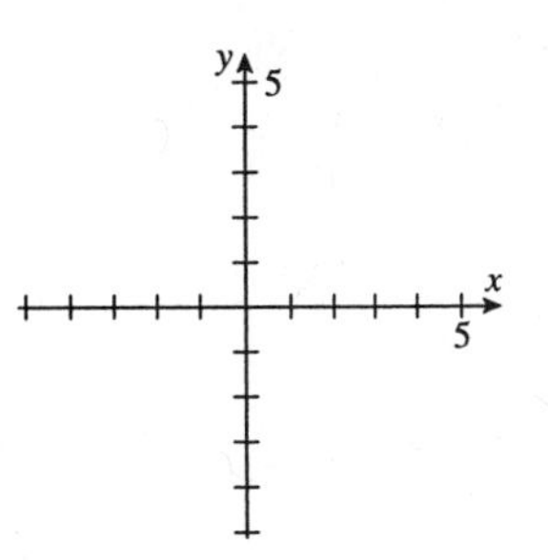

19. $h(x) = \left(\frac{1}{2}\right)^x$

19. ____________________

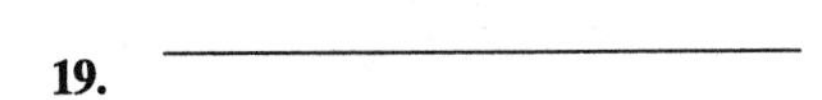

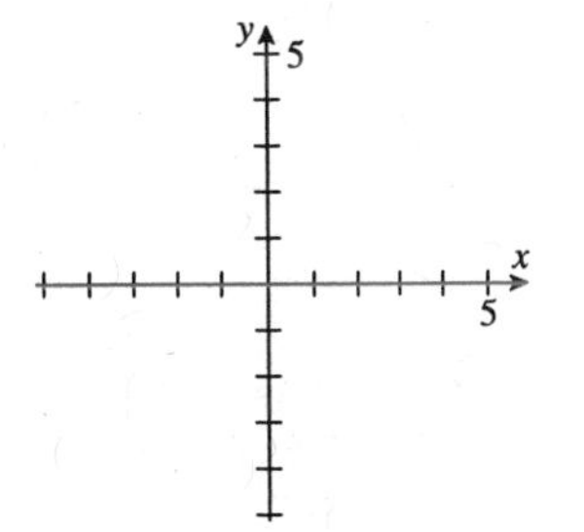

20. $k(x) = x^3 - 2$

20. ____________________

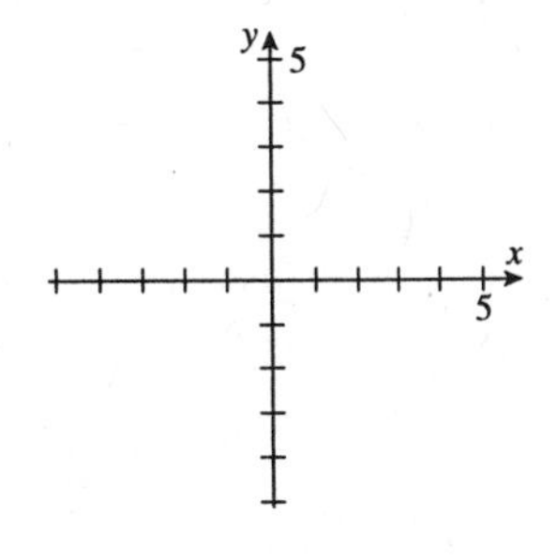

Chapter 9 Additional Exercises *(cont.)*

Name:____________________

In 21–24, find f^{-1} analytically, and graph f and f^{-1} on the same set of axes.

21. $f(x) = 0.5x - 3$

21.

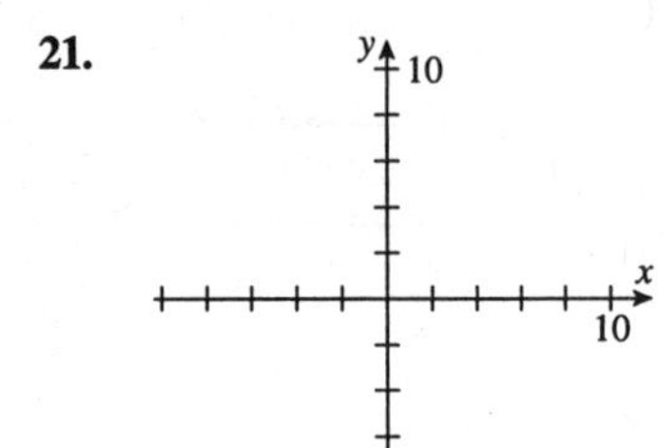

22. $f(x) = \dfrac{x^3 - 2}{3}$

22.

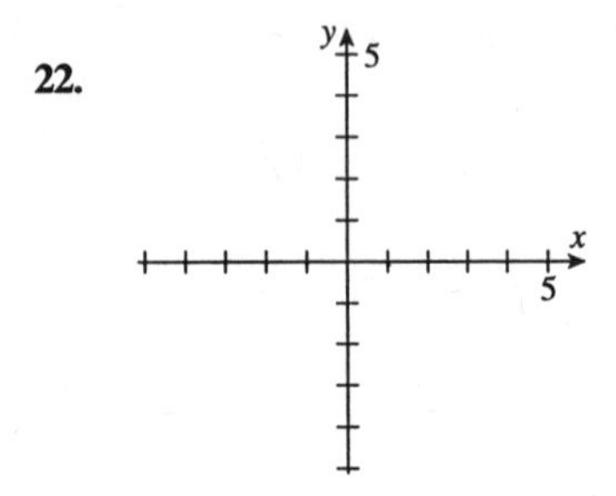

23. $f(x) = \dfrac{x}{x - 3}$

23.

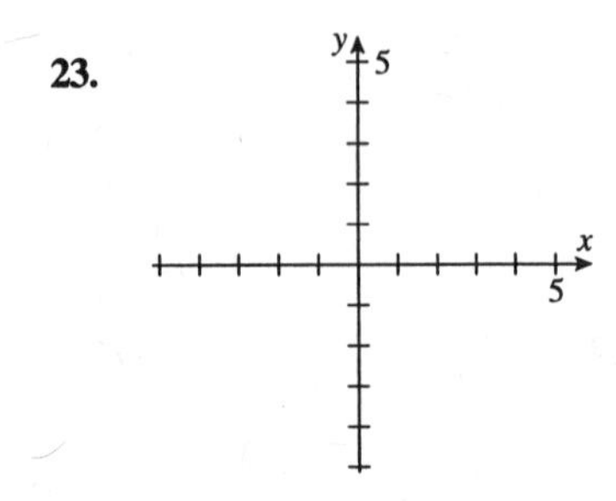

24. $f(x) = \sqrt{3 - x}$

24.

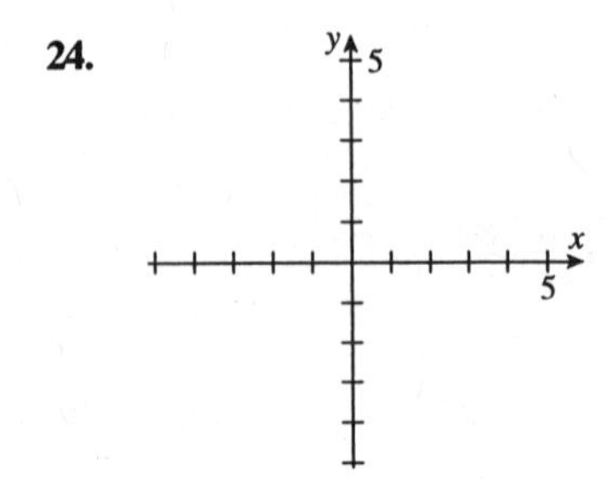

Chapter 9 Additional Exercises *(cont.)*

Name:____________________

In 25–26, decide whether the table fits a function of the form $y = \log_b x + k$.

25.

x	y
1.2	3
1.22	4
1.222	5
1.2222	6

25. ____________________

26.

x	y
0.3	2
3	4
30	6
300	8

26. ____________________

In 27–29, sketch the graph of each function by transforming an appropriate graph $y = \log_b x$. In each case, find the domain and range of the function, decide whether the function is increasing or decreasing and identify the points obtained by moving (1, 0).

27. $y = \log(x - 4)$

27. ____________________

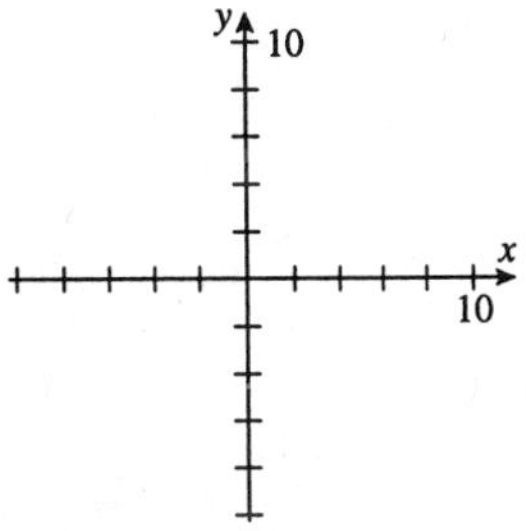

28. $y = \frac{1}{3}\ln x - 3$

28. ____________________

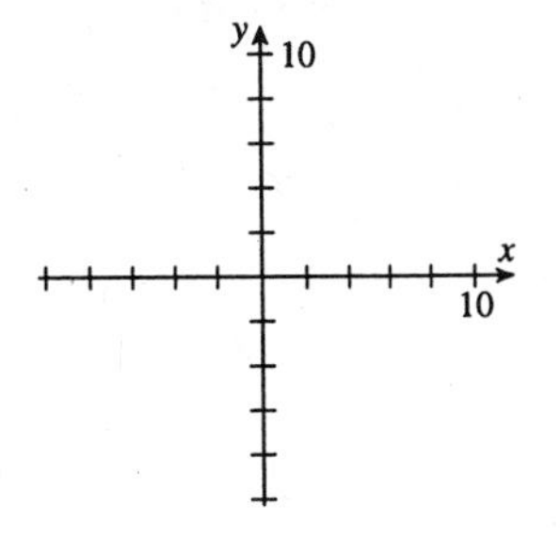

Name:______________________

29. $y = 3 - \log x$

29. ______________________

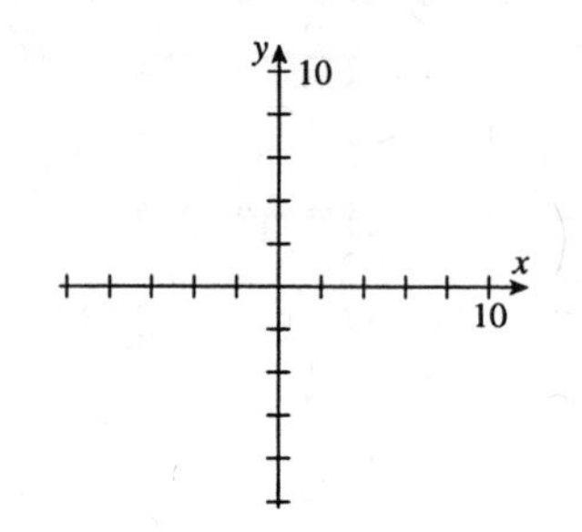

In 30–33, solve the equation.

30. $6\log_7(x+4) = 18$

30. ______________________

31. $\log_4 x + \log_4(x-6) = 2$

31. ______________________

32. $\log_x 8 + \log_x 125 = 3$

32. ______________________

33. $\ln\sqrt{5t} = 4$

33. ______________________

In 34–36, rewrite each expression to have no logarithms except $\log_7 x$.

34. $\log_7 \frac{1}{\sqrt[3]{x}}$

34. ______________________

35. $\log_7 \frac{x}{49}$

35. ______________________

36. $\log_7(x^3 + 9x^2) - \log_7(x+9)$

36. ______________________

In 37–38, find the equation of the regression line for the points (x, y) in the table.

37.

x	y
1	18
2	15
3	13
4	10

37. ______________________

38.

x	y
0.2	–6
0.5	2
0.7	8
1.0	16

38. ______________________

Chapter 9 Additional Exercises *(cont.)*

Name:______________________

In 39–40, plot the pairs $(\ln x, y)$ and assume that the values of x and y are subject to small errors. Decide whether x and y might be related by a logarithmic function.

39.

x	y
3	1
8	2
15	2.75
28	3

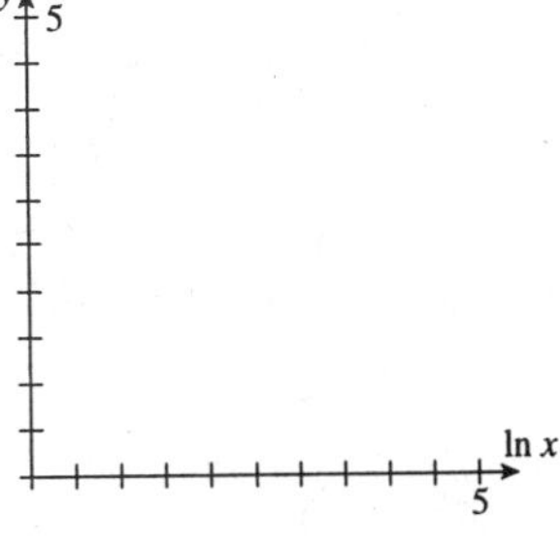

39. ______________________

40.

x	y
3	9
8	4.5
15	2
50	1

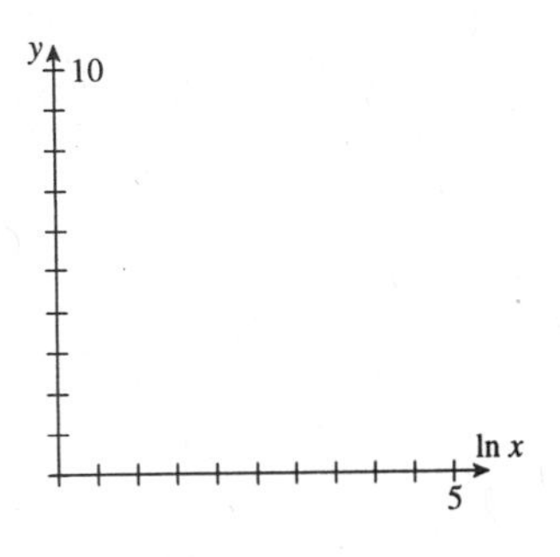

40. ______________________

41. Explain how to estimate $3^{\sqrt{5}}$ using rational exponents.

41. ______________________

42. You invest \$2000 at 4.75% interest. Then you find that another bank has a rate of 5.5%. Both have continuously compounded interest. How much money do you lose after 2 years?

42. ______________________

43. If interest is 7% compounded semiannually, explain why an investment of \$650 grows to $650\left(1+\frac{0.07}{2}\right)$ in 6 months.

43. ______________________

44. Whitney almost got hit in a crosswalk near the office where she works by a car driving around a curve. A building blocks drivers' line-of-sight until they are 40 feet from the crosswalk. The speed limit in the area is 25 miles per hour, but Whitney says this is too high.

a. Use the function for stopping distance (d) for cars traveling v miles per hour, $d = 0.06v^2 + 0.8v$, to argue Whitney's point.

44.a. ______________________

____

____

____

b. Use the inverse function, $v = \dfrac{5\sqrt{6d+16}-20}{3}$, to argue her point.

b. ______________________

c. Which function makes it easier for Whitney to argue her point? Why?

c. ______________________

Chapter 9 Additional Exercises *(cont.)*

Name:____________________

Use for 45–48: The intensity of earthquakes is often measured on the Richter scale. On this scale the magnitude M of an earthquake is related to its released energy E in ergs by $M = \dfrac{\log E - 11.8}{1.5}$.

45. A 1990 earthquake in Costa Rica had a magnitude of about 6.1. Find the amount of energy released by the earthquake.

45. ____________________

46. Another earthquake struck Costa Rica in 1991, with a magnitude of 7.4. Find the amount of anergy released by the 1991 earthquake.

46. ____________________

47. How many times more powerful was the 1991 earthquake than the 1990 one, in terms of energy released?

47. ____________________

48. Explain why an earthquake with a magnitude of $m+2$ is about 1000 times more powerful than one with a magnitude of m.

48. ____________________

49. Data on the amount of sugar and total carbohydrates in several foods are given in the table. Find the equation of the regression line for the points in the table and use it to predict the total carbohydrates in chicken noodle soup which has 1g of sugar.

49. ____________________

Food	Sugar (in grams)	Total Carbohydrates (in grams)
Peanut butter	3	7
Tomato soup	10	18
Children's cereal	13	25
Spaghetti sauce	15	23

50. Plot the pairs $(\ln x, y)$ where x is the amount of sugar and y is the total carbohydrates in the foods in the table. Decide whether x and y might be related by a logarithmic function.

50. 

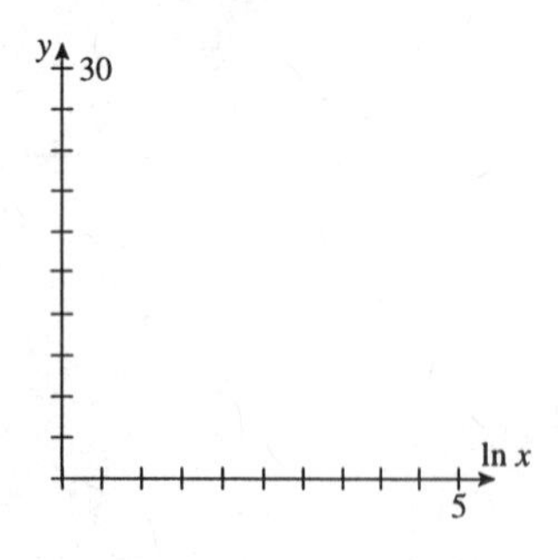

A Mathematical Looking Glass

Chocolate Surprise

(to follow Section 9-1)

In Rózsa Péter's book, *Playing with Infinity*, she tells of a well-known mathematician who, as a child, attempted to find the value of a chocolate candy bar. Ordinarily, this would be a simple matter of checking the price. However, this particular company put a coupon in the silver-paper wrapping of each bar, and ten coupons could be exchanged for another candy bar. Thus the value of a candy bar must be calculated to include the value of both the candy and the coupon.

For simplicity, suppose that the value of the candy without the coupon is \$1.00. Since the accompanying coupon is worth one tenth of a candy bar, the candy plus the coupon would appear to have a value of \$1.00 + \$0.10 = \$1.10. But wait! When we exchange the coupon for one tenth of a candy bar, we also receive another tenth of a coupon. This tenth of a coupon has a value of \$0.01, which must be added to the value of the original candy bar, making it worth \$1.00 + \$0.10 + \$0.01 = \$1.11. However, we are not done yet. The tenth of a coupon can purchase an additional one hundredth of a candy bar, which comes with an additional hundredth of a coupon, having a value of \$0.001. This in turn must be added to the value of the original candy bar, making it worth \$1.00 + \$0.10 + \$0.01 + \$0.001 = \$1.111. In Exercise 1, you can discover how to determine the candy bar's value in dollars by continuing this process.

In *Playing with Infinity*, Rózsa Péter explains the surprising conclusion of Exercise 1 as follows.

> It is enough to demonstrate that 9 coupons are worth one bar of chocolate, since then it is certain that one coupon is worth $\frac{1}{9}$th of this. Suppose that I have 9 coupons, then I can go into the shop and say: 'Please can I have a bar of chocolate? I should like to eat it here and now and I will pay afterwards.' I eat the chocolate, take out the accompanying coupon, and now I have 10 coupons, with which in fact I can actually pay and the whole business is concluded,

I have eaten the chocolate and I have no coupons left. So the exact value of 9 coupons is in fact one bar of chocolate, the value of one coupon is $\frac{1}{9}$th of a bar of chocolate, one bar of chocolate with a coupon is worth $1\frac{1}{9}$ bars of chocolate. So the sum of the infinite series

$$1 + \frac{1}{10} + \frac{1}{100} + \frac{1}{1000} + \frac{1}{10000} + \ldots$$

is exactly $1\frac{1}{9}$, quite tangibly, even edibly.

Exercise

1. (Problem Solving) The process described in **Chocolate Surprise** produces successive sums S_n of the terms in a geometric sequence. Find the value of the candy bar, in dollars, by finding the ceiling for the values of S_n.

A Mathematical Looking Glass
Fair Exchange?
(to follow Section 9-2)

One of the first recorded real estate transactions in North America was the purchase of Manhattan Island from the Algonquin tribe in 1626 by Peter Minuit, director-general of the Dutch West India Company. The exchange of the island for beads and cloth valued at 60 guilders (about $24) is widely regarded as having been a bargain for the Dutch, and a bad deal for the Algonquins. In fact, the Algonquins had many occasions to regret their association with the Dutch. For example, in 1638 one of Minuit's successors levied a tax on them for protection against the neighboring Iroquois, then refused assistance when the Iroquois attacked. The result of this act was four years of warfare between the Algonquins and the Dutch.

In strictly monetary terms, however, was the exchange a fair one? This question can be answered in several ways. For example, we could try to estimate the value of real estate around Manhattan in 1626. To take an alternate point of view, let's consider the present value of each side of the exchange. The value of all taxable real estate in New York City as of January, 1993, was $77.3 billion, and Manhattan accounts for a significant part of that total. On the other side of the exchange, the present value of the $24 depends on what was done with it. If the Algonquins had been able to take cash (instead of beads and cloth) and invest it, its value would have increased greatly over the intervening years. Let's see if we can estimate the potential amount of their investment as of 1994.

Of course, any estimate will be highly suspect, since we have no way of knowing the effects of wars, depressions, and other economic catastrophes on an investment over a period of 368 years. However, let's assume for simplicity that the interest rate over the entire time span was constant.

Exercises

1. (Interpreting Mathematics) If the Algonquins could have invested their $24 at 6%, find the value of their investment after 368 years.

   Surprise! At that rate, they could now buy back Manhattan plus the entire city of New York, and even afford to eat lunch there.

2. (Interpreting Mathematics) Repeat Exercise 1, using each of the following interest rates.

   a. 5% b. 7%

   Surprise again! Did you expect that the return on the investment would be so sensitive to the interest rate?

A Mathematical Looking Glass

The Gathering Field

(to follow Section 9-3)

Eric Riebling plays bass for a popular Pittsburgh rock group, The Gathering Field. Like most rock musicians, Eric is aware that feedback can either hinder or enhance the band's music. Unwanted feedback can cause microphones to screech unpleasantly, while controlled feedback can add harmonics and (intended) distortions to their guitarist's leads. In both cases, the word *feedback* refers to the fact that part of an amplifier's output produces additional amplification by being fed back into the system as input.

According to Eric, an important consideration in controlled feedback is the gain of the system. Informally, you may think of gain as the ratio of output volume to input volume. For example, in Exercise 1 you can verify that if Eric feeds 0.03% of the output back into an amplifier that magnifies sould by a factor of 60, the resulting gain is about 61.08. That is, the output volume is about 61.08 times the input volume.

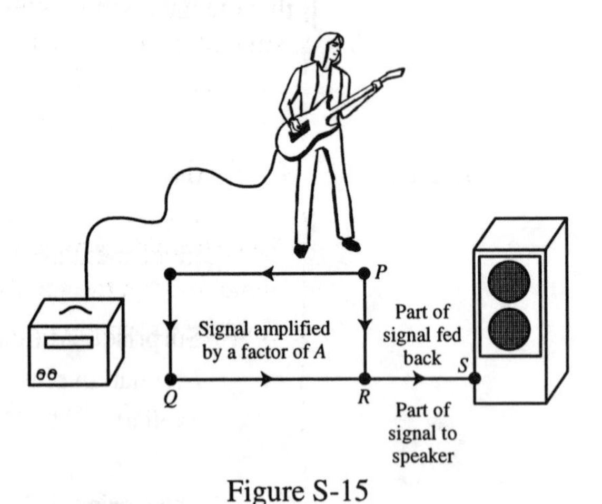

Figure S-15

In Figure S-15, Eric is feeding a fraction b of the output back into an amplifier that magnifies sound by a factor of A. For simplicity, let's suppose that the input signal has a strength of 1, measured in appropriate units. The resulting gain can be calculated from the following observations.

- Since the amplifier magnifies sound by a factor of A, the strength of the signal from the amplifier is A.

- When the signal reaches R, it splits into two parts. Since b is the fraction of the signal fed back into the amplifier, the strength of that part is Ab.

- The strength of the remaining part of the signal is $A(1 - b)$. It comes through the speaker at S, and forms part of what we hear.

Now the part of the signal with strength Ab is sent through Eric's bass again, and thus forms a new input. After passing through the amplifier, its strength is A^2b. When the new signal reaches R, it splits into two parts with strengths A^2b^2 and $A^2b(1 - b)$. As before, the second part comes through the speaker to form part of what we hear.

The part of the signal with strength A^2b^2 is sent through Eric's bass again. After passing through the amplifier, its strength is A^3b^2. When the new signal reaches R, it splits into two parts with strengths A^3b^3 and $A^3b^2(1 - b)$. The second part comes through the speaker to form part of what we hear.

Since the signal travels at the speed of light, the parts that come through the speaker are heard all at once, and have a combined strength of

$$A(1 - b) + A^2b(1 - b) + A^3b^2(1 - b) + \ldots$$

Exercises

1. (Making Observations)

   a. Verify that the strengths of the successive parts of the output signal form a geometric sequence.

   b. Verify that if $Ab < 1$, the combined strength of the signal (the gain) is

   $$G = \frac{A(1 - b)}{1 - A b}$$

   (If $Ab \geq 1$, the strength of the signal grows beyond all bounds, resulting in the unpleasant squealing we sometimes hear.)

   c. Verify that if $A = 60$ and $b = 0.0003$, the gain is $G = 61.08$.

2. (Making Observations) Eric controls the value of A by adjusting the volume control on his amplifier. In Exercise 1b, suppose that $A = 60$.

   a. Verify that G is a one-to-one function of b.

   b. Graph G as a function of b, and find the vertical and horizontal asymptotes on the graph.

   c. What are the domain and range of the function in the abstract? What are the domain and range in a physical context?

   d. What physical information is provided by each asymptote?

   e. Express b as a function of G.

f. Graph b as a function of G, and find the vertical and horizontal asymptotes on the graph. How do they relate to the asymptotes in part (b)?

3. (Making Observations) Eric controls the value of b by adjusting his distance from the amplifier. In Exercise 1b, suppose he stands so that $b = 0.0003$.

a. Verify that G is a one-to-one function of A.

b. Graph G as a function of A, and find the vertical and horizontal asymptotes on the graph.

c. What are the domain and range of the function in the abstract? What are the domain and range in a physical context?

d. What physical information is provided by each asymptote?

e. Express A as a function of G.

f. Graph A as a function of G, and find the vertical and horizontal asymptotes on the graph. How do they relate to the asymptotes in part (b)?

Chapter 9

Section 9.1

1. a. (Sample response) Each entry in the second column is obtained by multiplying the previous entry by 1.02. Thus the entry opposite $t = 4$ can be obtained by starting with the entry of 6 opposite $t = 0$ and multiplying by 1.02 four times. The final result is $6(1.02)^4$.

   b. (Sample response) If Table 29 is continued, the entry opposite any value of t can be obtained by starting with the entry of 6 opposite $t = 0$ and multiplying by 1.02 t times. The final result is $6(1.02)^t$.

3. The ratio of each y-value to the preceding one is 3, so the table fits an exponential function.

5. The ratio of each y-value to the preceding one is not constant, so the table does not fit an exponential function.

7. a.

x	y
5	243
6	729
7	2187
8	6561
9	19683
10	59049

When x is large, y is extremely large.

   b.

x	y (approx.)
−1	0.33
−2	0.11
−3	0.037
−4	0.012
−5	0.0041
−6	0.0014
−7	0.00046
−8	0.00015
−9	0.000051
−10	0.000017

When $x < 0$ and $|x|$ is large, y is extremely close to 0.

9. a. Since $b^x > 0$ for all values of x, the range of a function $y = b^x$ appears to be $(0, \infty)$.

   b. If a function $y = b^x$ is decreasing, the left tail of the graph points up and the right tail approaches the x-axis.

   c. If a function $y = b^x$ is increasing, the left tail of the graph approaches the x-axis and the right tail points up.

11. a. The table in Exercise 3 fits a function $y = Ab^x$.

   When $x = 0$, $y = \frac{2}{3}$, so $Ab^0 = \frac{2}{3}$

   When $x = 1, y = 2$, so $Ab^1 = 2$

   Solve the first equation for A: $A = \frac{2}{3}$

   Replace A by $\frac{2}{3}$ in the second equation:

   $\frac{2b}{3} = 2$

   $b = 3$

   Thus, $y = \frac{2}{3}3^x$.

b. The table in Exercise 4 fits a function
$y = Ab^x$.
When $x = 0, y = 64$: $Ab^0 = 64$
When $x = 2, y = 16$: $Ab^2 = 16$
From the first equation, $A = 64$. Replacing A by 64 in the second equation yields $64b^2 = 16$, and solving for b yields $b = \frac{1}{2}$.

Thus, $y = 64\left(\frac{1}{2}\right)^x$.

c. The table in Exercise 6 fits a function
$y = Ab^x$.
When $x = 1, y = 8$: $Ab^1 = 8$
When $x = 2, y = -12$: $Ab^2 = -12$
From the first equation, $A = \frac{8}{b}$. Replacing A by $\frac{8}{b}$ in the second equation yields
$\left(\frac{8}{b}\right)b^2 = -12$, and solving for b yields
$b = -\frac{3}{2}$. Then $A = -\frac{8}{3/2} = -\frac{16}{3}$. Thus,
$y = \left(-\frac{16}{3}\right)\left(-\frac{3}{2}\right)^x$.

13. (Sample response) Whenever x is increased by 1, the power of b is increased by 1, so the value of the function is multiplied by b. The ratio of the new value of f to the previous value is b.

15. a. $\sqrt{6} = 2.449489\ldots$

$3^2 = 9$	$3^3 = 27$
$3^{2.4} \cong 13.97$	$3^{2.5} \cong 15.59$
$3^{2.44} \cong 14.59$	$3^{2.45} \cong 14.76$
$3^{2.449} \cong 14.74$	$3^{2.450} \cong 14.76$
$3^{2.4494} \cong 14.75$	$3^{2.4495} \cong 14.75$

$3^{\sqrt{6}} \cong 14.75$, with an error of no more than 0.01.

b. $\pi = 3.141592\ldots$

$\pi^3 \cong 31.01$	$\pi^4 \cong 97.41$
$\pi^{3.1} \cong 34.77$	$\pi^{3.2} \cong 38.98$
$\pi^{3.14} \cong 36.40$	$\pi^{3.15} \cong 36.81$
$\pi^{3.141} \cong 36.44$	$\pi^{3.142} \cong 36.48$
$\pi^{3.1415} \cong 36.46$	$\pi^{3.1416} \cong 36.46$

$\pi^{\pi} \cong 36.46$, with an error of no more than 0.01.

17. a. (Sample response) Raising $\frac{1}{2}$ to a large positive power is equivalent to taking the reciprocal of 2 to a large positive power. Since the reciprocal of a large positive number is a small positive number, the value of $g(x)$ is positive and extremely near 0 when x is large.

b. (Sample response) Raising $\frac{1}{2}$ to a large negative power is equivalent to raising 2 to a large positive power. Thus when $x < 0$ and $|x|$ is large, $g(x)$ is extremely large.

c. (Sample response) The results taken together suggest that as the exponent x increases, the value of $\left(\frac{1}{2}\right)^x$ decreases. This in turn suggests that if $0 < b < 1$, $g(x) = b^x$ is a decreasing function, becoming arbitrarily large when $x < 0$ and $|x|$ is large and becoming arbitrarily close to 0 when x is large. The results also show that $\left(\frac{1}{2}\right)^x > 0$ for all values of x, and thus suggest that the range of g is $(0, \infty)$.

19. a.

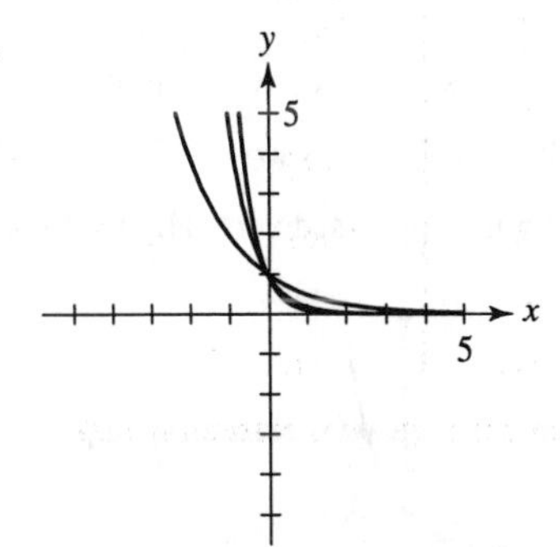

b.

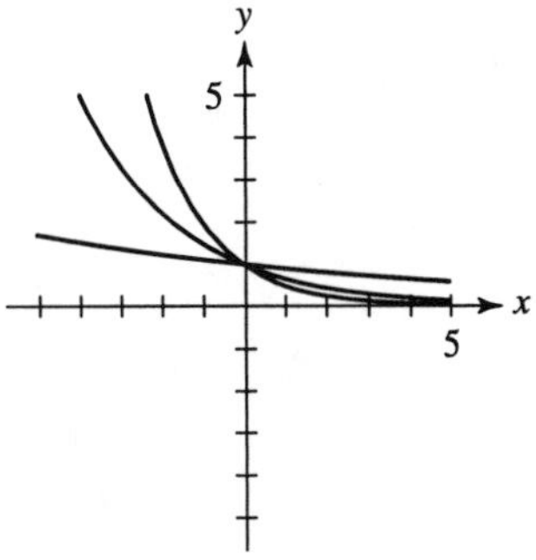

c. The domain is $(-\infty, \infty)$ and the range is $(0, \infty)$.

d. The y-intercept is (0, 1), and there are no x-intercepts.

e. $f(1) = b^1 = b$ and $f(-1) = b^{-1} = \frac{1}{b}$

f. The graph has a horizontal asymptote on the right side of the x-axis.

g. The function is decreasing on $(-\infty, \infty)$.

21. The graph is obtained from that of $y = 0.1^x$ by the transformations
$y = 0.1^x \to y = 0.1^{2x} \to y = 0.1^{2x} - 2$
The graph is compressed horizontally by a factor of 2, then shifted 2 units down.

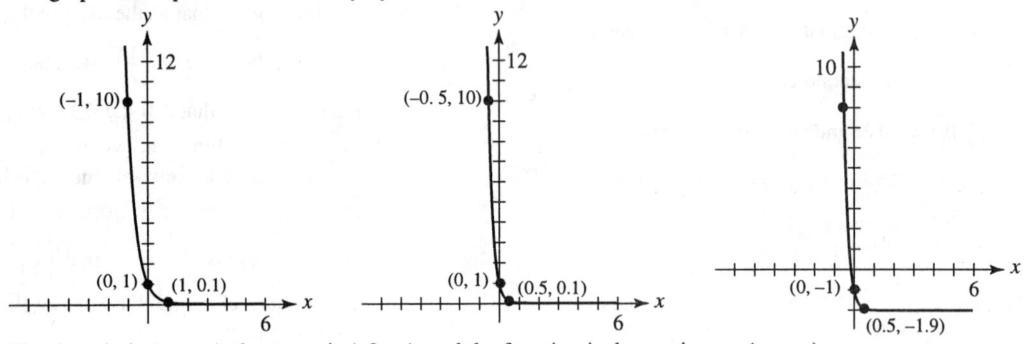

The domain is $(-\infty, \infty)$, the range is $(-2, \infty)$, and the function is decreasing on $(-\infty, \infty)$.

23. The graph is obtained from that of $y = 3^x$ by the transformations
$y = 3^x \to y = 3^{-x} \to y = -(3^{-x}) \to y = \left(-\frac{1}{2}\right)(3^{-x})$ The graph is reflected in the y-axis, then reflected in the x-axis, then compressed vertically by a factor of 2.

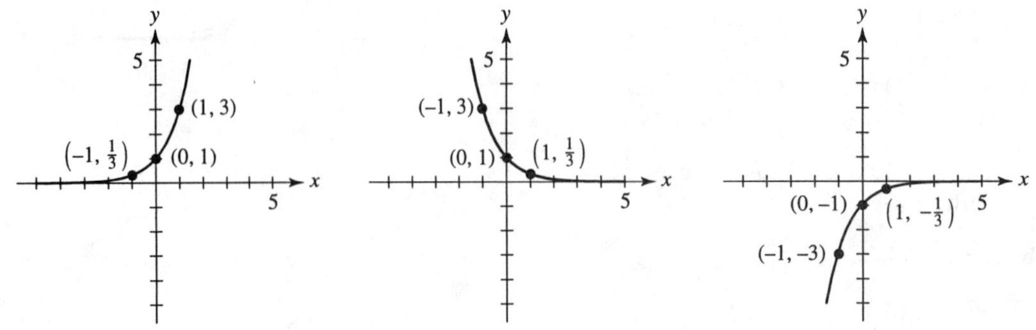

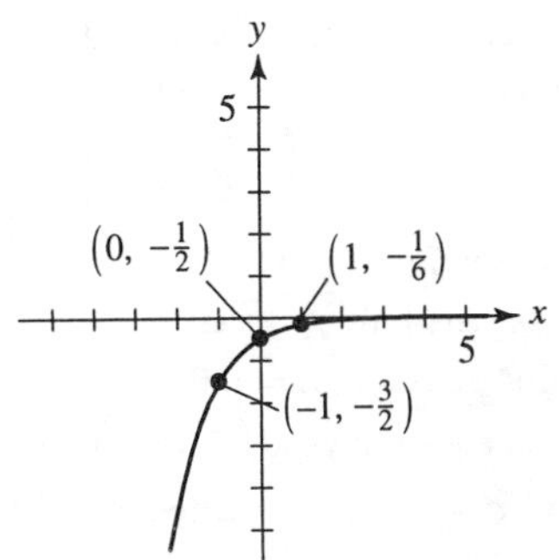

The domain is $(-\infty, \infty)$, the range is $(-\infty, 0)$, and the function is increasing on $(-\infty, \infty)$.

25. $f(0) = 1, f(1.71) \cong 2, f(3.42) \cong 4, f(5.13) \cong 8$
The doubling time is about 1.71.

27. $f(-2.32) \cong 1, f(-1.32) \cong 2, f(-0.32) \cong 4,$
$f(0.68) \cong 8$
The doubling time is 1.

29. The discussion in the Mathematical Looking Glass allows you to set up the following table, which fits a function $s = Ab^n$.

number of symbols (n)	size of symbol pool(s)
0	1
1	2
2	4
3	8

When $n = 0$, $s = 1$, so $Ab^0 = 1$
When $n = 1$, $s = 2$, so $Ab^1 = 2$
From the first equation, $A = 1$. Replacing A by 1 in the second equation yields $b = 2$. Thus, $s = 2^n$.

31. (Sample response) One reason is that by asking 20 yes/no questions, a child can distinguish among more than a million objects. Another reason is that most children choose familiar objects for their opponents to identify, so that the pool of likely objects is usually much smaller than a million.

33. $a = \left\{2, -\frac{2}{3}, \frac{2}{9}, -\frac{2}{27}, \frac{2}{81}, \ldots\right\}$
The ratio of each term to the preceding term is $-\frac{1}{3}$, so the sequence is geometric.

35. $a = \left\{\frac{1}{5}, \frac{2}{5}, \frac{4}{5}, \frac{8}{5}, \frac{16}{5}, \ldots\right\}$
The ratio of each term to the preceding term is 2, so the sequence is geometric.

37. The ratio of each number in the sequence to the preceding one is not constant, so the sequence is not geometric.

39. The ratio of each term to the preceding one is 1.1, and the first term is 1, so $a_n = 1.1^n$. In particular, $a_{10} = 1.1^{10} \cong 2.59$.

41. **a.** $rS_n = r(A + Ar + Ar^2 + \ldots + Ar^{n-1} + Ar^n)$
$= Ar + Ar^2 + Ar^3 + \ldots + Ar^n + Ar^{n+1}$

b. $(1-r)S_n$
$= S_n - rS_n$
$= (A + Ar + Ar^2 + \ldots + Ar^{n-1} + Ar^n)$
$\quad -(Ar + Ar^2 + Ar^3 + \ldots + Ar^n + Ar^{n+1})$
$= A - Ar^{n+1} = A(1 - r^{n+1})$

c. $S_n = \dfrac{A(1-r^{n+1})}{1-r}$

43. $A = 16$ and $r = -\frac{1}{4}$, so

$$S_5 = \frac{16\left(1-\left(-\frac{1}{4}\right)^6\right)}{1+\frac{1}{4}} \cong 12.80$$

$$S_{10} = \frac{16\left(1-\left(-\frac{1}{4}\right)^{11}\right)}{1+\frac{1}{4}} \cong 12.80$$

$$S_{20} = \frac{16\left(1-\left(-\frac{1}{4}\right)^{21}\right)}{1+\frac{1}{4}} \cong 12.80$$

45. $A = 2$ and $r = 1.1$, so

$$S_5 = \frac{2(1-1.1^6)}{1-1.1} \cong 15.43$$

$$S_{10} = \frac{2(1-1.1^{11})}{1-1.1} \cong 37.06$$

$$S_{20} = \frac{2(1-1.1^{21})}{1-1.1} \cong 128.00$$

47. For the sequence in Exercise 42,
$\frac{A}{1-r} = \frac{6}{1-\frac{1}{3}} = 9.$
For the sequence in Exercise 43,
$\frac{A}{1-r} = \frac{16}{1+\frac{1}{4}} = \frac{64}{5} = 12.8.$

49. The ratio of each y-value to the preceding one is 3, so the table fits a function $y = A \cdot b^x$.
When $x = -0.5$, $y = 0.007$, so $A \cdot b^{-0.5} = 0.007$
When $x = 0$, $y = 0.021$, so
$A \cdot b^0 = 0.021$
From the second equation, $A = 0.021$.
Replace A by 0.021 in the first equation:
$(0.021)b^{-0.5} = 0.007$
$b^{-0.5} = \frac{1}{3}$
$\frac{1}{\sqrt{b}} = \frac{1}{3}$
$b = 9$
The equation is $y = (0.021)(9^x)$.

51. The ratio of each y-value to the preceding one is not the same, so the table does not fit a function $y = Ab^x$.

53. When $x = 0$, $y = 7$, so $A \cdot b^0 = 7$
When $x = 2$, $y = 49$, so $A \cdot b^2 = 49$
From the first equation, $A = 7$. Replacing A by 7 in the second equation yields $7b^2 = 49$, so $b^2 = 7$, and $b = \sqrt{7}$. The equation is $y = 7\left(\sqrt{7}\right)^x$.

55. When $x = 3$, $y = 40$, so $A \cdot b^3 = 40$
When $x = 8$, $y = 30$, so $A \cdot b^8 = 30$
Solve the first equation for A: $A = \frac{40}{b^3}$
Replace A by $\frac{40}{b^3}$ in the second equation:
$\left(\frac{40}{b^3}\right)b^8 = 30$
$40b^5 = 30$
$b^5 = 0.75$
$b = \sqrt[5]{0.75} \cong 0.94$
$A = \frac{40}{\left(\sqrt[5]{0.75}\right)^3} \cong 47.54$
The equation is $y = (47.54)(0.94)^x$.

57. The graph is obtained from that of $y = 5^x$ by the transformations $y = 5^x \rightarrow y = 5^{x-2} \rightarrow y = 5^{x-2} + 1$
The graph is shifted 2 units to the right and 1 unit up.

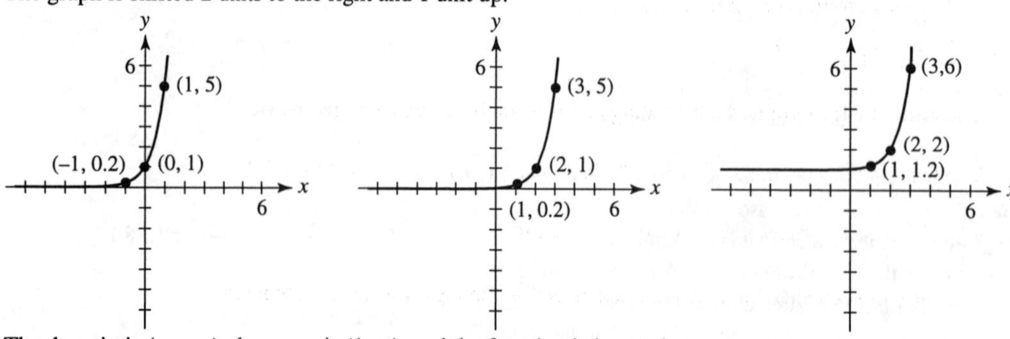

The domain is $(-\infty, \infty)$, the range is $(1, \infty)$, and the function is increasing.

59. The graph is obtained from that of $y = 2^x$ by the transformations $y = 2^x \rightarrow y = -2^x \rightarrow y = -2^x + 8$
The graph is reflected in the y-axis, then shifted 8 units up.

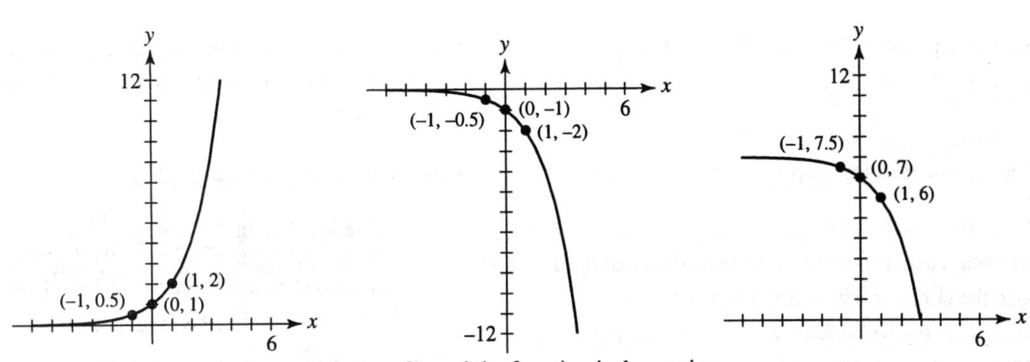

The domain is $(-\infty, \infty)$, the range is $(-\infty, 8)$, and the function is decreasing.

61. a. The graph of $y = 3^{2x}$ can be obtained by compressing the graph of $y = 3^x$ horizontally by a factor of 2.

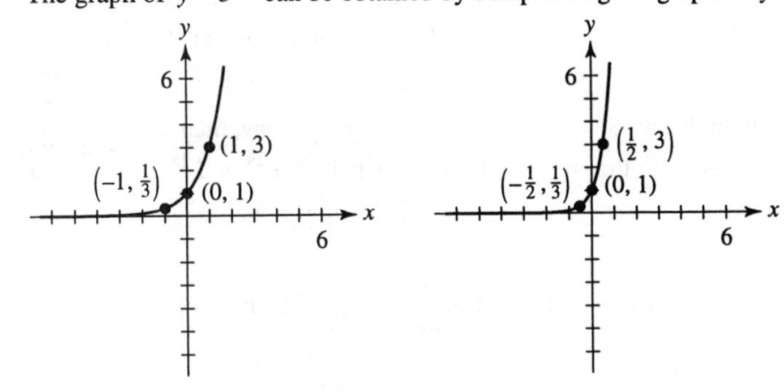

b. $4^{3x} = (4^3)^x = 64^x$, so $b_1 = 64$

63. a. The ratio of each term to the preceding one is 0.9, so the sequence is geometric.

b. The leading term is –1, so $a_n = -(0.9^n)$.
In particular, $a_8 = -(0.9^8) \cong -0.43$.

c. $|r| = 0.9 < 1$, so the sum of the series is
$\frac{A}{1-r} = \frac{-1}{1-0.9} = -10.$

65. a. The ratio of each term to the preceding one is 5, so the sequence is geometric.

b. The leading term is 10, so $a_n = 10 \cdot 5^n$. In particular, $a_8 = 10 \cdot 5^8 = 3{,}906{,}250$.

c. $|r| = 5 > 1$, so the series does not have a finite sum.

67. a. The ratio of each term to the preceding one is $-\frac{3}{4}$, so the sequence is geometric.

b. The leading term is 1, so $a_n = \left(-\frac{3}{4}\right)^n$. In particular, $a_8 = \left(-\frac{3}{4}\right)^8 \cong 0.10$.

c. $|r| = \frac{3}{4} < 1$, so the sum of the series is
$\frac{A}{1-r} = \frac{1}{1+\frac{3}{4}} = \frac{4}{7}$.

69. The ratio of each term to the preceding one is not constant, so the sequence is not geometric.

71. An increase of 1 in the fret number always produces the same percentage increase in frequency, so the relationship between n and F can be described by an equation $F = Ab^n$ for some values of A and b. Since the frequencies at frets 0 and 12 are 220 and 440, respectively,
$f(0) = 220$, so $Ab^0 = 220$
$f(12) = 440$, so $Ab^{12} = 440$
From the first equation, $A = 220$. Substitute 220 for A in the second equation to obtain $b = \sqrt[12]{2} \cong 1.06$. Thus,
$f(n) = 220\left(\sqrt[12]{2}\right)^n$.

73. Solution using exponential functions:
On the first day, there is 1 $(= 2^0)$ fly in the jar. Since the number doubles each day, there are 2^{n-1} flies in the jar after n days. Therefore, it takes 2^{29} flies to fill the jar in 30 days. Since half of 2^{29} is 2^{28}, the jar is half full after 29 days.

Easier solution:
Since the number of flies in the jar doubles each day, it is half full one day before it is completely full, that is after 29 days.

Section 9-2

1. **a.** After one year you owe the original \$10,000, plus interest amounting to 5.9% of \$10,000. The total amount you owe, in dollars, is
10,000 + 0.059(10,000) = 10,000(1 + 0.059).

b.

interest periods	amount owed
0	10,000
1	10,000(1 + 0.059) = 10,590
2	10,590(1 + 0.059) = 11,214.81
3	11,214.81(1 + 0.059) ≅ 11,876.48
4	11,876.48(1 + 0.059) ≅ 12,577.20
5	12,577.20(1 + 0.059) ≅ 13,319.25

c. In the second column of the table, the ratio of each term to the preceding one is 1.059, and the initial term is 10,000. The table defines the sequence $a_n = 10,000(1.059)^n$. The amount owed after t years is $10,000(1.059)^t$ dollars.

3. a. After one third of a year you owe the original \$10,000, plus interest amounting to one third of 5.9% of \$10,000. The total amount you owe, in dollars, is $10,000\left(1+\frac{0.059}{3}\right)$.

b.

interest periods	amount owed
0	10,000
1	$10,000\left(1+\frac{0.059}{3}\right) \cong 10,196.67$
2	$10,196.67\left(1+\frac{0.059}{3}\right) \cong 10,397.20$
3	$10,397.20\left(1+\frac{0.059}{3}\right) \cong 10,601.68$
4	$10,601.68\left(1+\frac{0.059}{3}\right) \cong 10,810.18$
5	$10,810.18\left(1+\frac{0.059}{3}\right) \cong 11,022.78$
6	$11,022.78\left(1+\frac{0.059}{3}\right) \cong 11,239.56$
7	$11,239.56\left(1+\frac{0.059}{3}\right) \cong 11,460.61$
8	$11,460.61\left(1+\frac{0.059}{3}\right) \cong 11,686.00$
9	$11,686.00\left(1+\frac{0.059}{3}\right) \cong 11,915.82$
10	$11,915.82\left(1+\frac{0.059}{3}\right) \cong 12,150.17$
11	$12,150.17\left(1+\frac{0.059}{3}\right) \cong 12,389.12$
12	$12,389.12\left(1+\frac{0.059}{3}\right) \cong 12,632.77$
13	$12,632.77\left(1+\frac{0.059}{3}\right) \cong 12,881.22$
14	$12,881.22\left(1+\frac{0.059}{3}\right) \cong 13,134.55$
15	$13,134.55\left(1+\frac{0.059}{3}\right) \cong 13,392.86$

c. After n interest periods, you owe $10,000\left(1+\frac{0.059}{3}\right)^n$ dollars. After t years ($3t$ interest periods) you owe $10,000\left(1+\frac{0.059}{3}\right)^{3t}$ dollars.

5. The ceiling appears to be about \$13,431.

7.

n	$\left(1+\frac{1}{n}\right)^n$
1,000	2.717
1,000,000	2.718280
1,000,000,000	2.718281827

9. a. $4000(1.149)^3 \cong 6067.64$

$4000\left(1+\frac{0.149}{12}\right)^{36} \cong 6237.27$

$4000e^{(0.149)(3)} \cong 6254.46$

b. $50,000(1.086)^{30} \cong 594,107.13$

$50,000\left(1+\frac{0.086}{12}\right)^{360} \cong 653,813.37$

$50,000e^{(0.086)(30)} \cong 659,856.91$

c. $500(1+2)^{0.5} \cong 866.03$

$500\left(1+\frac{2}{12}\right)^6 \cong 1260.81$

$500e^{(2)(0.5)} \cong 1359.14$

11.

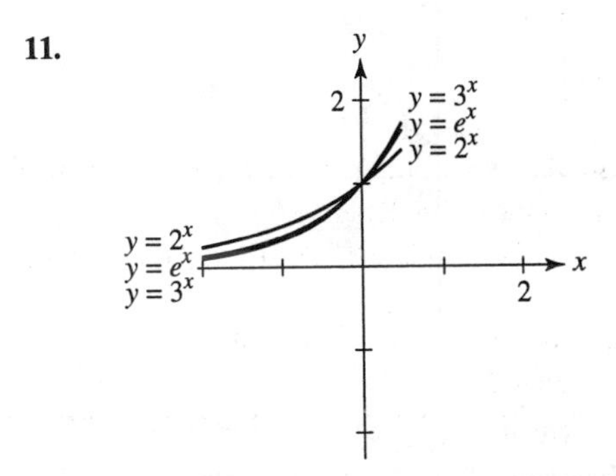

13. (Sample response) If the independent variable is time, then the average rate of change in f over any time interval is the average rate at which f grows over that interval. If the average rate of change is always equal to the value of f at some point within the interval, then f grows at a rate equal to its size.

15. $f(6700) = 2^{-6700/5730} \cong 0.44$, so about 44% of the carbon-14 found in living wood was in the sample.

17. Zoom in on the graph of $y = e^x$ to conclude that $y = 3$ when $x \cong 1.10$. Therefore $f(t) \cong e^{1.10t}$.

19. Zoom in on the graph of $y = e^x$ to conclude that $y = \frac{1}{2}$ when $x \cong -0.69$. Therefore $f(t) \cong 3e^{-0.69t}$.

21. The equation can be obtained from $y = e^x$ through the transformation
$y = e^x \rightarrow y = e^{0.5x}$
The graph is stretched horizontally by a factor of 2.

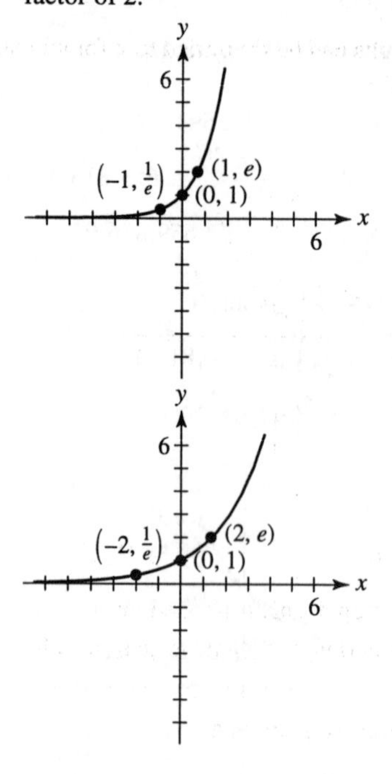

23. The equation can be obtained from $y = e^{-0.5x}$ (Exercise 22) by the transformation
$y = e^{-0.5x} \rightarrow y = 2e^{-0.5x}$
The graph is stretched vertically by a factor of 2.

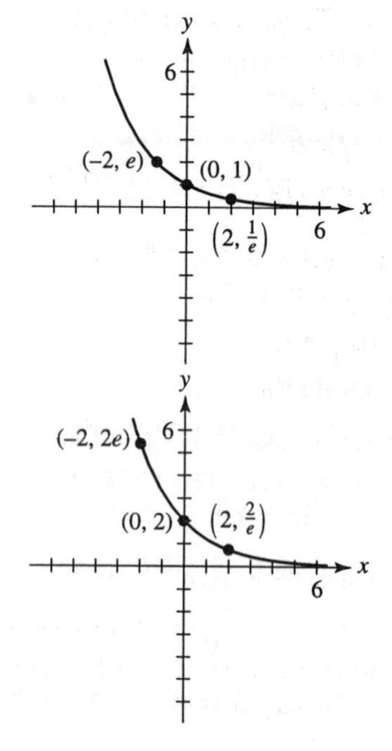

25. See the descriptions in the solutions to Exercises 21–24.

27. a. The amount of money in her account after t months will be
$f(t) = (10{,}114.32)e^{0.0485t/12}$.

b. $f(3.8) = (10{,}114.32)e^{(0.0485)(3.8)/12}$
$\cong \$10{,}270.86$

29. a. At 6% interest, you had $10{,}000e^{(0.06)(0.5)} \cong \$10{,}304.55$ in your account after 6 months. At 6.2% interest, you would have had $10{,}000e^{(0.062)(0.5)} \cong \$10{,}314.86$. You lost about \$10.31.

b. At 6% interest, you will have $10{,}000e^{(0.06)(50)} \cong \$200{,}855.37$ after 50 years. At 6.2%, you will have $10{,}000e^{(0.062)(50)} \cong \$221{,}979.51$. The difference is about \$21,124.14.

31. a. $f(t) = 2^{-kt}$, and $f(4.5\times 10^9) = 2^{-1}$:

$2^{(-4.5\times10^9)k} = 2^{-1}$

$(-4.5\times 10^9)k = -1$

$k \cong 2.22\times 10^{-10}$

The function is $f(t) = 2^{(-2.22\times10^{-10})t}$.

b. From Exercise 16, $2 \cong e^{0.69}$, so

$f(t) = 2^{(-2.22\times10^{-10})t}$

$= (e^{0.69})^{(-2.22\times10^{-10})t}$

$= e^{(-1.54\times10^{-10})t}$

$f(2\times10^8) = e^{(-1.54\times10^{-10})(2\times10^8)} \cong 0.97$,

so the rock contains about 97% of its original uranium 238.

c. Zoom in on the graph of $f(t) = e^{(-1.54\times10^{-10})t}$ to conclude that $y =$ 0.73 when $t \cong 2.04\times 10^9$, so the rock is about 2 billion years old.

33. $6! = 720$; $7! = 5040$; $10! = 3{,}628{,}800$; $25! \cong 1.55\times 10^{25}$

Section 9-3

1. $(F\circ G)(x) = F[G(x)]$

$= F(x-2) = (x-2)+2 = x$

$(G\circ F)(x) = G[F(x)]$

$= G(x+2) = (x+2) - 2 = x$

3. The first and last rows have the same second entry and different first entries, so the function is not one-to-one.

5. All rows have the same second entry and different first entries, so the function is not one-to-one.

7. No horizontal line intersects the graph more than once, so the function is one-to-one.

9. The horizontal line which comprises the graph intersects the graph more than once, so the function is not one-to-one.

11. No horizontal line intersects the graph more than once, so the function is one-to-one.

13. No horizontal line intersects the graph more than once, so the function is one-to-one.

15. $(f\circ g)(x) = f[g(x)]$

$= f\left[\frac{5}{9}(x-32)\right] = \frac{9}{5}\left[\frac{5}{9}(x-32)\right] + 32$

$= (x-32) + 32 = x$

$(g\circ f)(x) = g[f(x)]$

$= g\left(\frac{9}{5}x + 32\right) = \frac{5}{9}\left[\left(\frac{9}{5}x+32\right) - 32\right]$

$= \frac{5}{9}\left(\frac{9}{5}x\right) = x$

Both results can be simplified to x for all real values of x.

17. a.

b. The points $\left(-1, \frac{1}{2}\right)$, (0, 1), (1, 2), and (2, 4) make up a table generated by $f(x) = 2^x$. The corresponding table generated by $y = f^{-1}(x)$ contains those points with their coordinates reversed.

c. The plotted points suggest that the points (a, b) and (b, a) are always reflections of each other in the line $y = x$.

d.

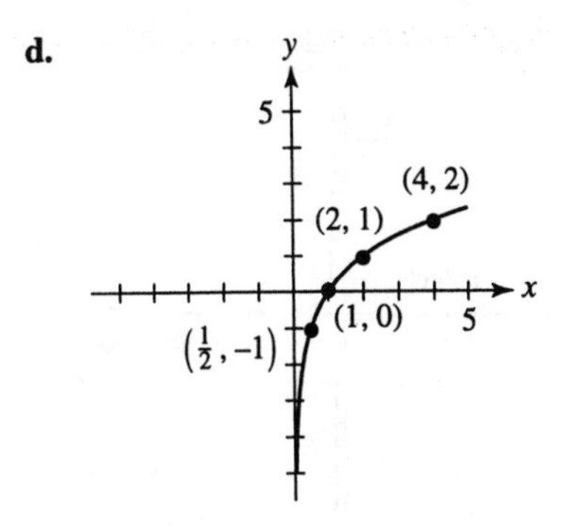

19. a. The domain of g is $(-\infty, 1) \cup (1, \infty)$.

b. $f[g(x)] = f\left(-\frac{2x}{x-1}\right)$

$= \frac{-\frac{2x}{x-1}}{-\frac{2x}{x-1}+2} = \frac{-\frac{2x}{x-1}}{-\frac{2x}{x-1}+2} \cdot \frac{x-1}{x-1}$

$= \frac{-2x}{-2x+2(x-1)}$ (if $x \neq 1$) $= \frac{-2x}{-2} = x$ for all x in the domain of g.

21. a.

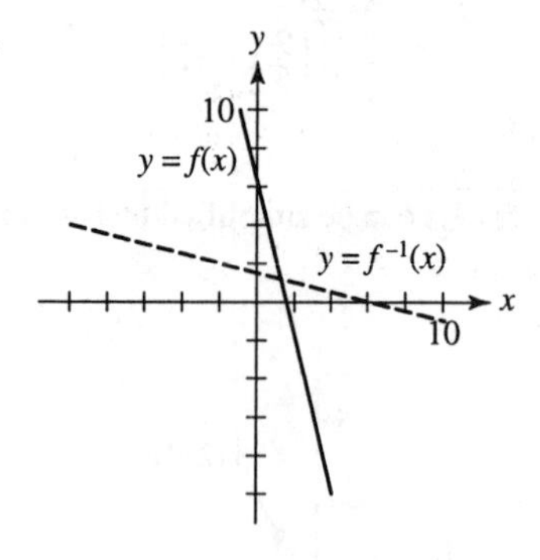

b. First, write f using x and y:

$y = 6 - 4x$

Next, interchange x and y, and solve for y:

$x = 6 - 4y$

$4y = 6 - x$

$y = \frac{6-x}{4}$

Now define $g(x) = \frac{6-x}{4}$, and check to see whether $g = f^{-1}$:

$g[f(x)] = g(6-4x)$

$= \frac{6-(6-4x)}{4} = \frac{4x}{4} = x$

$f[g(x)] = f\left(\frac{6-x}{4}\right)$

$= 6 - 4\left(\frac{6-x}{4}\right) = 6-(6-x) = x$

Therefore $g = f^{-1}$.

23. a.

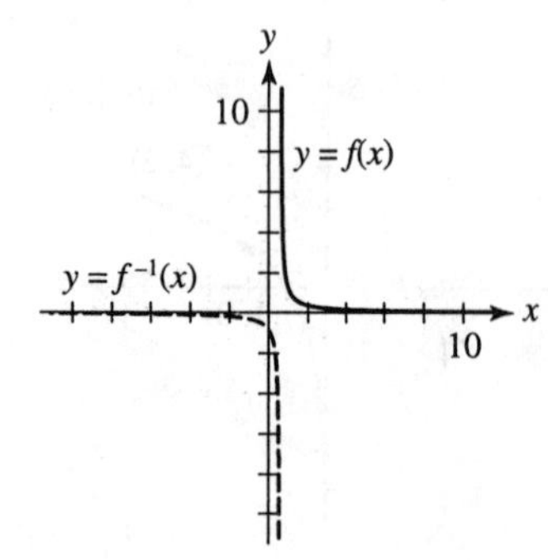

b. First, write f using x and y:

$y = \frac{1}{2x-1}$

Next, interchange x and y, and solve for y:

$x = \frac{1}{2y-1}$

$x(2y-1) = 1$

$2xy - x = 1$

$2xy = x + 1$

$y = \frac{x+1}{2x}$

Now define $g(x) = \frac{x+1}{2x}$, and check to see whether $g = f^{-1}$:

$g[f(x)] = g\left(\frac{1}{2x-1}\right)$

$= \frac{\frac{1}{2x-1}+1}{2\left(\frac{1}{2x-1}\right)} = \frac{\frac{1}{2x-1}+1}{2\left(\frac{1}{2x-1}\right)} \cdot \frac{2x-1}{2x-1}$

$= \frac{(2x-1)+1}{2}$ $\left(\text{if } x \neq \frac{1}{2}\right) = \frac{2x}{x} = x$ for all x in the domain of f

$f[g(x)] = f\left(\frac{x+1}{2x}\right)$

$= \frac{1}{2\left(\frac{x+1}{2x}\right)-1} = \frac{1}{2\left(\frac{x+1}{2x}\right)-1} \cdot \frac{2x}{2x}$

$= \frac{2x}{2(x+1)-2x}$ (if $x \neq 0$) $= \frac{2x}{2} = x$ for all x in the domain of g.

Therefore $g = f^{-1}$.

25. a.

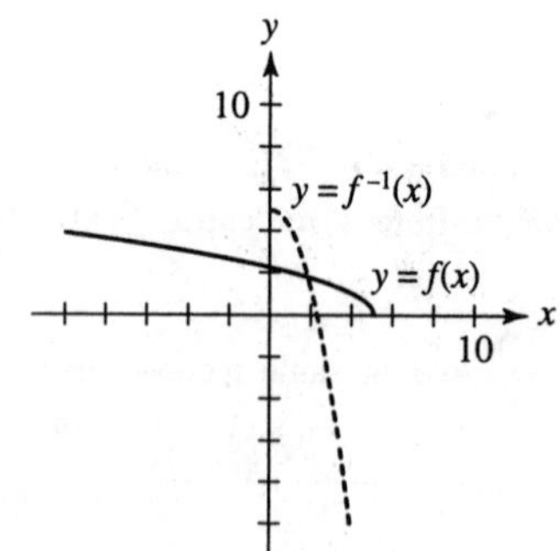

b. To find f^{-1} analytically, write f using x and y: $y = \sqrt{5-x}$

Interchange x and y, and solve for y:

$x = \sqrt{5-y}$

$x^2 = 5 - y$

$y = 5 - x^2$

Define $g(x) = 5 - x^2$, and check to see

whether $g = f^{-1}$:

$g[f(x)] = g\left(\sqrt{5-x}\right) = 5 - \left(\sqrt{5-x}\right)^2$

$= 5 - (5 - x) = x$ for all x in the domain of f.

$f[g(x)] = f(5 - x^2)$

$= \sqrt{5-(5-x^2)} = \sqrt{x^2} = |x| = x$ if $x \ge 0$.

Thus $f^{-1}(x) = 5 - x^2$ for $x \ge 0$.

27. a.

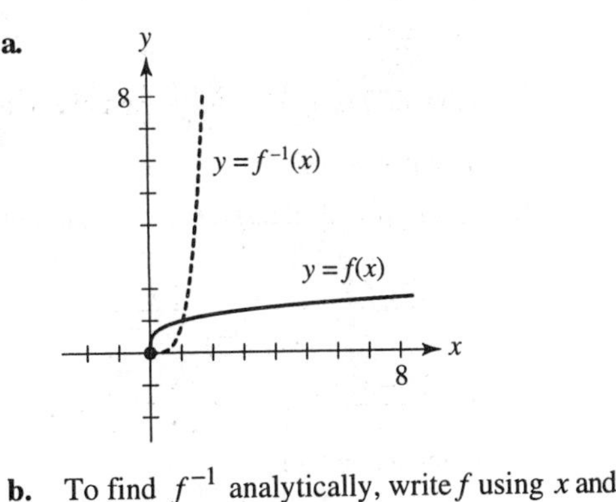

b. To find f^{-1} analytically, write f using x and y: $y = \sqrt[4]{x}$

Interchange x and y, and solve for y:

$x = \sqrt[4]{y}$

$y = x^4$

Define $g(x) = x^4$, and check to see whether $g = f^{-1}$:

$g[f(x)] = g\left(\sqrt[4]{x}\right) = \left(\sqrt[4]{x}\right)^4 = x$ for all x in the domain of f.

$f[g(x)] = f(x^4) = \sqrt[4]{x^4} = |x| = x$ if $x \ge 0$.

Thus $f^{-1}(x) = x^4$ for $x \ge 0$.

29. a. $v = \dfrac{-0.8 \pm \sqrt{0.8^2 - 4(0.06)(-d)}}{2(0.06)}$

$v = \dfrac{-0.8 \pm \sqrt{0.64 + 0.24d}}{0.12}$

$v = \dfrac{-0.8 \pm \sqrt{0.04(16 + 6d)}}{0.12}$

$v = \dfrac{-0.8 \pm 0.2\sqrt{16 + 6d}}{0.12} \cdot \dfrac{25}{25}$

$v = \dfrac{-20 \pm 5\sqrt{16 + 6d}}{3}$

b. The function $d = 0.06v^2 + 0.8v$ is one-to-one on $[0, \infty)$. Its inverse is found by reversing the roles of the independent and dependent variables, as was done in part (a).

31. Since the vertex of the graph is (0, 10), the two one-to-one functions are $f_1(x) = 10 - 2x^2$ for $x \le 0$ and $f_2(x) = 10 - 2x^2$ for $x \ge 0$.

To find $f_1^{-1}(x)$, rewrite $f_1(x)$ as

$y = 10 - 2x^2$ for $x \le 0$

Interchange x and y, and solve for y:

$x = 10 - 2y^2$ for $y \le 0$

$2y^2 = 10 - x$ for $y \le 0$

$y^2 = \dfrac{10 - x}{2}$ for $y \le 0$

$y = \pm\sqrt{\dfrac{10 - x}{2}}$ for $y \le 0$

Since $y \le 0$, $y = -\sqrt{\dfrac{10 - x}{2}}$. Thus

$f_1^{-1}(x) = -\sqrt{\dfrac{10 - x}{2}}$. Similarly

$f_2^{-1}(x) = \sqrt{\dfrac{10 - x}{2}}$.

33. Since the vertex of the graph is (1, 2), the two one-to-one functions are $f_1(x) = 3(x-1)^2 + 2$ for $x \le 1$ and $f_2(x) = 3(x-1)^2 + 2$ for $x \ge 1$.

To find $f_1^{-1}(x)$, rewrite $f_1(x)$ as

$y = 3(x-1)^2 + 2$ for $x \le 1$

Interchange x and y, and solve for y:

$x = 3(y-1)^2 + 2$ for $y \le 1$

$3(y-1)^2 = x - 2$ for $y \le 1$

$(y-1)^2 = \dfrac{x-2}{3}$ for $y \le 1$

$y - 1 = \pm\sqrt{\dfrac{x-2}{3}}$ for $y \le 1$

Since $y \le 1$, $y - 1$ is negative, so

$y - 1 = -\sqrt{\dfrac{x-2}{3}}$

$y = 1 - \sqrt{\dfrac{x-2}{3}}$

Thus $f_1^{-1}(x) = 1 - \sqrt{\dfrac{x-2}{3}}$. Similarly

$f_2^{-1}(x) = 1 + \sqrt{\dfrac{x-2}{3}}$.

35. Answers will vary.

37. No two rows in the table have the same second entry, so the function is one-to-one.

39. No two rows in the table have the same second entry, so the function is one-to-one.

41.

x	$f^{-1}(x)$
$\sqrt{3}$	1
12,946	2.8
0	7π
e	$\frac{2}{3}$

$f^{-1}[f(1)] = f^{-1}(\sqrt{3}) = 1$ $\qquad$ $f[f^{-1}(\sqrt{3})] = f(1) = \sqrt{3}$

$f^{-1}[f(2.8)] = f^{-1}(12{,}946) = 2.8$ $\qquad$ $f[f^{-1}(12{,}946)] = f(2.8) = 12{,}946$

$f^{-1}[f(7\pi)] = f^{-1}(0) = 7\pi$ $\qquad$ $f[f^{-1}(0)] = f(7\pi) = 0$

$f^{-1}[f(\frac{2}{3})] = f^{-1}(e) = \frac{2}{3}$ $\qquad$ $f[f^{-1}(e)] = f(\frac{2}{3}) = e$

43. Many vertical lines intersect the graph more than once, so y is not a function of x at all. (No horizontal line intersects the graph more than once, so x is a function of y, but it is not one-to-one.)

45. No horizontal line intersects the graph more than once, so y is a one-to-one function of x.

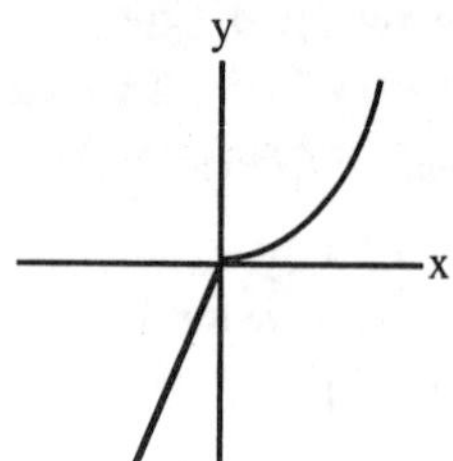

47. No horizontal line intersects the graph more than once, so y is a one-to-one function of x.

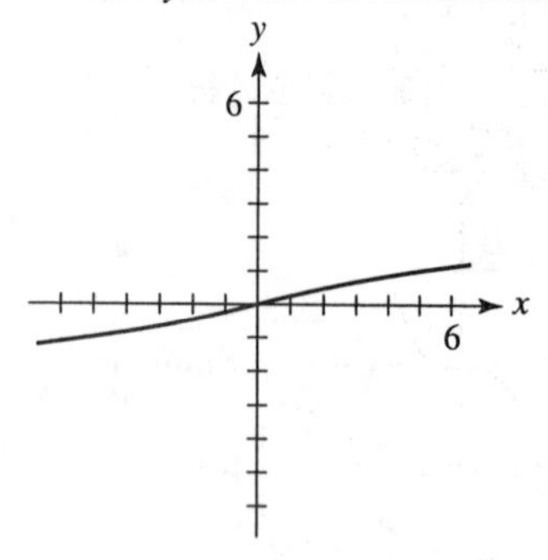

49. No horizontal line intersects the graph more than once, so y is a one-to-one function of x.

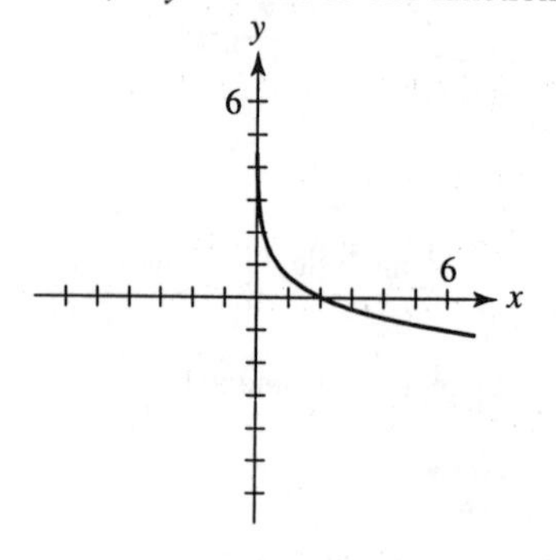

51. Many horizontal lines intersect the graph more than once, so y is not a one-to-one function of x.

53. No horizontal line intersects the graph more than once, so y is a one-to-one function of x.

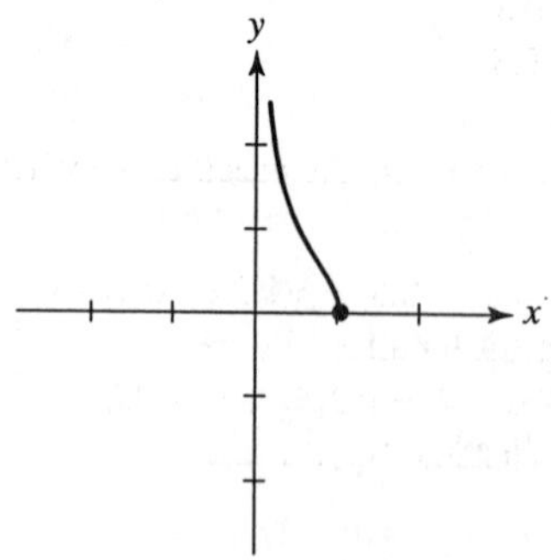

55. $g[f(x)] = g(3 - x) = 3 - (3 - x) = x$ for all x
$g[g(x)] = f(3 - x) = 3 - (3 - x) = x$ for all x
Therefore f and g are inverses.

57. $f[g(x)] = f\left(\frac{6x}{1-x}\right) = \frac{\frac{6x}{1-x}}{\frac{6x}{1-x}+6} \cdot \frac{1-x}{1-x}$

$= \frac{6x}{6x+6(1-x)} = \frac{6x}{6x+6-6x}$

$= \frac{6x}{6} = x$ for $x \neq 1$

$g[f(x)] = g\left(\frac{x}{x+6}\right) = \frac{6\left(\frac{x}{x+6}\right)}{1-\frac{x}{x+6}} \cdot \frac{x+6}{x+6}$

$= \frac{6x}{(x+6)-x} = \frac{6x}{6} = x$ for $x \neq -6$

Therefore f and g are inverses.

59. $g[f(x)] = g\left[(x-7)^4\right]$

$\sqrt[4]{(x-7)^4} + 7 = |x-7| + 7 = x$ if $x \geq 7$

$f[g(x)] = f\left(\sqrt[4]{x}+7\right) = \left[\left(\sqrt[4]{x}+7\right)-7\right]^4$

$= \left(\sqrt[4]{x}\right)^4 = x$ for all $x \geq 0$

Therefore f and g are not inverses.

(If $f_1(x) = (x-7)^4$ for $x \geq 7$, then $g = f_1^{-1}$.)

61. $g[f(x)] = g(4) = \frac{1}{4}$ for all x

$f[g(x)] = f\left(\frac{1}{4}\right) = 4$ for all x

Therefore f and g are inverses.

63. To find f^{-1} analytically, rewrite f using x and y:

$y = 0.2x - 0.8$

Interchange x and y, and solve for y:

$x = 0.2y - 0.8$

$0.2y = x + 0.8$

$y = 5x + 4$

Define $g(x) = 5x + 4$, and check to see whether $g = f^{-1}$:

$g[f(x)] = g(0.2x - 0.8) = 5(0.2x - 0.8) + 4$

$= x - 4 + 4 = x$ for all x

$f[g(x)] = f(5x + 4) = 0.2(5x + 4) - 0.8$

$= x + 0.8 - 0.8 = x$ for all x

Therefore $f^{-1}(x) = 5x + 4$

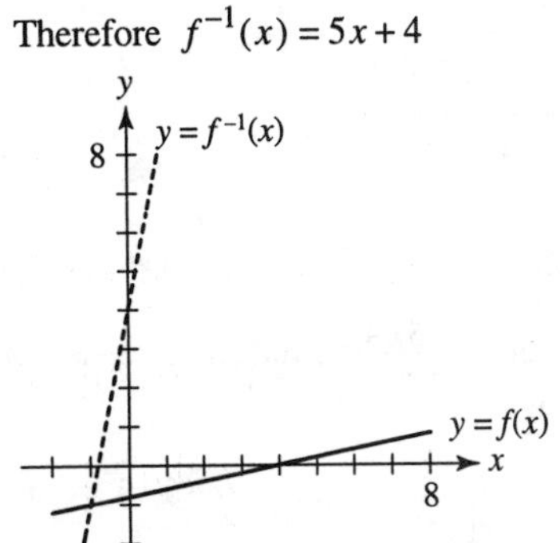

65. To find f^{-1} analytically, rewrite f using x and y:

$y = \frac{x^3+8}{3}$

Interchange x and y, and solve for y:

$x = \frac{y^3+8}{3}$

$3x = y^3 + 8$

$y^3 = 3x - 8$

$y = \sqrt[3]{3x-8}$

Define $g(x) = \sqrt[3]{3x-8}$, and check to see whether $g = f^{-1}$:

$g[f(x)] = g\left(\frac{x^3+8}{3}\right) = \sqrt[3]{3\left(\frac{x^3+8}{3}\right) - 8}$

$= \sqrt[3]{(x^3+8)-8} = \sqrt[3]{x^3} = x$ for all x.

$f[g(x)] = f\left(\sqrt[3]{3x-8}\right) = \frac{\left(\sqrt[3]{3x-8}\right)^3 + 8}{3}$

$= \frac{(3x-8)+8}{3} = \frac{3x}{3} = x$ for all x.

Therefore $f^{-1}(x) = \sqrt[3]{3x-8}$.

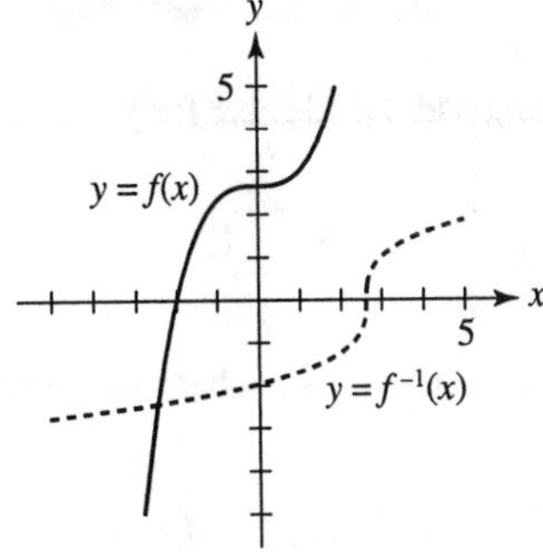

67. To find f^{-1} analytically, rewrite f using x and y:

$y = \frac{5}{x-5}$

Interchange x and y, and solve for y:

$x = \frac{5}{y-5}$
$x(y-5) = 5$
$xy - 5x = 5$
$xy = 5x + 5$
$y = \frac{5x+5}{x}$

Define $g(x) = \frac{5x+5}{x}$, and check to see whether $g = f^{-1}$:

$g[f(x)] = g\left(\frac{5}{x-5}\right) = \frac{5\left(\frac{5}{x-5}\right)+5}{\frac{5}{x-5}}$

$= \frac{5\left(\frac{5}{x-5}\right)+5}{\frac{5}{x-5}} \cdot \frac{x-5}{x-5} = \frac{5(5)+5(x-5)}{5} = \frac{5x}{5} = x$

for all $x \neq 5$.

$f[g(x)] = f\left(\frac{5x+5}{x}\right) = \frac{5}{\frac{5x+5}{x}-5}$

$= \frac{5}{\frac{5x+5}{x}-5} \cdot \frac{x}{x} = \frac{5x}{(5x+5)-5x} = \frac{5x}{5} = x$

for all $x \neq 0$.

Therefore $f^{-1}(x) = \frac{5x+5}{x}$.

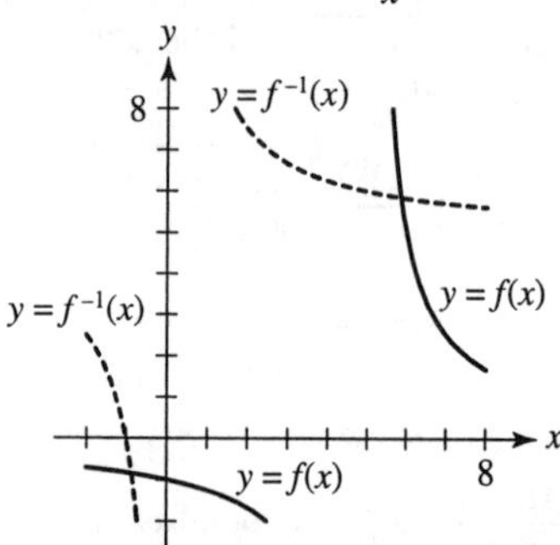

69. To find f^{-1} analytically, rewrite f using x and y:

$y = \sqrt{10-x}$

Interchange x and y, and solve for y:

$x = \sqrt{10-y}$
$x^2 = 10 - y$
$y = 10 - x^2$

Define $g(x) = 10 - x^2$, and check to see whether $g = f^{-1}$:

$g[f(x)] = g\left(\sqrt{10-x}\right) = 10 - \left(\sqrt{10-x}\right)^2$
$= 10 - (10 - x) = x$ for all x in the domain of f.

$f[g(x)] = f(10 - x^2) = \sqrt{10-(10-x^2)}$
$= \sqrt{x^2} = |x| = x$ if $x \geq 0$.

Therefore $f^{-1}(x) = 10 - x^2$ for $x \geq 0$.

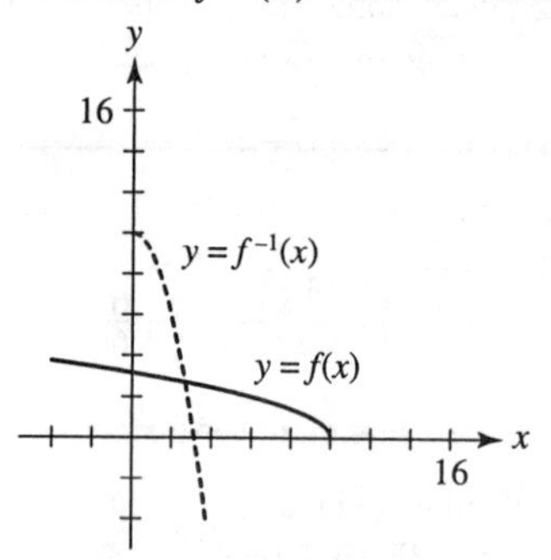

71. Since the vertex of the graph is (0, 7), the two one-to-one functions are $f_1(x) = 7 - 3x^2$ for $x \leq 0$, and $f_2(x) = 7 - 3x^2$ for $x \geq 0$.

To find the inverse of $f_1(x)$, rewrite it as

$y = 7 - 3x^2$ for $x \leq 0$

Interchange x and y, and solve for y:

$x = 7 - 3y^2$ for $y \leq 0$
$3y^2 = 7 - x$ for $y \leq 0$
$y^2 = \frac{7-x}{3}$ for $y \leq 0$
$y = \pm\sqrt{\frac{7-x}{3}}$ for $y \leq 0$

Since $y \leq 0$, we know that $y = -\sqrt{\frac{7-x}{3}}$

Thus $f_1^{-1}(x) = -\frac{\sqrt{7-x}}{3}$. Similarly,

$f_2^{-1}(x) = \sqrt{\frac{7-x}{3}}$.

73. Since the vertex of the graph is (–2, 1), the two one-to-one functions are $f_1(x) = (x+2)^2 + 1$ for $x \leq -2$, and $f_2(x) = (x+2)^2 + 1$ for $x \geq -2$.

To find the inverse of $f_1(x)$, rewrite it as

$y = (x+2)^2 + 1$ for $x \leq -2$

Interchange x and y, and solve for y:

$x = (y+2)^2 + 1$ for $y \leq -2$
$(y+2)^2 = x - 1$ for $y \leq -2$
$y + 2 = \pm\sqrt{x-1}$ for $y \leq -2$

Since $y \leq -2$, we know that $y + 2 \leq 0$, so

$y + 2 = -\sqrt{x-1}$
$y = -\sqrt{x-1} - 2$

Thus $f_1^{-1}(x) = -\sqrt{x-1} - 2$. Similarly,

$f_2^{-1}(x) = \sqrt{x-1} - 2$.

75. The function is increasing on $(-\infty, -1)$ and $(-1, 0)$, and decreasing on $(0, 1)$ and $(1, \infty)$. It has an inverse on each of these intervals.

77. The function is decreasing on $(-\infty, -3)$ and $(0, 3)$, and increasing on $(-3, 0)$ and $(3, \infty)$. It has an inverse on each of these intervals.

Section 9-4

1. First obtain the equation in the form
$y = 20 + Ab^t$.
When $t = 0$, $y = 37$, so $20 + Ab^0 = 37$
When $t = 7$, $y = 37 - 5.2 = 31.8$, so
$20 + Ab^7 = 31.8$
From the first equation, $A = 17$. Replacing A by 17 in the second equation yields
$20 + 17b^7 = 31.8$
$17b^7 = 11.8$
$b^7 \cong 0.69$
$b = \sqrt[7]{0.69} \cong 0.95$
Thus, $y = 20 + 17(0.95)^t$. From the graph of $y = e^x$, estimate that $e^x = 0.95$ when $x \cong -0.052$. Thus $y = 20 + 17e^{-0.052t}$.

3. a. No horizontal line intersects the graph of $y = 2^x$ more than once.

b. (Sample response) Solving $2^x = 7$ is equivalent to finding the point $(x_0, 7)$ on the graph of $y = g(x)$. This in turn is equivalent to finding the point $(7, x_0)$ on the graph of $y = g^{-1}(x)$, so that $x_0 = g^{-1}(7)$.

5.

x	$y = \log_5 x$
5	1 = power to which 5 is raised to obtain 5
25	2 = power to which 5 is raised to obtain 25
125	3 = power to which 5 is raised to obtain 125
625	4 = power to which 5 is raised to obtain 625

7. It must be true that $\log_b 1 = 0$ for any base $b > 0$ because $b^0 = 1$ for any $b > 0$. Similarly, $\log_b b = 1$ because $b^1 = b$ for any $b > 0$.

9. The ratio of each x-value to the previous one is $\frac{2}{3}$, so the table fits a function $y = \log_b x + k$.

11. The ratio of each x-value to the previous one is 1.1, so the table fits a function $y = \log_b x + k$.

13. a.

x	$h(x)$
–2	1
–1	3
0	9
1	27
2	81

b.

x	$h^{-1}(x)$
1	–2
3	–1
9	0
27	1
81	2

c. The table for the inverse of $h(x) = 9 \cdot 3^x$ in Exercise 13b and the table for $y = \log_3 x - 2$ in Exercise 12c are identical.

15.

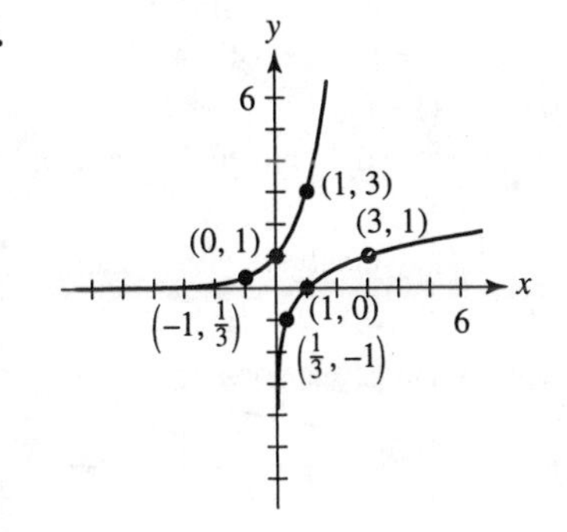

17. **a.** If $c = \log_{1/b} x$, then $x = \left(\frac{1}{b}\right)^c$, so it is also true that $x = b^{-c}$, which means that $-c = \log_b x$. Thus $\log_{1/b} x = -\log_b x$.

b.

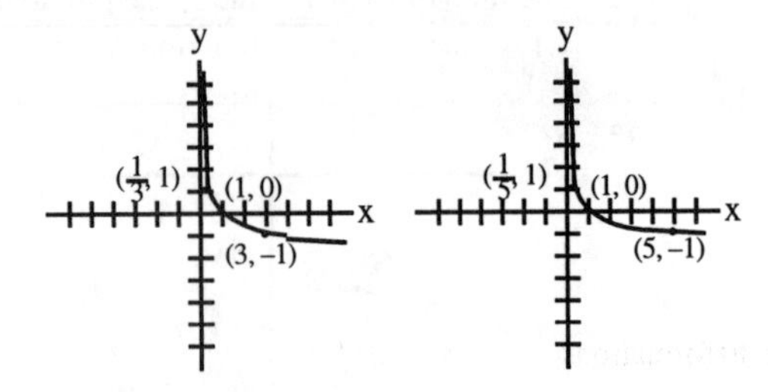

19. The graph is obtained from that of $y = \log_2 x$ by the transformations
$y = \log_2 x \rightarrow y = 0.5\log_2 x \rightarrow$
$y = 0.5\log_2(x+2)$
The graph is compressed vertically by a factor of 2, then shifted 2 units to the left.

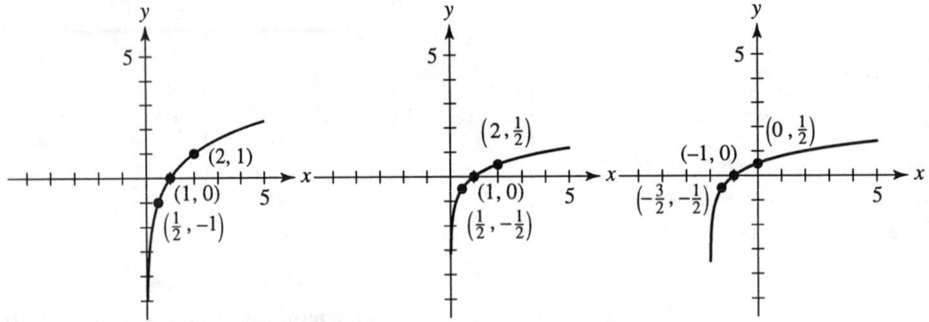

The domain is (–2, ∞) and the range is (–∞, ∞). The function is increasing.

21. The graph is obtained from that of $y = \log_3 x$ by the transformations
$y = \log_3 x \rightarrow y = \log_3(x-4) \rightarrow$
$y = \log_3(x-4)-1$
The graph is shifted 4 units to the right and 1 unit down.

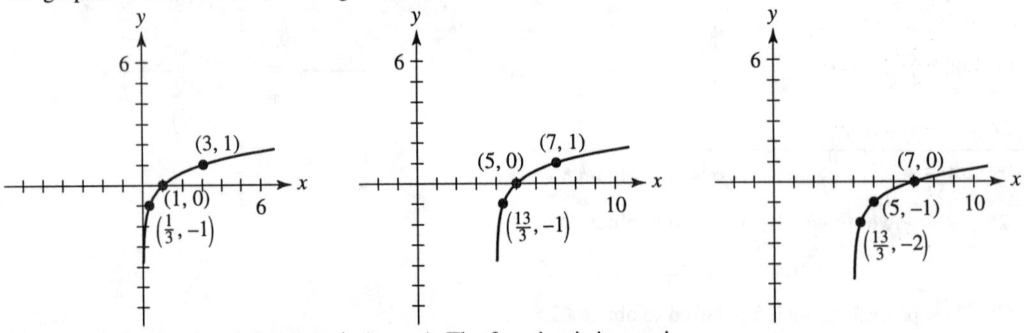

The domain is (4, ∞) and the range is (–∞, ∞). The function is increasing.

23.

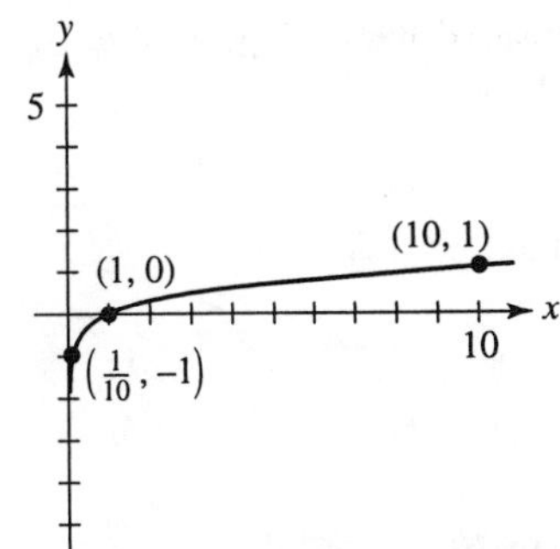

25. The graph is obtained from that of $y = \log x$ by the transformations

$y = \log x \rightarrow y = \log 2x \rightarrow y = 3\log 2x$

The graph is compressed horizontally by a factor of 2, then stretched vertically by a factor of 3.

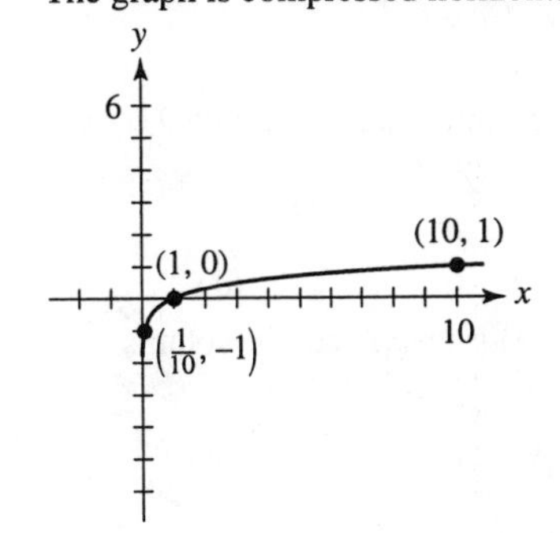

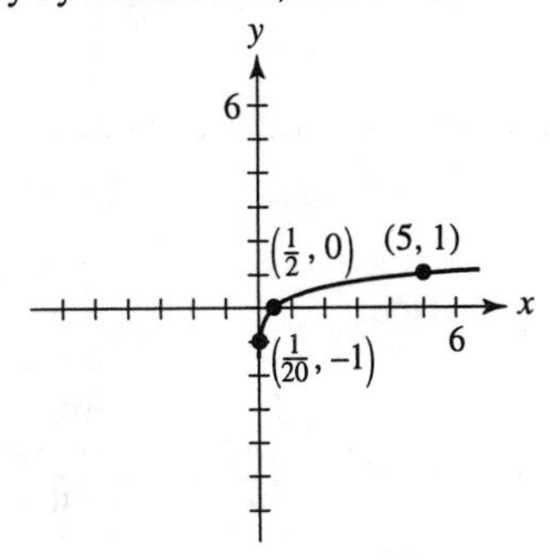

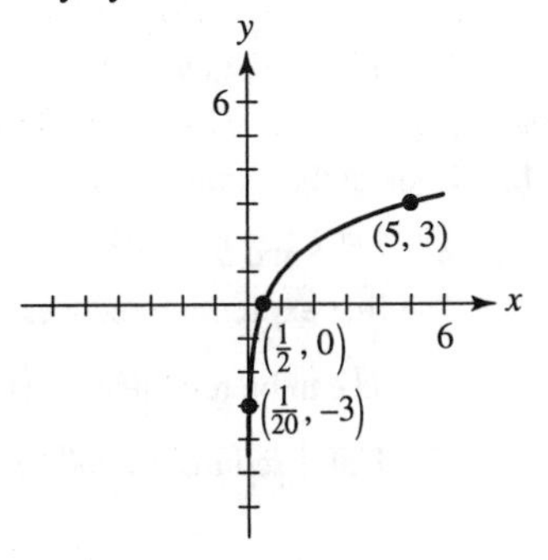

27. $4e^{5t} = 102.8$

$e^{5t} = 25.7$

$5t = \ln 25.7$

$t = \dfrac{\ln 25.7}{5} \cong 0.65$

29. $2 \cdot 10^x + 0.42 = 3.29$

$2 \cdot 10^x = 2.87$

$10^x = \dfrac{2.87}{2}$

$x = \log\left(\dfrac{2.87}{2}\right) \cong 0.16$

31. $\ln(4t - 100) = 0.57$

$4t - 100 = e^{0.57}$

$4t = e^{0.57} - 100$

$t = \dfrac{e^{0.57} - 100}{4} \cong -24.56$

33. $e^{-0.00012t} = 0.45$

$-0.00012t = \ln 0.45$

$t = -\dfrac{\ln 0.45}{0.00012} \cong 6654,$

so Crater Lake is about 6700 years old.

35. **a.** To see why $\log_b\left(\frac{m}{n}\right) = \log_b m - \log_b n$, let's look at the expressions

$b^{\log_b(m/n)}$ and $b^{(\log_b m - \log_b n)}$.

If these two powers of b are equal, then the exponents must be equal. (This is another way of saying that the function $y = b^x$ is one-to-one.)

The first expression reduces to $\frac{m}{n}$, and the second can be rewritten as $\frac{b^{\log_b m}}{b^{\log_b n}}$, which also reduces to $\frac{m}{n}$. This establishes that

$\log_b\left(\frac{m}{n}\right) = \log_b m - \log_b n.$

b. Look at the expressions

$b^{\log_b(m^a)}$ and $b^{a \log_b m}$.

The first expression reduces to m^a, and the second can be rewritten as $\left(b^{\log_b m}\right)^a$, which also reduces to m^a. This establishes that

$\log_b\left(m^a\right) = a \log_b m.$

37. $\log \frac{a}{bc} = \log a - \log bc = \log a - (\log b + \log c)$

39. $\log a^2 b = \log a^2 + \log b = 2\log a + \log b$

41. $x = Ab^y$

$\log_b x = \log_b\left(Ab^y\right)$

$\log_b x = \log_b A + \log_b b^y$

$\log_b x = \log_b A + y$

$y = \log_b x - \log_b A$

This equation has the form

$y = \log_b x + k$, with $k = -\log_b A$.

43. $2\log_2 x + \log_2 2x = 1$

$\log_2 x^2 + \log_2 2x = 1$

$\log_2[\left(x^2\right)(2x)] = 1$

$\log_2\left(2x^3\right) = 1$

$2x^3 = 2$

$x = 1$

Check in the original equation:

$2\log_2 1 + \log_2 2(1) = 1$

$2(0) + 1 = 1$

$1 = 1$

The solution is valid.

45. $\frac{\ln x}{\ln(x-12)} = 2$

$\ln x = 2\ln(x-12)$

$\ln x = \ln(x-12)^2$

$x = (x-12)^2$

$x = x^2 - 24x + 144$

$x^2 - 25x + 144 = 0$

$(x-9)(x-16) = 0$

$x = 9, 16$

Check each solution in the original equation:

$\frac{\ln 9}{\ln(9-12)} = 2$

$\frac{\ln 9}{\ln(-3)} = 2$

Since $\ln(-3)$ is undefined, the solution $x = 9$ is extraneous.

$\frac{\ln 16}{\ln(16-12)} = 2$

$\frac{\ln 16}{\ln 4} = 2$

$2 = 2$

The only valid solution is $x = 16$.

47. **a.** $\log_4 7 = \frac{\log 7}{\log 4} \cong 1.40$

b. $\log_7 4 = \frac{\log 4}{\log 7} \cong 0.71$

c. $\log_{2/3} 13 = \frac{\log 13}{\log \frac{2}{3}} \cong -6.33$

d. $\log_{29} 29 = 1$ (You don't need the change of base formula for this one!)

e. $\log_5 \sqrt{11} = \frac{\log \sqrt{11}}{\log 5} \cong 0.74$

f. $\log_{0.4} 16 = \frac{\log 16}{\log 0.4} \cong -3.03$

g. $\log_3 17.6 = \frac{\log 17.6}{\log 3} \cong 2.61$

h. $\log_{12} \frac{16}{9} = \frac{\log \frac{16}{9}}{\log 12} \cong 0.23$

49.

x	y
1	0
10	1
100	2
1000	3

51. a. $36^{1/2} = 6$, so $\log_{36} 6 = \frac{1}{2}$

b. $6^2 = 36$, so $\log_6 36 = 2$

c. $11^{1/2} = \sqrt{11}$, so $\log_{11} \sqrt{11} = \frac{1}{2}$

d. $3^{-4} = \frac{1}{81}$, so $\log_3 \frac{1}{81} = -4$

e. $\log_5 5^{-7} = -7$

f. $\left(\frac{1}{2}\right)^{-4} = 16$, so $\log_{1/2} 16 = -4$

g. $\left(\frac{2}{3}\right)^2 = \frac{4}{9}$, so $\log_{2/3} \frac{4}{9} = 2$

h. $\left(\sqrt{2}\right)^4 = (2^{1/2})^4 = 2^2 = 4$, so $\log_2 4 = 4$

53. The y-values are equally spaced, and the ratio of each x-value to the preceding one is 2, so the table fits a function $y = A \log_b x + k$.

55. The y-values are equally spaced, and the ratio of each x-value to the preceding one is 0.1, so the table fits a function $y = A \log_b x + k$.

57. The graph is obtained from that of $y = \ln x$ by the transformation

$y = \ln x \rightarrow y = 4 \ln x$

The graph is stretched vertically by a factor of 4.

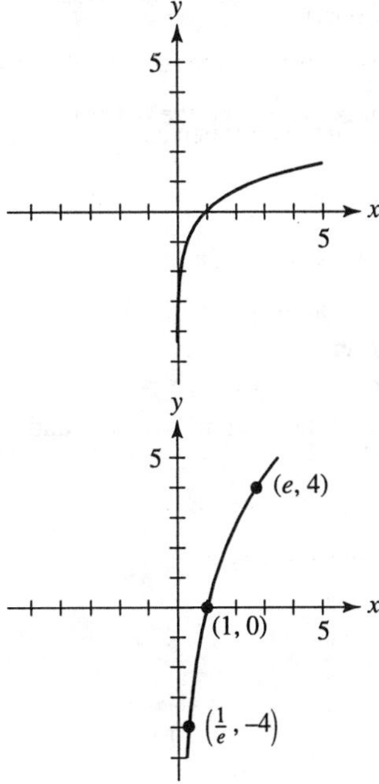

The domain is $(0, \infty)$, the range is $(-\infty, \infty)$, and the function is increasing.

59. The graph is obtained from that of $y = \log_3 x$ by the transformations

$y = \log_3 x \rightarrow y = \frac{1}{3}\log_3 x \rightarrow y = \frac{1}{3}\log_3 x + 3$

The graph is compressed vertically by a factor of 3, then shifted 3 units up.

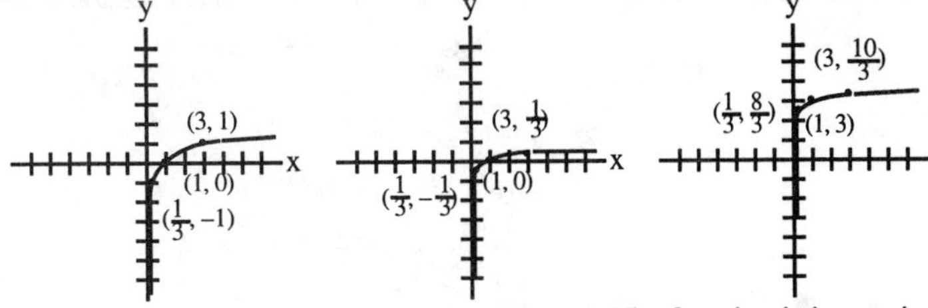

The domain is $(0, \infty)$, and the range is $(-\infty, \infty)$. The function is increasing.

61. The graph is obtained from that of $y = \log x$ by the transformations
$y = \log x \rightarrow y = \log 3x \rightarrow y = 6\log 3x$
The graph is compressed horizontally by a factor of 3, and stretched vertically by a factor of 6.

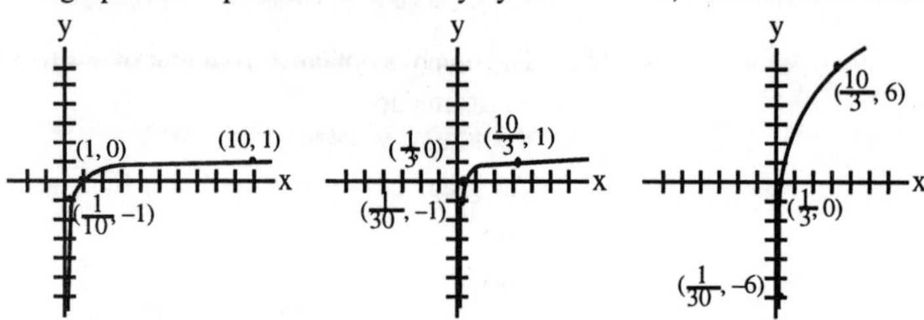

The domain is $(0, \infty)$, and the range is $(-\infty, \infty)$. The function is increasing.

63. The graph is obtained from that of $y = \ln x$ by the transformations
$y = \ln x \rightarrow y = \ln(x+1) \rightarrow y = \ln(x+1) - 2$
The graph is shifted 1 unit to the left and 2 units down.

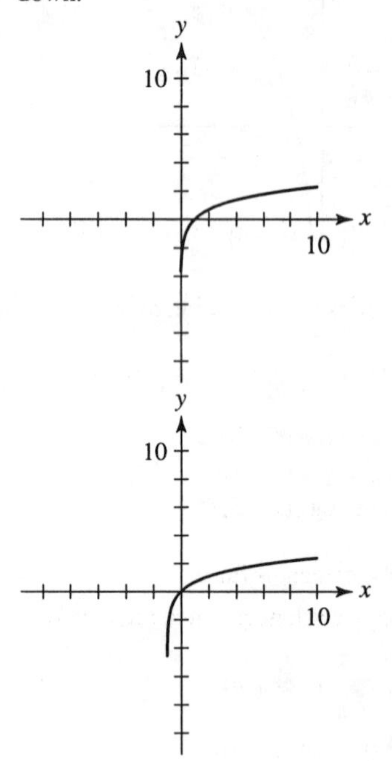

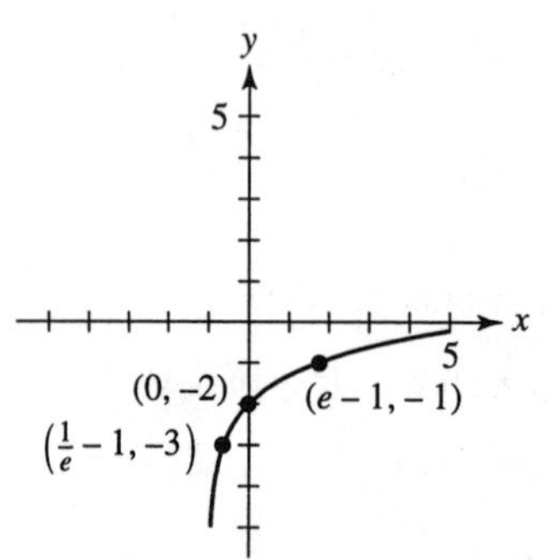

The domain is $(-1, \infty)$, and the range is $(-\infty, \infty)$. The function is increasing.

65. $5^{3t-4} = 9$
$3t - 4 = \log_5 9$
$t = \dfrac{\log_5 9 + 4}{3} \cong 1.79$

67. $3 \cdot 4^{v+5} - 2 = 0.21$
$3 \cdot 4^{v+5} = 2.21$
$4^{v+5} = \dfrac{2.21}{3}$
$v + 5 = \log_4\left(\dfrac{2.21}{3}\right)$
$v = \log_4\left(\dfrac{2.21}{3}\right) - 5 \cong -5.22$

69. $e^{4x+5} = 2e^{3x-1}$
$\ln(e^{4x+5}) = \ln(2e^{3x-1})$
$\ln(e^{4x+5}) = \ln 2 + \ln(e^{3x-1})$
$4x + 5 = \ln 2 + 3x - 1$
$x = \ln 2 - 6 \cong -5.31$

71. $12\log_8(x-1)=4$
$\log_8(x-1)=\frac{1}{3}$
$x-1=8^{1/3}=2$
$x=3$
Check: $12\log_8(3-1)=4$
$12\log_8 2=4$
$12\left(\frac{1}{3}\right)=4$
$4=4$
The solution is valid.

73. $\ln\sqrt{t}=5$
$\sqrt{t}=e^5$
$t=(e^5)^2=e^{10}\cong 22{,}026.47$
Check:
$\ln\sqrt{e^{10}}=5$
$\ln e^5=5$
$5=5$ The solution is valid.

75. $2\ln x=\ln(6-x)$
$\ln x^2=\ln(6-x)$
$x^2=6-x$
$x^2+x-6=0$
$(x+3)(x-2)=0$
$x=-3, 2$
Check:
$2\ln(-3)=\ln[6-(-3)]$
Since $\ln(-3)$ is not defined, the solution $x=-3$ is extraneous.
$2\ln 2=\ln(6-2)$
$2\ln 2=\ln 4$
$\frac{\ln 4}{\ln 2}=2$
$2=2$
The solution $x=2$ is valid.

77. $\log(x-3)+\log(x^2+3x+9)=\log 37$
$\log[(x-3)(x^2+3x+9)]=\log 37$
$\log(x^3-27)=\log 37$
$x^3-27=37$
$x^3=64$
$x=4$
Check:
$\log(4-3)+\log[4^2+3(4)+9]=\log 37$
$\log 1+\log 37=\log 37$
$\log 37=\log 37$ The solution is valid.

79. $\log_4(x+5)+\log_4\left(\frac{1}{x}\right)=\frac{1}{2}$
$\log_4\left[(x+5)\left(\frac{1}{x}\right)\right]=\frac{1}{2}$
$(x+5)\left(\frac{1}{x}\right)=4^{1/2}$
$\frac{x+5}{x}=2$
$x+5=2x$
$x=5$
Check:
$\log_4(5+5)+\log_4\left(\frac{1}{5}\right)=\frac{1}{2}$
$\log_4 10+\log_4 0.2=0.5$
$\log_4 2=0.5$
$0.5=0.5$ The solution is valid.

81. $\log_4\sqrt{x}=\log_4 x^{1/2}=\frac{1}{2}\log_4 x$

83. $\log_4\frac{x}{16}=\log_4 x-\log_4 16$
$=\log_4 x-\log_4 4^2=\log_4 x-2$

85. $\log_4\frac{1}{x^3}=\log_4(x^{-3})=-3\log_4 x$

87. $\log_4(x^2+x)-\log_4(x+1)$
$=\log_4\frac{x^2+x}{x+1}=\log_4\frac{x(x+1)}{x+1}=\log_4 x$

89. **a.** The statement is true, because
$\log_2(32x)=\log_2 32+\log_2 x$
$=\log_2 2^5+\log_2 x=5+\log_2 x$

b. The statement is false
$[\log_b x+\log_b 3=\log_b(3x)$, not $\log_b(x+3)]$.

c. The statement is false
$[3\log_2 x=\log_2 x^3$, not $\log_2(3x)]$.

d. The statement is true.

e. The statement is true.

f. The statement is true, because
$\log_b(bx)=\log_b b+\log_b x=\log_b x+1$.

91. **a.** $\frac{I_1}{I_2}=(2.512)^{M_2-M_1}$
$\log\left(\frac{I_1}{I_2}\right)=\log(2.512)^{M_2-M_1}$

$$\log I_1 - \log I_2 = (M_2 - M_1)\log 2.512$$
$$M_2 - M_1 = \frac{1}{\log 2.512}(\log I_1 - \log I_2)$$
$$M_2 - M_1 \cong 2.50(\log I_1 - \log I_2)$$

b. If $M_1 = 6$ and $M_2 = 29$, then
$$\frac{I_1}{I_2} = (2.512)^{23} \cong 1.59 \text{ billion}$$

c. If $\frac{I_1}{I_2} = 10$, then
$$(2.512)^{M_2 - M_1} = 10$$
$$\log(2.512)^{M_2 - M_1} = 10$$
$$\log(2.512)^{M_2 - M_1} = \log 10$$
$$(M_2 - M_1)\log 2.512 = 1$$
$$M_2 - M_1 = \frac{1}{\log 2.512} \cong 2.50$$

93. a. If the weaker and stronger earthquakes release E_1 and E_2 ergs of energy, then
$$\frac{\log E_2 - 11.8}{1.5} - \frac{\log E_1 - 11.8}{1.5} = 1$$
$$\log E_2 - 11.8 - (\log E_1 - 11.8) = 1.5$$
$$\log E_2 - \log E_1 = 1.5$$
$$\log\left(\frac{E_2}{E_1}\right) = 1.5$$
$$\frac{E_2}{E_1} = 10^{1.5} \cong 31.62$$

b. From part (a) each increase of 1 on the Richter scale multiplies the released energy by a factor of $10^{1.5}$. An increase of 4 multiplies the released energy by a factor of $(10^{1.5})^4 = 10^6 = 1{,}000{,}000$.

c. For the 1906 San Francisco earthquake,
$$\frac{\log E - 11.8}{1.5} = 8.25$$
$$\log E - 11.8 = 12.38$$
$$\log E = 24.18$$
$$E = 10^{24.18} \cong 1.51 \times 10^{24}$$
For the 1964 Alaskan earthquake,
$$\frac{\log E - 11.8}{1.5} = 8.40$$
$$\log E - 11.8 = 12.60$$
$$\log E = 24.40$$
$$E = 10^{24.40} \cong 2.51 \times 10^{24}$$

95. The 2 on the lower ruler and the 3 on the upper ruler represent lengths of log 2 and log 3, respectively. The position of the slide rule demonstrates that the sum of these two lengths is log 6, since the 3 on the upper ruler is opposite the 6 on the lower ruler. Thus
$\log 6 = \log 2 + \log 3 = \log(2 \cdot 3)$, so $6 = 2 \cdot 3$.

Section 9-5

1. a.

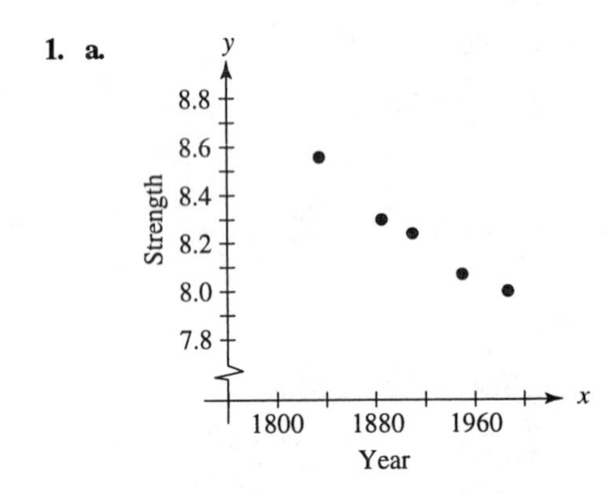

b. Lines will vary.

3. Answers will vary.

5. a.

Lines and equations will vary, depending in part on whether time is expressed in minutes or seconds.

b. Estimates will vary, but should be around 3:47.

c. Answers will vary, but should be unrealistically large for the year 1, and negative for the year 3000.

d. Answers will vary, but the time interval for which the equation is approximately valid should clearly exclude the years 1 and 3000.

7. a. $n = 4$

$\sum x = 0.5 + 0.9 + 1.5 + 1.9 = 4.8$

$\sum y = 7 + 6 + 5 + 4 = 22$

$\sum x^2 = 0.25 + 0.81 + 2.25 + 3.61 = 6.92$

$\sum xy = 3.5 + 5.4 + 7.5 + 7.6 = 24.0$

$\bar{x} = \frac{4.8}{4} = 1.2$

$\bar{y} = \frac{22}{4} = 5.5$

$m = \frac{4(24) - (4.8)(22)}{4(6.92) - (4.8)^2} \cong -2.07$

The regression line is $y - 5.5 = -2.07(x - 1.2)$, or $y = -2.07x + 7.98$.

b. $n = 4$

$\sum x = -4 - 2.5 + 0 + 1 = -5.5$

$\sum y = 230 + 250 + 350 + 460 = 1290$

$\sum x^2 = 16 + 6.25 + 0 + 1 = 23.25$

$\sum xy = -920 - 625 + 0 + 460 = -1085$

$\bar{x} = -\frac{5.5}{4} = -1.38$

$\bar{y} = \frac{1290}{4} = 322.5$

$m = \frac{4(-1085) - (-5.5)(1290)}{4(23.25) - (-5.5)^2} \cong 43.90$

The regression line is
$y - 322.5 = 43.90(x + 1.38)$, or
$y = 43.90x + 383.08$.

c. $n = 4$

$\sum x = 1000 + 1500 + 1800 + 2300 = 66000$

$\sum y = 10 + 16 + 19 + 23 = 68$

$\sum x^2 = 1{,}000{,}000 + 2{,}250{,}000 + 3{,}240{,}000 + 5{,}290{,}000 = 11{,}780{,}000$

$\sum xy = 10{,}000 + 24{,}000 + 34{,}200 + 52{,}900 = 121{,}100$

$\bar{x} = \frac{66\ 000}{4} = 16{,}500$

$\bar{y} = \frac{68}{4} = 17$

$m = \frac{4(121{,}100) - (66\ 000)(68)}{4(11\ 780\ 000) - (66\ 000)^2} \cong 0.00093$

The regression line is
$y - 17 = 0.00093(x - 16{,}500)$ or
$y = 0.00093x + 1.66$.

9. Answers will vary.

11. Estimates will vary, but should be around 3000 years.

13. a.

t	y	$Y = \ln y$
0	335	5.81
1	1285	7.16
2	3933	8.28
3	9015	9.11
4	18,195	9.81
5	31,452	10.36
6	46,053	10.74
7	55,388	10.92
8	78,215	11.27

b. Equations will vary, but should be close to $\ln y = 6 + 0.6t$.

c. Calculations will vary, according to the equation obtained in part b. Here is a sample calculation beginning with $\ln y = 6 + 0.6t$:

$\ln y = 6 + 0.6t$

$e^{\ln y} = e^{6+0.6t}$

$y = e^6 e^{0.6t}$

$y \cong 403.43e^{0.6t}$

d. Graphs will vary.

15. a. $n = 9$

$\sum t = 36$

$\sum Y = 83.46$

$\sum t^2 = 204$

$\sum tY = 373.13$

$\bar{t} = 4$

$\bar{Y} = 9.27$

$m = \frac{9(373.13) - (36)(83.46)}{9(204) - (36)^2} \cong 0.65$

The regression line is
$Y - 9.27 = 0.65(t - 4)$, or
$Y = 0.65t + 6.67$.

b. $Y = 0.65t + 6.67$
$\ln y = 0.65t + 6.67$
$y = e^{0.65t+6.67}$
$y = e^{0.65t}e^{6.67}$
$y = 788.40e^{0.65t}$

c.

t	y
0	788
1	1510
2	2893
3	5541
4	10,615
5	20,333
6	38,949
7	74,608
8	142,915

17. The exponential model constructed in Exercise 15 was not an accurate predictor of the number of deaths from AIDS in 1991. This indicates that the actual relationship between t and y may not be exponential.

19. a. If $y = Ax^m$, then
$\ln y = \ln(Ax^m) = \ln A + \ln x^m$.

b. Because $\ln x^m = m \ln x$, the equation can be written $y = \ln A + m \ln x$.

c. The equation is $\ln y = m \ln x + \ln A$. This has the form $\ln y =$ (constant) $\ln x$ + (constant), with m and $\ln A$ as the two constants.

21. Tables will vary.

23. Kepler's third law states that y^2 is proportional to x^3, that is, $y^2 = kx^3$ for some k. You can find k by using any data point from the table.

For example, $x = 149.6$ and $y = 1$ for the earth, so
$1^2 = k(149.6)^3$
$k = \dfrac{1}{149.6^3}$
Therefore $y^2 = \dfrac{1}{149.6^3}x^3$, so
$y = \sqrt{\dfrac{1}{149.6^3}x^3} \cong 0.00055x^{3/2}$. This agrees with the function obtained in Exercise 22b.

25. a. The linear equation is $y \cong 0.0034x - 1.94$, and $r \cong 0.99986$.

b. The logarithmic equation is $y \cong 39.78x - 306.17$, and $r \cong 0.92$.

c. The exponential equation is $y \cong 6.59(1.000077)^x$, and $r \cong 0.92$.

d. The power equation is $y \cong 0.0016x^{1.071}$, and $r \cong 0.99998$.

The power curve gives the best fit.

27. Answers will vary.

29.

x	$\ln x$	y
1	0	–5
7	1.95	–1
12	2.48	0
20	3.00	1

Equations will vary, but should be close to $y = 2 \ln x - 5$.

31.

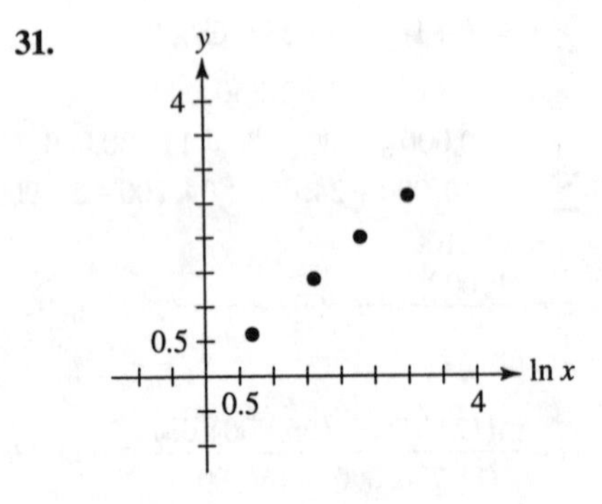

The table is approximately logarithmic.

33.

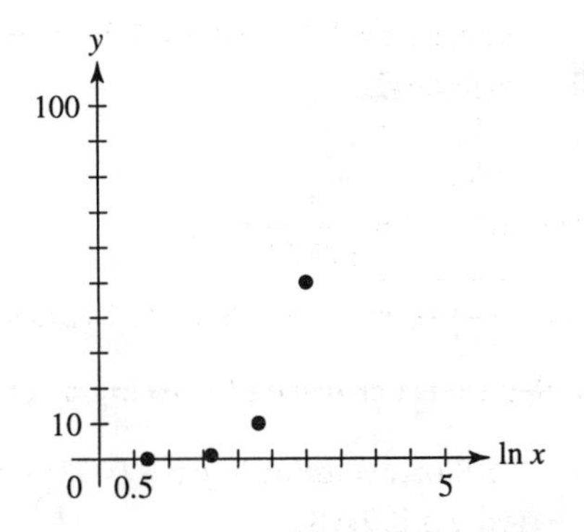

The table is not approximately logarithmic.

35. a.

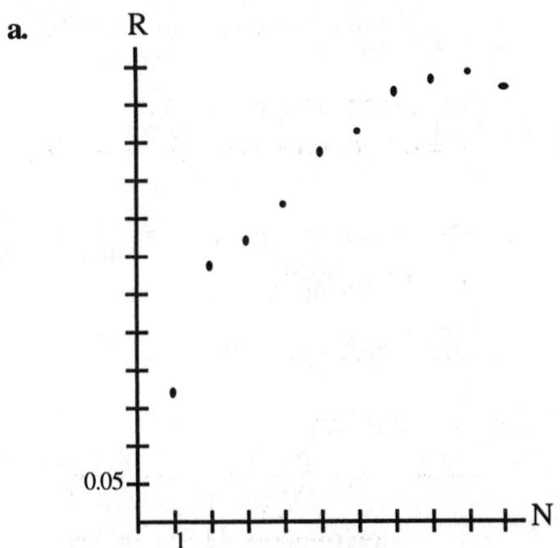

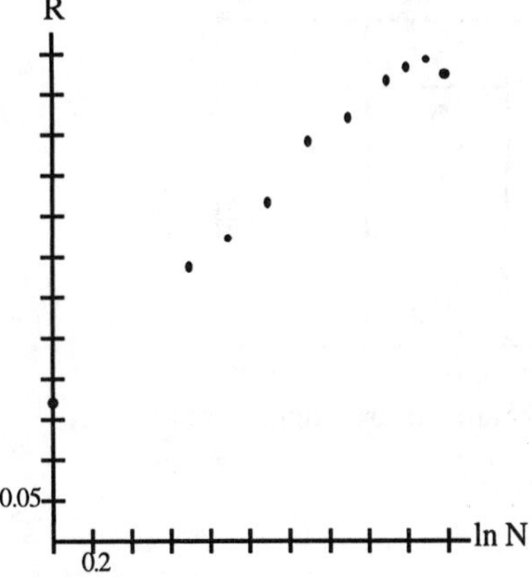

The data looks more nearly logarithmic.

b. Equations will vary, but should be close to $R = 0.2 + 0.2 \ln N$.

37.

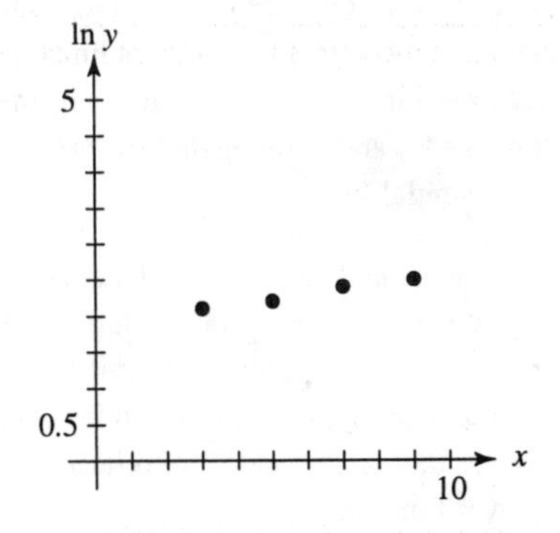

The table is approximately exponential.

39.

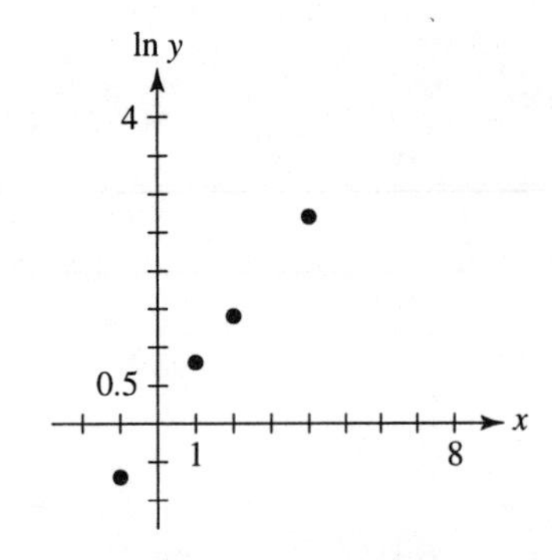

The table is approximately exponential.

41.

x	$\ln x$	y	$\ln y$
1	0	2	0.69
7	1.95	90	4.50
12	2.48	300	5.70
20	3.00	750	6.62

Lines will vary. Equations should be close to $\ln y = 2 \ln x + 0.7$, leading to something close to $y = 2x^2$.

43.

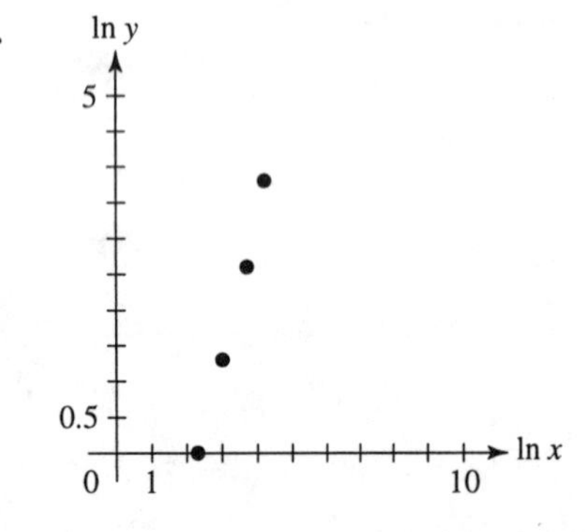

The table approximately fits a power function.

45.

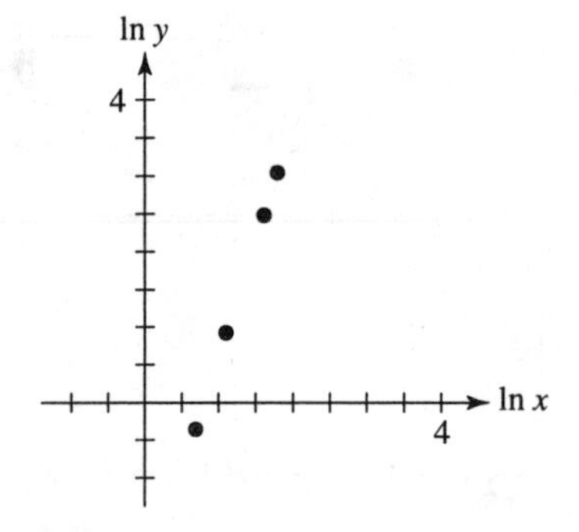

The table approximately fits a power function.

47. The best fit is given by the linear equation $y \cong 11.51 - 1.05x$ ($r \cong -0.99994$).

49. The best fit is given by the power equation $y \cong 1.04x^{0.49}$ ($r \cong 0.999988$).

51. **a.** If t represents time in years since 1780, and y represents population in millions, then the best fit is given by the exponential equation $y \cong 2.94(1.03)^t$.

b. The equation predicts an 1860 population of about 31,392,000 people, and an 1870 population of about 42,211,000 people.

c. The Civil War would account for some of the shortfall.

53. **a.** The horizontal and vertical distances on semilog paper are proportional to x and $\log y$, respectively. The plot is linear if the relationship between x and $\log y$ is linear, that is, if x and y are related by an exponential function.

b. The horizontal and vertical distances on log-log paper are proportional to $\log x$ and $\log y$, respectively. The plot is linear if the relationship between $\log x$ and $\log y$ is linear, that is, if x and y are related by a power function.